STRUCTURAL MECHANICS IN REACTOR TECHNOLOGY

TRANSACTIONS OF THE 9TH INTERNATIONAL CONFERENCE ON
STRUCTURAL MECHANICS IN REACTOR TECHNOLOGY / LAUSANNE
17-21 AUGUST 1987

Structural Mechanics in Reactor Technology

VOLUME A
Indexes, Abbreviations, Supplement

Editor
FOLKER H.WITTMANN

A.A.BALKEMA / ROTTERDAM / BOSTON / 1987

SPONSORING COMPANIES AND
ORGANIZATIONS

Ecole Polytechnique Fédérale de
  Lausanne, Switzerland
Eidgenössisches Institut für
  Reaktorforschung, Würenlingen,
  Switzerland
Municipalité de Lausanne, Lausanne,
  Switzerland

CO-SPONSORING COMPANIES AND
ORGANIZATIONS

Nuclear Regulatory Commission (NRC),
  Washington, D.C., USA
Société Coopérative Nationale pour
  l'Entreposage des Déchets Radioactifs
  (CEDRA/NAGRA), Baden, Switzerland
Société des Chaux et Ciments de la
  Suisse Romande, Lausanne, Switzerland
Kernkraftwerk Leibstadt AG, Leibstadt,
  Switzerland
Kernkraftwerk Gösgen-Däniken AG,
  Däniken, Switzerland
S.A. L'Energie de l'Ouest-Suisse (EOS),
  Lausanne, Switzerland
Bernische Kraftwerke AG, Bern,
  Switzerland

Kernkraftwerk Kaiseraugst AG, Olten,
  Switzerland
Société des Forces Motrices de
  Chancy-Pougny, Geneva, Switzerland
Sprecher & Schuh AG Niederspannung,
  Aarau, Switzerland
Gesellschaft des Aare-und-Emmen-
  kanals, Solothurn, Switzerland
BBC Brown Boveri & Cie, Baden,
  Switzerland
Electricité Neuchâteloise SA, Corcelles,
  Switzerland
SWISSAIR, Zürich, Switzerland
Control Data, Zürich, Switzerland
Banque Cantonale du Valais, Sion,
  Switzerland

CO-SPONSORING ASSOCIATIONS

American Nuclear Society (ANS),
  LaGrange Park, Ill., USA
European Nuclear Society (ENS), Bern,
  Switzerland
Swiss Nuclear Society (SNS),
  Würenlingen, Switzerland
International Union of Testing and
  Research Laboratories for Materials
  and Structures (RILEM), Paris,
  France

A special grant for publication of the Transactions of SMIRT-9 and special volume
Structural mechanics in reactor technology: Advances 1987, from Nuclear Regulatory
Commission (NRC) Washington D.C., USA, is gratefully acknowledged.

*The text of the various papers in this volume were set individually
by typists under the supervision of each of the authors concerned.*

For the complete set of volumes, ISBN 90 6191 762 X
For Volume A, ISBN 90 6191 761 1
For volume B, ISBN 90 6191 763 8
For volume C, ISBN 90 6191 764 6
For volume D, ISBN 90 6191 765 4
For volume E, ISBN 90 6191 766 2
For volume F, ISBN 90 6191 767 0
For volume G, ISBN 90 6191 768 9
For volume H, ISBN 90 6191 769 7
For volume J, ISBN 90 6191 770 0
For volume K1, ISBN 90 6191 771 9
For volume K2, ISBN 90 6191 772 7
For volume L, ISBN 90 6191 773 5
For volume M, ISBN 90 6191 774 3
For volume N, ISBN 90 6191 775 1

Preface

More than 850 papers have been submitted for presentation during SMIRT-9. Most of these contributions will be published as usual in Volumes B to N of the Transactions. Volume A contains a comprehensive index covering all contributions included in the Transactions of SMIRT-9. The reader will thus find easily volume and page number of all contributions in this volume A.

To further facilitate the use of the Transactions of SMIRT-9 during and after the conference an authors index has been prepared and is included in volume A. The more than 2000 authors are listed in alphabetical order.

In many papers more or less commonly used abbreviations are used. We found that a certain number of abbreviations such as JGSCC or MSBR are widely used but not necessarily understood by all those interested in the field of structural mechanics in reactor technology. Sometimes the exact meaning of an abbreviation is familiar to a small group of experts exclusively. Therefore we prepared a list of frequently used abbreviations. SMIRT-Conferences traditionally bring together specialists coming from all different branches of mechanics in reactor technology. Hopefully this compilation will be useful for a more efficient interdisciplinary exchange of ideas.

Finally fourty-three papers have arrived here after volumes B to N had been sent already for printing. These contributions are also included in volume A and they are regrouped according to the Division of the scientific program of SMIRT-9 in which they have been accepted for presentation.

It may be mentioned at this point that contributions to the opening session and the Principal Division Lectures will also be published by A.A.Balkema Publishers, the Netherlands, in a special volume entitled 'Structural Mechanics in Reactor Technology – Advances 1987'. This volume will be made available automatically to all registered participants of SMIRT-9.

It is my great pleasure to thank all collaborators of the Laboratory for Building Materials of Swiss Federal Institute of Technology Lausanne. Without their continuous efforts it would not have been possible to prepare more than 850 papers for publication in the Transactions of SMIRT-9 and to include an authors index, a subject index and a list of abbreviations.

Lausanne, June 1987

F.H.Wittmann

Scheme of the complete work

VOLUME A
Indexes, Abbreviations, Supplement

VOLUME B
Computational Mechanics and Computer-Aided Engineering
Coordinators
T.BELYTSCHKO, Northwestern University, Evanston, Ill., USA
J.DONEA, C.E.C., Ispra, Italy

VOLUME C
Fuel Elements and Assemblies
Coordinators
H.HOLTBECKER, C.E.C., Ispra, Italy
Y.R.RASHID., ANATECH International Corporation, La Jolla, Calif., USA

VOLUME D
Experience with Structures and Components in Operating Reactors
Coordinators
D.G.EISENHUT, NUS Corporation, Gaithersburg, Md., USA
R.NOEL, Electricité de France, Paris

VOLUME E
Fast Reactor Core Coolant Circuit Structures
Coordinators
Y.CHANG, Argonne National Laboratory, Ill., USA
M.LIVOLANT, CEA-CEN, Saclay, France

VOLUME F
LWR Pressure Components
Coordinators
D.G.H.LATZKO, Delft University of Technology, Netherlands
B.TOMKINS, United Kingdom Atomic Energy Authority, Risley

VOLUME G
Fracture Mechanics and NDE
Coordinators
R.NICHOLS, Nichols Consultancies, Culcheth, Warrington, UK
G.YAGAWA, University of Tokyo, Japan

VOLUME H
Concrete and Concrete Structures
Coordinators
Z.P.BAZANT, Northwestern University, Evanston, Ill., USA
T.ZIMMERMANN, Swiss Federal Institute of Technology, Lausanne

VOLUME J
Extreme Loading and Response of Reactor Containment
Coordinators
W.J.AMMANN, Swiss Federal Institute of Technology, Zürich
A.HUBER, Motor-Columbus Consulting Engineering, Baden, Switzerland
J.D.STEVENSON, Stevenson & Associates, Cleveland, Ohio, USA

VOLUMES K1 and K2
Seismic Response Analysis of Nuclear Power Plant Systems
Coordinators
A.H.HADJIAN, Bechtel Western Power Corporation, Los Angeles, Calif., USA
H.SHIBATA, University of Tokyo, Japan
J.P.WOLF, Elektrowatt Engineering Services, Ltd, Zürich, Switzerland

VOLUME L
Inelastic Behaviour of Metals and Constitutive Equations
Coordinators
J.-P.BOEHLER, Institut de Mécanique de Grenoble, St-Martin d'Hères, France
E.KREMPL, Rensselaer Polytechnic Institute, Troy, N.Y., USA

VOLUME M
Structural Reliability – Probabilistic Safety Assessment
Coordinators
A.H.S.ANG, University of Illinois, Urbana, USA
G.APOSTOLAKIS, University of California, Los Angeles, USA
P.KAFKA, Gesellschaft für Reaktorsicherheit (GRS)mbH, Garching, FR Germany

VOLUME N
Mechanical and Thermal Problems of Fusion Reactors
Coordinators
C.C.BAKER, Argonne National Laboratory, Argonne, Illinois, USA
P.KOMAREK, Kernforschungszentrum Karlsruhe, ITP, FR Germany

Table of contents

B Computational mechanics and computer-aided engineering

Contact with friction modeling for the study of a bolted junction

F.Lebon & M.Raous
CNRS, Laboratoire de Mécanique et d'Acoustique et GRECO G.D.E., Marseille, France
D.Boulegues
Technicatome, Les Milles, France

1 INTRODUCTION

Many structural analysis problems are concerned by contact phenomena. A good knowledge of the contact displacements and the contact forces between the different parts of the structure is generally essential in structure assembling. The special boundary behaviour has a strong influence on the distribution of the stresses in the whole structure and on his total fiability. The contact behaviour is strongly non linear because of the non penetration conditions on the one hand, and because of the friction on the other. On such problems the real contact zone and the contact forces are unknown "a priori" and have to be determined during the resolution. The non penetration will be caracterized by unilateral conditions and the friction will be described by a constitutive law (the Coulomb friction law in this paper).

The application presented here concerns the assembling of the three parts of a bolted junction using a pressing ring. There are three contact zones in this problem. A good description of the contact phenomena is essential to ensure tightness.

From a theoretical point of view, this class of problems leads to variational or quasi-variational inequations containing undifferentiable terms (Raous-Latil). For frictionless problems, formulation and mathematical results have been given by Lions-Stampacchia, Stampacchia,... For problems including friction, one can refer to Duvaut-Lions, Duvaut, Campos-Oden-Kikuchi, Cocu,...

From a numerical point of view, different classes of algorithms are used: penalisation (Campos-Oden-Kikuchi,...), non linear programming methods (Karbling,...), Newton-Raphson (Curnier,...), contact finite elements (Bathe,...). Our methods are based on projection techniques coupled with overrelaxed Gauss-Seidel methods including condensation procedures (reduction of the number of variables). Non linear programming methods and iterative procedures on special boundary conditions are however also used.

2 SIGNORINI PROBLEM INCLUDING FRICTION

The problem of the contact of several solids is a generalization of the contact between a solid and a rigid obstacle, which in turn if one takes friction into account is an extension of the Signorini problem.

3

2.1 Formulation

Under the small deformation hypothesis the kinematic relations are written :

$$(1) \quad e_{ij} = \frac{1}{2} (u_{i,j} + u_{j,i})$$

where e_{ij} denotes the strain tensor components and u_i the displacement components. For elastic material, we set :

$$(2) \quad \sigma_{ij} = K_{ijkl} \, e_{kl}$$

where σ_{ij} are the stress tensor components and K_{ijkl} the elasticity tensor. The equilibrium equations are written :

$$(3) \quad \sigma_{ij,j} = - \phi_i^1 \quad (\phi_i^1 \text{ are the components of the volumic forces })$$

$$(4) \quad \sigma_{ij} \, n_j = \phi_i^2 \quad (\phi_i^2 \text{ are the components of the surfacic forces })$$

On the contact boundary we use a local coordinate referential $(\vec{n}, \vec{t})$ where $\vec{n}$ is the exterior normal vector to the surface. We denote u^n, u^t, F^n, F^t respectively the normal and tangential components of the displacements and of the forces on the contact aera. The unilateral conditions are written :

$$u^n \leq 0$$

$$(5) \qquad F^n \leq 0$$

$$u^n . \, F^n = 0$$

The Coulomb friction law on the tangential components can be written (see Duvaut-Lions) :

$$(6) \quad | F^t | \leq \mu | F^n | \quad (\mu \text{ is the friction coefficient})$$

$$\text{with if } | F^t | < \mu | F^n | \text{ then } u^t = 0$$

$$\text{if } | F^t | = \mu | F^n | \text{ then } u^t = -\lambda \, F^t \, , \, \lambda > 0 \text{ if } | F^n | \neq 0$$

The variational form of the equilibrium equations leads in this case to a quasi variational inequation. We formulate now the problem as a sequence of Tresca problems (where the sliding limit g is given) associated with a fixed point method on this sliding limit $g_{k+1} = \mu | F^n (u_k) |$ (see Raous-Latil).

The Tresca problem can be formulated as a variational inequation containing undifferentiable terms. Because of the symetry of the elasticity mapping, this problem is equivalent to the minimization under constraints of the following functional :

$$(7) \quad J(v) = \frac{1}{2} a(v,v) - (f,v) + \int g | v^t | \, dl$$

where :   - $a(v,v)$ is a coercive and symetrical bilinear form associated
   to the elasticity mapping
         - (f,v) is a linear form associated to the given loads
       - the last term is the work of the friction force in the tangential displacement v^t.

4

2.2 Numerical methods

We have extended the Cryer-Christopherson method to the friction case including the undifferentiable term. It is an overrelaxed Gauss-Seidel method with projection (see Raous-Latil). Using a condensation procedure (see Taallah), we reduce the problem to the only variables concerned by the contact non linearities. A preliminary preparation for the reduced problem has to be done once and only once : it is a partial inversion of the system. This is specially efficient in the case of several loading cases. Others ameliorations and performances are presented in Raous-Latil.

We also use a direct non linear programming method (Lemke-Cottle-Dantzig algorithm) for frictionless problems. It is very efficient on the condensed problem but it cannot be extended to the friction case.

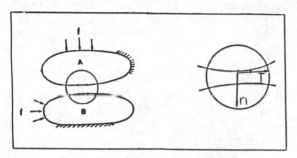

Figure 1 : Contact of two bodies

3 EXTENSION TO A TWO BODY CONTACT CASE

Denoting with the indices A et B the variables concerning respectively the solids A and B, we can write the unilateral conditions and the friction law under the following forms :

$$u_A^n + u_B^n \leq 0$$

$$F_A^n = F_B^n$$

(8)

$$F_A^n \leq 0$$

$$(u_A^n + u_B^n) F_A^n = 0$$

n_A and n_B are the outside normal to the contact aeras Γ_A^C and Γ_B^C.

$$F_A^t = F_B^t$$

$$|F_A^t| \leq \mu |F_A^n|$$

(9)

if $|F_A^t| < \mu |F_A^n|$ then $u_A^t + u_B^t = 0$

if $|F_A^t| = \mu |F_A^n|$ then $u_A^t + u_B^t = -\lambda F_A^t$, $\lambda > 0$, if $F_A^n \neq 0$

Using the change of variables $d^n = u_A^n + u_B^n$ and $d^t = u_A^t + u_B^t$, we give a new formulation of (7) with the variables u_A^t and d^t. The variable d^n is used in the definition of the convex K which is the space of the

5

contraints concerning the minimization. This is essential for the projection procedure.

For frictionless problem, the Lemke-Cottle-Dantzig has been modified and extended to this case : it it still efficient.

For the problem including friction, we use the Cryer-Christopherson method associated again with the condensation procedure.

4 APPLICATION

4.1 Contact with a rigid obstacle

We present here an example used as a validation test in a workshop of the GRECO "Grandes Déformations et Endommagement". Computations on this example have been done by different laboratories in this group. It concerns the compression of a long bar on a plane (plane strain problem). The geometry is given on figure 2 :

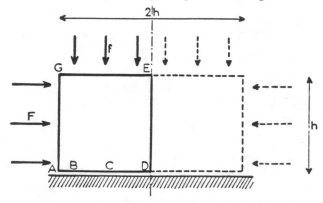

Figure 2 : Compression of a bar on a plane

The values of the parameters are : h = 40 mm, E = 13000 daN/mm^2, ν = 0.2, F = 10 daN/mm, f = −5 dan/mm, μ = 1.

Table 1. Displacements and forces on the contact aera.

NODE	u^n	u^t	F^n	F^t
1	-0.571E-03	0.147E-01	0.333E-03	-0.214E-03
2	-0.271E-03	0.137E-01	0.195E-03	-0.595E-03
3	-0.487E-04	0.128E-01	0.138E-03	-0.504E-03
4	0.000E+00	0.118E-01	-1.16	-1.16
5	0.000E+00	0.107E-01	-2.16	-2.16
6	0.000E+00	0.96BE-02	-2.84	-2.83
7	0.000E+00	0.865E-02	-3.40	-3.40
8	0.000E+00	0.765E-02	-3.87	-3.87
9	0.000E+00	0.669E-02	-4.29	-4.29
10	0.000E+00	0.576E-02	-4.66	-4.66
11	0.000E+00	0.487E-02	-5.01	-5.01
12	0.000E+00	0.403E-02	-5.33	-5.33
13	0.000E+00	0.324E-02	-5.64	-5.64
14	0.000E+00	0.250E-02	-5.95	-5.95
15	0.000E+00	0.182E-02	-6.27	-6.27
16	0.000E+00	0.120E-02	-6.61	-6.61
17	0.000E+00	0.667E-03	-6.99	-6.99
18	0.000E+00	0.244E-03	-7.50	-7.50
19	0.000E+00	-0.527E-25	-8.16	-7.61
20	0.000E+00	0.345E-26	-8.59	-5.07
21	0.000E+00	0.113E-26	-8.68	-4.33
22	0.000E+00	-0.73BE-28	-8.75	-3.73
23	0.000E+00	-0.122E-27	-8.80	-3.23
24	0.000E+00	0.181E-28	-8.87	-2.79
25	0.000E+00	-0.701E-28	-8.90	-2.41
26	0.000E+00	-0.529E-28	-8.94	-2.06
27	0.000E+00	0.653E-28	-8.96	-1.73
28	0.000E+00	0.262E-28	-8.99	-1.41
29	0.000E+00	-0.156E-27	-9.01	-1.11
30	0.000E+00	0.596E-28	-9.04	-0.827
31	0.000E+00	0.27BE-27	-9.04	-0.542
32	0.000E+00	0.15BE-27	-9.07	-0.260

6

We have a unilateral contact with friction on the part AD. The displacements and the forces on the contact zone are given in the table 1.
We observe that the relations (5) and (6) are verified with a very good precision on the three different zones of the contact boundary:
- a non contact zone (nodes 1 to 3): $u^n < 0$ and $F^n = F^t = 0$,
- a sliding zone (nodes 4 to 18): $u^n = 0$, $F^n < 0$ (contact), $u^t \neq 0$ and $|F^t| = |F^n|$ (the friction coefficient $\mu = 1.$),
- a sticked zone (nodes 19 to 33): $u^n = 0$, $F^n < 0$ (contact), $u^t = 0$ and $|F^t| < |F^n|$.

4.2 Study of a bolted junction

The structure is the assembling of the three parts of bolted junction by the use of a pressing ring. Results given on figure 3 concern the application of the closing pressure on the ring before introducing the pressure. The figure 3 gives the deformation of the structure (with an amplification coefficient equal to 100.) and the level curves of the component σ_{zz} of the stress tensor.

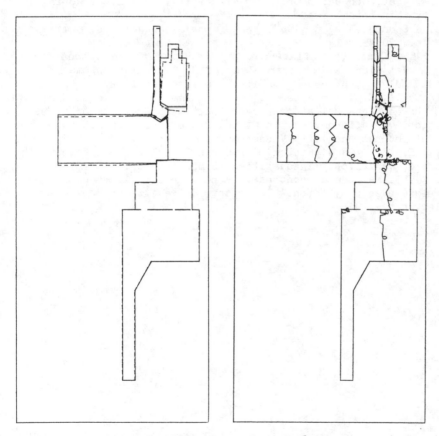

Figure 3 : deformation and component σ_{zz} of the stress tensor

A complete study with different loading cases and different friction coefficients has been done in the context of a contract with Technicatome.

REFERENCES

Bathe, K.J. & Chaudhary, A. 1985. A solution method for planar and axisymetric contact problems. Int. J. Num. Meth. Engn. 21:65-88.

Campos, L.T. & Oden, J.T. & Kikuchi, N. 1982. A numerical analysis of a class of contact problems with friction in elastostatics. Computer Meth. Appl. Mech. Engn. 34:821-847.

Cocu, M. 1984. Existence of solutions of Signorini problems with friction. Int. J. Engn. Sci. 22: 567-575.

Curnier, A. 1984. A theory of friction. Int. j. Solids Struct. 20:637-659.

Duvaut, G. 1980. Equilibre d'un solide élastique avec contact unilatéral et frottement de Coulomb. CRAS- Série A. 290:263-265.

Duvaut, G. & Lions, J.L. 1972. Les inequations en mecanique et en physique. Paris: Dunod.

Klarbling, A. 1984. Contact problem in elasticity. Thesis. Linkoping University. Linkoping, Sweden.

Lions, J.L. & Stampacchia, G. 1967. Variational inequalities. Com. Pure Appl. Math.. 20:493-519.

Raous, M. 1985. Contacts unilatéraux avec frottement en viscoélasticité. In Del Piero & F. Maceri (eds.). Unilateral problems in structural analysis, p. 269-297. Vienne:Springer-Verlag.

Raous, M. & Latil, J.C. (to appear). Code d'éléments finis pour des problémes de contacts unilatéraux avec frottement formulés en termes d'inéquations variationnelles. In G. Del Piero (ed.). Unilateral problems in structural analysis. Vienne:Springer-Verlag.

Stampacchia, G. 1972. On a problem of numerical analysis connected with the theory of variational inequalities. Symposia Math.. 10:281-293.

Taallah, F. 1985. Méthode de condensation pour un algorithme de Gauss-Seidel relaxé avec projection. DEA report, University of Provence, Marseille, France.

The general solutions of toroidal shells and curved tubes

Z.Xia

Shanghai Ship & Shipping Research Institute, People's Republic of China

W.Zhang

Department of Engineering Mechanics, Tsinghua University, Beijing, People's Republic of China

1 INTRODUCTION

Because of the wide application of thin-walled toroidal shells and curved tubes in industry the stress and deformation analysis for them attracts the attention of many engineers and researchers. For example, there are more than 10,000 bends used in one nuclear reactor unit and more than million bellows per year are produced in some industrial countries. In the present paper the term toroidal shell is used for those which are closed in ϕ-direction but unclosed in x-direction while the curved tubes are closed in x-direction and unclosed in ϕ-direction. (Fig.1) For the toroidal shells Zhang Wei (1948), Clark (1950) gave asymptotic solutions of symmetric loadings. Chernykh (1972) obtained an asymptotic solution of wind-type loadings. For the curved tubes a detailed survey was given in AMR by Axelrad and Emmerling (1984). But the solutions of nonsymmetric loadings and with complex boundary conditions are relatively less studied.

In the present paper the general solution of toroidal shells and curved tubes has been obtained based on Novozhilov's thin shell equations.(1962) They can be used in stress and deformation analysis of those structures with various boundary conditions and subjected to symmetric or nonsymmetric loadings.

2 BASIC EQUATIONS

The governing equations for shells of revolution in the form of function of complex variables are

$$G(\bar{U})-[1-ic(\frac{1}{R_1}-\frac{1}{R_2})\frac{1}{\cos^2 x}]\frac{\partial^2 \bar{T}}{\partial \phi^2}=\bar{f}(x,\phi)$$

$$-icG(\bar{T})+\bar{T}+(\frac{1}{R_1}-\frac{1}{R_2})\frac{1}{\cos^2 x}\bar{U}=R_2\bar{q}_n' \tag{1}$$

For the toroidal shells or curved tubes we have

$$R_1=r_0 \quad R_2=R_0\eta/\cos x \quad \eta=1+a\cos x \quad a=r_0/R_0 \tag{2}$$

Eliminating $\bar{U}$ from eqns.(1) a partial differential eqn. of fourth

9

order for function $\bar{T}$ is obtained.

$$\eta^4\frac{\partial^4\bar{T}}{\partial x^4}-8a\eta^3\sin x\frac{\partial^3\bar{T}}{\partial x^3}+[i\frac{ar_0}{c}\eta^3\cos x+$$

$$\eta^2(1+14a^2-4a\cos x-19a^2\cos^2 x)]\frac{\partial^2\bar{T}}{\partial x^2}-[i\frac{ar_0}{c}\eta^2\sin x(3+7a\cos x)$$

$$+a\eta\sin x(2+4a^2-6a\cos x-12a^2\cos^2 x)]\frac{\partial\bar{T}}{\partial x}+a^4\frac{\partial^4\bar{T}}{\partial\phi^4}+2a^2\eta^2\frac{\partial^4\bar{T}}{\partial x^2\partial\phi^2}$$

$$-4a^3\eta\sin x\frac{\partial^3\bar{T}}{\partial x\partial\phi^2}+[i\frac{a^2r_0}{c}\eta^2+a^3(a-2\cos x-2a\cos^2 x)]\frac{\partial^2\bar{T}}{\partial\phi^2}+$$

$$i\frac{ar_0}{c}\eta[7a+(9a^2-2)\cos x-12a\cos^2 x-12a^2\cos^3 x]\bar{T}=\bar{F}(x,\phi) \qquad (3)$$

All internal force components can be expressed by complex function $\bar{T}$

$$T_1=-\frac{c}{r_0}\eta\,Im\frac{\partial^2\bar{T}}{\partial x^2}+\frac{ac}{r_0}\sin x\,Im\frac{\partial\bar{T}}{\partial x}-a\cos x Re\bar{T}-\frac{a^2c}{r_0\eta}Im\frac{\partial^2\bar{T}}{\partial\phi^2}+q_n r_0\eta \qquad (4)$$

...

Then the displacement components can be derived by integrating the elastic relations.

$$\epsilon_1=\frac{1}{r_0}(\frac{\partial u}{\partial x}+w)=\frac{1}{Eh}(T_1-\nu T_2) \qquad (5)$$

...

3 THE GENERAL SOLUTION OF TOROIDAL SHELLS

Note that the toroidal shells are closed in ϕ-direction with period of 2π and all the coefficients of eqn.(3) are merely functions of x, so it is appropriate to assume

$$\bar{T}(x,\phi)=\sum_{m=0}^{\infty}[\sum_{k=0}^{\infty}(A_k+iB_k)x^k]e^{im\phi} \qquad (6)$$

All hamonics of $\bar{T}$ are uncoupled and the recursion formula for the constants A_k, B_k can be obtained after substituting (6) into (3). For every harmonic of $\bar{T}$ there are eight arbitrary constants A_1, A_2, A_3, A_4, B_1, B_2, B_3, B_4 which will be determined from the boundary conditions.
    Three cases may be distinguished:

1.1 m=0, symmetric loadings. There are only 5 independent boundary forces at both boundaries of a toroidal shell. To decide the 8 arbitaray constants in the solution 3 additional conditions are required. When integrating the elastic relations an arbitrary constant representing rigid-body motion will be obtained.

10

1.2 m=1, wind-type loadings. There are only 6 independent forces at both boundaries and 2 additional conditions are required. Two arbitrary constants representing rigid-body motion will be appeared when integrating the elastic relations.

1.3 m≥2. The 8 boundary forces at both ends are all in self-equilibrium. No additional condition is required and there is no rigid-body motion in the case.

4 THE GENERAL SOLUTION OF CURVED TUBES

The curved tubes are closed in x-direction with period of 2π, so let

$$\bar{T}(x,\phi)=\sum_{m=0}^{\infty} \bar{T}_m(\phi)\cos mx+\sum_{m=1}^{\infty} \bar{T}_m(\phi)\sin mx \qquad (7)$$

The two parts of the Fourier series in (7) are called symmetric (or in-plane) part and antisymmetric (or out-of-plane) part, respectively. They are uncoupled after substituting (7) into (3) and an ordinary differential matrix equation of biquadratic order is obtained

$$a^4(I)\{\bar{T}^{(4)}(\phi)\}+(\bar{B})\{\bar{T}^{(2)}(\phi)\}+(\bar{C})\{\bar{T}(\phi)\}=\{\bar{F}(\phi)\} \qquad (8)$$

where

$$\{\bar{T}(\phi)\}=[\bar{T}_0(\phi),\bar{T}_1(\phi),\ldots]^T \quad \{\bar{F}(\phi)\}=[\bar{F}_0(\phi),\bar{F}_1(\phi),\ldots]^T \qquad (9)$$

Let

$$\{\bar{T}(\phi)\}=\{\bar{A}\}e^{\bar{k}\phi} \qquad (10)$$

The $\bar{K}$ and $\{\bar{A}\}$ are eigenvalues and corresponding eigenvectors of eqn.(11)

$$[\bar{K}^4a^4(I)+\bar{K}^2(\bar{B})+(\bar{C})]\{\bar{A}\}=\{0\} \qquad (11)$$

If the first L iterms are taken in the Fourier expansion, 2L eigenvalues and corresponding eigenvectors will then be solved. There is a zero eigenvalue and a negative unit eigenvalue but all other eigenvalues are complex and the solution can be written as:

$$\{\bar{T}(\phi)\}=\sum_{j=1}^{2L} [\bar{G}_j e^{a_j(\phi-\phi_2)}(\cos\beta_j\phi+i\sin\beta_j\phi)+$$

$$\bar{H}_j e^{-a_j(\phi-\phi_1)}(\cos\beta_j\phi-i\sin\beta_j\phi)]\{\bar{A}\}_j \qquad (12)$$

where ϕ_1 and ϕ_2 are end values of ϕ, G_j and H_j are complex constants which will be determined from the boundary conditions while a_j and β_j are real and imaginal part of the eigenvalue $\bar{K}_j$.

5 SOME NUMERICAL AND EXPERIMENTAL RESULTS

5.1 A model as shown in Fig.2 was skillfully made by special process. The shell is made of epoxy resin and the cylinders stuck to the shell at the boundaries of x=π are made of steel. Strains of the shell were measured with the resistance gauges and the displacements were measured by laser interferometry. Comparision

11

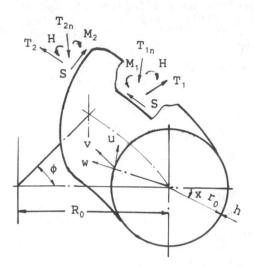

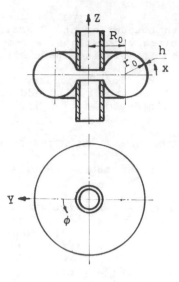

Fig.1 Geometry & internal forces

Fig.2 Toroidal shell
subjected to
bending moment

between the test and the calculated results is illustrated in Fig.3
and Fig.4, where scattered symbols are test values. It is clear
that the two results are in good agreement.

5.2 A closed toroidal shell subjected to local linear distributed
load is shown in Fig.5. The $q(\phi)$ should be expanded into Fourier
series and then the solution obtained by adding all harmonic
solutions together. It is found that the error would be less than
5% by taking 18 terms in the Fourier expansion.

5.3 A bend subjected to an in-plane bending moment with warping
ringframes at both ends is shown in Fig.6. The same problem has
been calculated by Ory and Wilczek (1983) based on the semimembrane
shell theory. The distribution of longitudinal stress versus x at
$\phi=L/2R$ is shown in Fig.6. The semimembrane theory disregards the
longitudinal internal moments M_2 and H, so it presents only one
curve of σ_2. But our result has revealed that there are
considerable longitudinal moments and the stress curves of the
external and inner surfaces of the bend differ from eath other.

6 CONCLUSION

The general solution of toroidal shells and curved tubes is
presented. A series of calculated results and experimental results
have justified their reliability. It is expected that the method
presented in the paper will provide reliable solutions for
versatile boundary-value problems with more complex loading
conditions.

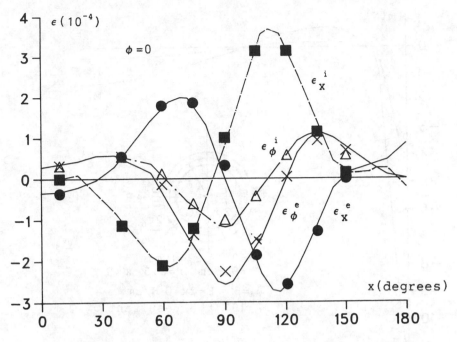

Fig.3 Strains vs x at external & inner surfaces

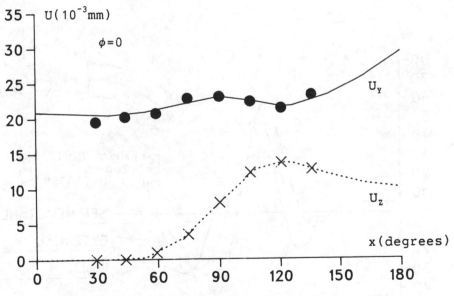

Fig.4 Displacements vs x at external surface

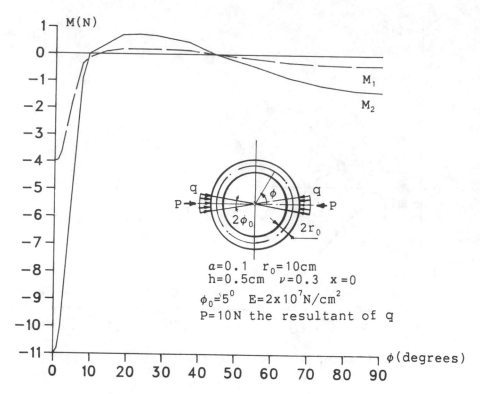

Fig.5 Closed toroidal shell subjected to
local linear distributed load

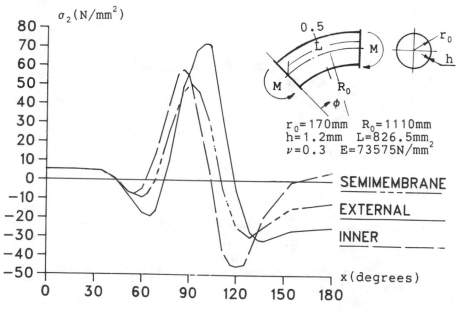

Fig.6 Bend with warping ringframes at both ends

REFERENCES

Zhang Wei 1948. Der Spannungszustand in Kreisringschale und
   anlichen Schalen mit Scheitelkreisringen unter dresymmetrischer
   Belastung. Science Report of Tsinghua University. 5A:289-349.
Clark,R.A. 1950. On the Theory of Thin-walled Toroidal Shells.
   J. Math. & Phys. 29:146-178.
Chernykh,K.F. 1972. Lineinaia Teoriia Obolochek. Leningrad.
   Izd. Lening. Univ.
Axelrad,E.L. & Emmerling,F.A. 1984. Elastic Tubes.
   Applied Mechanics Reviews. 37:891-897
Hovozhilov,V.V. 1962. Teoriia Tonkikh Obolochek. Lening.
   Izd. L. Cudpromgiz.
Ory,H. and Wilczek,E. 1983. Stress and Stiffness Calculation
   of Thin-walled curved pipes with realistic boundary conditions
   being loaded in the plane of curvature.
   Int. J. Pres. & Piping. 12(3): 167-189.

Numerical simulation of stable growing cracks controlled by incremental path-independent integrals

W.Brocks

Bundesanstalt für Materialforschung und -prüfung (BAM), Berlin, FR Germany

H.-H.Erbe & H.Yuan

Technische Universität Berlin, FR Germany

1 INTRODUCTION

In a contribution to ECF6, Erbe (1986) discussed the advantages of path-area integrals proposed by S. N. Atluri and coworkers, see Brust et al. (1985).

With respect to this theoretical basis we present results of FE calculations on growing mode-I cracks. The numerically simulated crack growth is controlled by experimental data of load line displacement, V_{LL}, vs. crack growth, Δa. The J-resistance curves obtained from the experimental load vs. load line displacement curves are compared with the numerical values of the contour integral calculated along various paths around the crack tip.

Two CT 25-specimens have been investigated with different initial and final crack lengths and different thicknesses.

The results are part of a broader study with specimens of different geometries and stable growing cracks until instability.

2 THEORETICAL CONSIDERATIONS

The formally calculated increment of J_1

$$\Delta J_1 = \int_{\Gamma_\varepsilon} [\Delta W n_1 - (t_i + \Delta t_i)\Delta u_{i,1} - \Delta t_i u_{i,1}]ds \qquad (1)$$

is equivalent to $(\Delta T_p^*)_1$, see Brust et al. (1985).

With respect to incremental plasticity, Γ_ε being the near field path (nf) and Γ some far field path (ff), we can state that

$$(\Delta T_p^*)_{1_{nf}} = (\Delta T_p^*)_{1_{ff}} + \text{some area-term} \qquad (2)$$

The area-term vanishes in deformation theory of plasticity.

The summation of these incremental terms over the load history yields J_1 and $(T_p^*)_1$, respectively. If $W = \Sigma \Delta W$ is the total accumulated stress work density then equation (1) can be used in incremental plasticity. Otherwise, i. e. by requiring W to be the strain energy density, equation (1) can be used within the limits of the deformation theory of plasticity (Moran & Shih 1986).

Choosing Γ_ε, the near field path, to evaluate $\Delta J_1 = (\Delta T_p^*)_1$ for growing cracks care must be taken because Γ_ε must be fixed with respect to the same material points. We come back to this point later.

On the other hand the Δa-step value of the FE analysis (Figure 1) is not arbitrary as has been discussed by several investigators following theoretical studies by Rice in 1966, see Nakagaki, Chen and Atluri (1979).

With these considerations in mind we try to give hints in our short study supporting the presumption that the near field values of J_1 or $(T_p^*)_1$ become constant on a level independent of the FE-mesh and the specimen geometry.

3 NUMERICAL INVESTIGATIONS

Our numerical calculations are based on experimental data got from compact tension specimens (Table 1). Incremental plasticity is used with v. Mises yield condition and small deformations are assumed (material nonlinearity only). The calculation is controlled by the experimental loadline displacement vs. Δa-curve and can be verified by comparison of the numerically with the experimentally obtained J vs. Δa-curve.

We used six different paths for the evaluation of the contour integral (Figure 2). Two of them are near field paths. Path 1 runs around the crack tip moving with the growing crack and, hence, passes different material points for each Δa. Path 2 surrounds the crack tip for the whole process of crack extension and is fixed to the same material points. Path 6 lies in the far field whereas the paths 3, 4 and 5 lie somewhere in between. The results of the numerical values together with the experimental J-resistance curves are shown in Figure 4a, b for both specimens.

The calculated J-values of path 1 tend to zero after initiation of crack growth supporting the presumption that J_{nf} levels against a constant value with further stable growing of the crack (results of path 2). The plane stress results of specimen #1 (Figure 4a) can be interpreted in this way showing the same qualitative tendency of the plane stress calculations of Brust et al. (1985). The results of our plane strain calculations for specimen #2 (Figure 4b) are somewhat different. The J-values of paths 2, 3 and 4 do not remain constant but start to decrease after some small amount of crack growth.

On the other hand, the stress field in the ligament is not altered significantly within the course of stable crack growth after initiation but just shifted by Δa (Figures 5a, b), in a similar way as the crack tip opening displacement is (Shih, deLorenzi and Andrews 1979 and Figure 3).

Clearly, these few calculations are not sufficient to confirm or disprove the presumption that the near-field values of J_1 and $(T_p^*)_1$, respectively, tend to a constant value with the growing crack. We cannot yet exclude the influence of the specimen geometry and FE-mesh size on the results. This is the aim of further investigations.

ACKNOWLEDGEMENT

The authors thank Dr. D. Hellmann of GKSS and Dr. K. Wobst of BAM for the experimental data.

18

REFERENCES

Aurich, D. et al. 1986. Analyse und Weiterentwicklung bruchmechanischer Versagenskonzepte auf der Grundlage von Forschungsergebnissen auf dem Gebiet der Komponentensicherheit, Teilvorhaben: Werkstoffmechanik. Förderkennzeichen: 150 490 6, Berlin: Bundesanstalt für Materialforschung und -prüfung.

Brust, F. W. et al. 1985. Further studies on elastic-plastic stable fracture utilizing the T* integral. Engineering Fracture Mechanics 22: 1079 - 1103.

Erbe, H.-H. 1986. On incremental path-independent integrals charakterizing stable crack growth. Presented at 6th European Conference on Fracture ECF6, Amsterdam.

Moran, B. & C. F. Shih 1986. Crack tip and associated domain integrals from momentum and energy balance. Report ONR 0365/2 Office of Naval Research.

Nakagaki, M., W. H. Chen & S. N. Atluri 1979. A finite element analysis of stable crack growth - 1. ASTM STP 668: 195 - 213.

Schwalbe, K.-H. & D. Hellmann 1984. Correlation of stable crack growth with the J-integral and the crack tip opening displacement, effects of geometry, size, and material. Report GKSS 84/E/37, Geesthacht (FRG): GKSS-Forschungszentrum.

Shih, C. F., H. G. deLorenzi & W. R. Andrews 1979. Studies on crack initiation and stable crack growth. ASTM STP 668: 65 - 120.

Table 1: Geometries and material data of the investigated compact specimens CT 25

No.	experimental results	W mm	B mm	a_0 mm	Δa_{max} mm
1	GKSS	50	5	25.3	6.7
2	BAM	50	25,sg 20%	29.4	2.8

	material	σ_{yield} MPa	J_i N/mm	FEM calculation
1	35 NiCrMo 16	431	22	plane stress
2	20 MnMoNi 55	459	156	plane strain

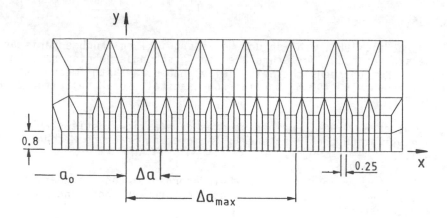

Figure 1. Detail of the FE mesh in the vicinity of the growing crack

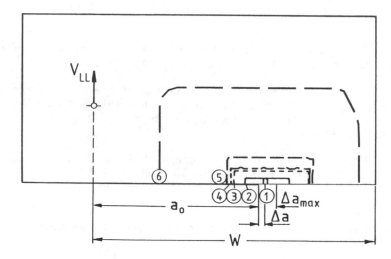

Figure 2. Upper half of the compact specimen with integration paths for the evaluation of J-integral

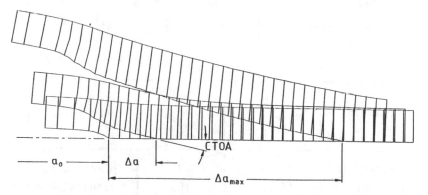

Figure 3. Deformed FE mesh in the vicinity of the growing crack at initiation, some Δa and Δa_{max}, respectively.

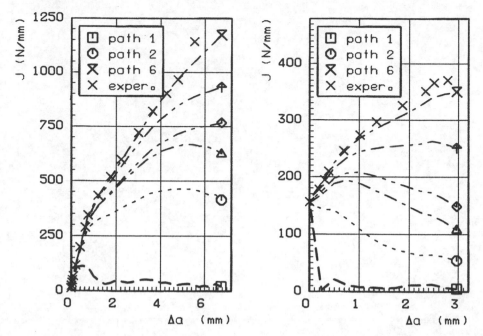

Figure 4. J-integral vs. crack growth, numerical and experimental results for

a) specimen #1   b) specimen #2

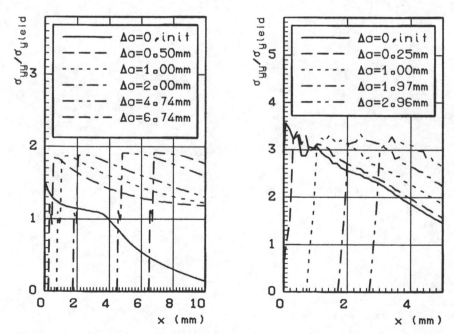

Figure 5. Normal stresses σ_{yy} in the ligament ahead of the growing crack for

a) specimen #1   b) specimen #2

21

D Experience with structures and components in operating reactors

Evaluation of pipe failure incidents

R.H.Vollmer

TENERA Master Limited Partnership, Bethesda, Md., USA, formerly USNRC

L.C.Shao & A.Taboada

USNRC, Washington, D.C., USA

ABSTRACT

Catastrophic failures of large pipes have occurred in reheat
systems of fossil plants and in a main feedwater line of a
nuclear plant. This paper describes the incidents and
compares the details of design and operation with those of
safety related piping in nuclear plants. Probable failure
mechanisms are discussed as is the relevance and implication
of such failures to the safety of nuclear plants.

1. Introduction

During 1984 and 1985 the NRC Piping Review Committee performed
a comprehensive review of requirements for safety related,
nuclear power plant piping. As part of this review the
Committee completed in-depth evaluations in the areas of:
stress corrosion cracking in piping of boiling water reactors,
seismic design, potential for pipe breaks, and other dynamic
loads and load combinations.[1] These evaluations resulted
in recommendations for appropriate changes to NRC requirement
for nuclear power plant piping and for research needed to
resolve outstanding piping issues. In part, this review and
related NRC research provide a basis to support the leak-before-
break concept for certain pipe systems which, when applied
would enhance maintainability and permit the removal of
massive pipe whip restraints and jet impingement barriers.

Subsequent to this review, catastrophic failures of large welded pipes occurred in reheat systems at two major fossil power plants and in the main feedwater line of a nuclear plant. These pipe failures resulting in damage to the plants as well as injuries and, at two plants, fatalities to personnel. Because of the concerns with pipe breaks, members of the Committee visited these plants to review the incidents, observe the pipe failures and determine any information that might relate to Committee recommendations or be applicable to safe operation of nuclear plants.

This paper describes the incidents and compares details of pipe design and operation with the design, and operating conditions of safety-related pipe in light water reactor nuclear plants. Probable failure mechanisms are discussed, as is the relevance of these failures to the safety of nuclear reactor piping.

2. Pipe Failure at The Mohave Generating Station

On June 9, 1985, a 20-foot section of a 30-inch diameter high pressure steam pipe of the Mohave Generating Station, ruptured without warning, releasing high pressure steam that resulted in the death of six plant personnel and injuries to 10 others. This plant consists of two coal-fired, steam turbine-generator units, each rated at 790 MWe. It was designed in the late 1960's to deliver superheated steam to high pressure turbines at 1000 °F and 3500 psig, and began commercial operation in 1971.

Failure occurred in a "hot reheat" line that operates at 1000 °F with loads of up to 580 psig. The piping system was designed to the ANSI/ASME B31.1 power piping code. The failed pipe was 30-inch in diameter and 1.5-inch thick and was fabricated by seam welding alloy steel plate containing 1¼% chromium and ½% molybdenum.

26

Pipe rupture occurred in a straight horizontal spool piece along the seam weld which unfortunately was oriented to release escaping steam towards the personnel work area. The fracture was described as a "fish mouth" rupture approximately 20-feet long and approximately 6 feet at its widest point.

A metallurical analysis of the failed pipe reported that the crack initiated in a midwall location, at the seam weld-base metal interface, and apparently grew in size over a period of time along this interface due to the occurrence of creep rupture. During this crack growth period the inner pipe surface was penetrated at several locations exposing the crack surface to steam environment. Final rupture occurred when the crack grew so large that the degraded pipe could no longer contain the pressure loads.

3. Pipe Failure at the Monroe Power Plant

On January 30, 1986 a pipe rupture very similar to the one at the Mohave Power Plant, occurred at the Monroe Power Plant located at Monroe, Michigan injuring several of the plant personnel. This facility is a coal fired plant with four turbine-generator units, each rated at 750 MWe, and has steam conditions at $1000°F$ and 3600 psig. The plant began commercial operations in June 1971 and had operated for 97,000 hrs. at the time of failure.

The pipe failure at the Monroe Power Plant occurred also in a "hot reheat" line which was designed to operate at $1000°F$ and 750 psig applying the ANSI/ASME B31.1 power piping code. The failed pipe was a 32-in in diameter and 1.5-inch thick seam welded pipe made of alloy steel plate, containing $2\frac{1}{4}\%$ chromium and 1% molybdenum.

As in the Mohave failure the pipe rupture occurred in a straight, horizontal spool piece along the seam weld which although causing significant damage to the plant equipment

27

did not result in serious injuries. The fracture was a
"fish mouth" rupture approximately 20 feet long and 6 ft. at
its wides point.

A metallurgical analysis of the failed pipe reported numerous
cracks initiating on the outside diameter as well as at
mid-wall that linked up along the toe of the weld at locations
of weld repairs. It appeared, that at the time of failure,
a crack pre-existed that was over 10 feet long and as much
as 80% through the wall.

4. General Observations

The pipe failures at Monroe and Mohave had several similarities.
Both failures occurred explosively in large diameter horizontal
reheat lines after long term operation at 1000°F and relatively
similar pressures (750 psig and 580 psig). Both failures
occurred primarily along a seam weld and opened up fish
mouth fractures that were approximately 20 feet long and 6
feet wide. Both failed pipes were of the same vintage, made
to similar material specification and designed to ASME/ANSI
B31.1. However, the Mohave pipe was made of 1¼ Cr-½ Mo
material and the Monroe pipe was made of improved 2¼ Cr-1Mo
material. Although, as yet, no formal failure report on
either incident has been released by the utilities, it is
generally understood that the failures were brought on by
creep rupture in the fusion zone of weldment at locations
containing manufacturing flaws. As a result of these
findings the power industry has instituted periodic inspections
to mitigate this problem.

5. Comparisons with Safety Related Piping in Nuclear Reactors

The failed Mohave and Monroe piping and their operating
conditions are not typical of the safety related piping in
light water reactor nuclear systems reviewed by the NRC

Piping Review Committee. In particular, the 1000°F operating
temperature is significantly higher and makes the alloy
steel material susceptable to creep rupture and creep
fatigue problems as well as to further metallurgical changes
not expected at the lower operating temperatures of light
water reactors. The Chromium, molybdenum materials used in
the reheat lines are not used in safety-related systems of
light water reactors.

In safety-related systems, light water reactors use austenitic
stainless steel and carbon steel for piping materials which
tend to be tougher and more weldable than to alloy steels.
In addition to material and operating temperature differences,
there are other factors which mitigate, against such a
failure in nuclear piping. These include:

o   Improved design and analysis
o   Improved welder qualification
o   Better control of pre- and post-weld heat treatments

o   A higher quality nondestructive examination during
      construction
o   Inservice inspection--now being instituted in fossil
      plants
o   Leak detection requirements
o   Generally more and improved quality assurance

6. Pipe Failure at Surry Power Station

On December 9, 1986, an 18-inch pipe of the nonnuclear part
of the Surry Power Station, Surry County, Virginia, failed
catastrophically releasing high pressure steam and resulted
in burn injuries to eight plant personnel that proved fatal
in four cases. The Surry Power Station is a two-unit plant,
each of which is a three-loop pressurized water reactor
rated at 775 MWe. The plant began commercial operation in

1972 and has operated for approximately 76,000 hours.

Failure occurred as a complete separation of a carbon steel suction line to a main feedwater pump in an elbow at a transition to a 24-inch condensate supply header pipe. In addition to the double ended break, pipe whip reactions twisted and rotated the pipe causing distortion and buckling away from the break and displacing the broken pipe end about 10 feet. As a result, a large fragment was ripped off of the broken pipe end and thrown a distance of approximately 20 feet.

The primary cause of the pipe failure was reported to be gross thinning of the pipe wall over a large area caused by erosion/corrosion. The pipe thickness had decreased from 0.5-inch to as thin as 0.04-inch, making the pipe unable to withstand minor pressure surges.

The rapid wastage of carbon steel pipe by an erosion/corrosion mechanism has been well documented by several European investigators.[2][3][4] Based on the data reported, it is apparent that a combination of the turbulance inducing pipe design, and fluid conditions were ideal for erosion/corrosion to take place in the carbon steel pipe at Surry.

A comprehensive inspection of the remainder of the piping in the Surry plant found significant wall thinning in other pipes at both Surry Units indicating the potential for a generic problem. As a result the NRC and the nuclear industry have initiated programs to evaluate (1) the extent of the problem, (2) the safety significance and risk aspects and (3) the existing state of knowledge on the erosion/corrosion mechanisms and variables that influence this mechanism. The results of these programs should result in guidelines and possible changes to code rules and safety requirements to avoid such an occurrence in the future.

7. <u>References</u>

(1) NUREG-1061, Volumes 1-5, Report of the U.S. Nuclear Regulatory Commission Piping Review Committee, April 1985.

(2) G.J. Bignold et al., Erosion-corrosion in nuclear steam generators, in Proceedings of the Second Meeting on Water Chemistry of Nuclear Reactors, British Nuclear Engineering Society, 1980.

(3) H.G. Heitmann and P. Schub, Initial experience gained with a high pH value in the secondary system of PWRs, in Proceedings of the Second Meeting on Water Chemistry of Nuclear Reactors, British Nuclear Engineering Society, 1980.

(4) Berge, Ph., Ducreaux, J., Saint-Paul, P., "Effects of Chemistry on Erosion-Corrosion of Steels in Water and Wet Steam," pg. 19, <u>Water Chemistry of Nuclear Reactor Systems 2</u>, British Nuclear Energy Society, London, 1980.

Pipe dynamic loading caused by water slug formation due to steam condensation

C.A.Hengge, C.E.Rupp & D.V.Wilczynski
Nuclear Engineering Division, Toledo Edison, Ohio, USA

ABSTRACT

The Davis-Besse Nuclear Power Station, owned by the Toledo Edison Company, was designed with two steam turbine driven auxiliary (emergency) feedwater pumps (AFP). On June 9, 1985, following a loss of main feedwater, both AFP turbines tripped on overspeed (4500 RPM) approximately 25 seconds after initiation. The root cause of the overspeed trips was determined to be the injection of water into the AFP turbines formed by the condensation of steam in the turbine inlet piping. Prior to June 9, 1985, Toledo Edison discovered damage to pipe supports on the steam inlet lines. This damage was also attributed to water slugs created by steam condensation in the lines. A comprehensive test program had been developed to determine the transient loads generated by these water slugs and to identify potential corrective actions. Modifications implemented following the June 9, 1985 event resolved the water slug formation problem and eliminated the need to perform this test program.

INTRODUCTION

Prior to June 9, 1985, auxiliary feedwater system actuation was accomplished by opening a normally closed motor operated steam isolation valve which admitted steam at approximately 535°F into pipe at ambient (70°F) temperature. The system is designed such that each auxiliary feedpump turbine (AFPT) can be supplied with steam from either of two steam generators (see Figure 1). With both steam generators intact (i.e. normal pressure), the original actuation logic would open only the isolation valve for the associated steam generator. The isolation valves located in the crossover lines (from the alternate steam generator) remain closed. The relative lengths of pipe initially at ambient temperature exposed to steam for this actuation configuration were 369 feet for AFPT #1 and 124 feet for AFPT #2.

To improve the reliability of the auxiliary feedwater system, an actuation logic modification was implemented in January 1985 to open both steam isolation valves for each AFPT. This would require the failure of two valves to result in the failure of an AFPT to receive steam. This modification increased the length of ambient temperature pipe exposed to steam to 686 feet for AFPT #1 and 415 feet for AFPT #2.

Plant inspections performed subsequent to the implementation of this actuation logic modification discovered damage to numerous piping hangers on these steam lines. This damage was determined to be caused by water slug forces created by steam condensation in the cold piping during the initiation of steam flow into the AFPT piping system.

The AFPT steam crossover lines are comprised of numerous short horizontal runs with 90 degree elbows to allow for thermal expansion. Initial calculations conservatively determined the mass of water condensed as a result of heating the AFPT crossover lines from 70°F to 535°F to be approximately 700 pounds of water for each line. This was in addition to the mass of water condensed during the heating of the common steam line between the crossover line check valve and the AFPT. A review was performed of the design and installation of steam traps on the AFPT steam inlet lines. This review determined that the installed steam traps were not adequate to remove the mass of water generated during the heating of the steam piping from 70°F to 535°F following system actuation.

The additional mass of water generated in the crossover line and the impact forces generated when this mass of water was injected into the common AFPT steam inlet line, which consists of long vertical and horizontal runs, resulted in pipe movement and dynamic loads sufficient to cause the pipe hanger damage discovered.

In addition to the resulting pipe hanger damage, the injection of this water into the AFPT was determined to be the primary cause of the overspeed trip of both AFPT's during the June 9, 1985 loss of main and auxiliary feedwater event which occurred at Davis-Besse. This determination was based upon the results of past testing of the turbines at the vendor and Toledo Edison experience of AFPT starts on cold steam supply lines versus hot steam supply lines.

The resolution to this problem was accomplished by the installation of a new normally closed AFPT steam inlet valve located approximately 20 feet from the AFPT. The existing steam isolation valves associated with the AFPT crossover lines (MS 106A and MS 107A) were aligned normally open. This alignment maintains the AFPT steam inlet piping in a hot pressurized condition which will prevent the formation of large amounts of water during system actuation.

METHODOLOGY AND FINDINGS

The investigation conducted prior to June 9, 1985 was initiated when a snubber was found damaged on the steam line to the Auxiliary Feedwater Pump Turbines (AFPT). The snubber was located at a point where the steam supply piping turned 90 degrees. A walkdown of the entire steam line revealed several other snubbers and supports that were damaged and studs that had pulled out of the walls.

It was well known that water was condensing in the steam lines upstream of the turbine. The turbine manufacturer had been contacted to verify that no problem would exist with the turbine due to the water slugs. The turbine manufacturer stated that no detrimental effects to the turbine would occur based on testing that was performed during the turbine design phase. It was then decided that the most cost effective way to solve the problem was to size the supports to be able to withstand the water slug formation.

34

To determine the required sizing of the supports, a testing program was initiated. The objectives of the testing were to:

(1) Verify the cause and magnitude of the high transient loads on the Auxiliary Feedwater Turbine steam supply piping.

(2) Identify potential long term corrective actions.

The testing was to be done in a progression such that the shortest run of pipe would be tested first (steam generator 2 to AFPT 2). The longest run of pipe would be run second (steam generator 1 to AFPT 2) and both runs of pipe consecutively would be run last. Due to the requirements for cold piping to initiate each phase of the test, a cooldown period of the piping would be necessary. This cooldown period would allow analysis of the data from the previous test to ensure that no further damage would result from the progressively more rigorous testing.

The test program consisted of instrumenting certain hanger locations on the steam line. This instrumentation would consist of strain gages on snubbers and sway struts, position detectors on the same snubbers and sway struts, pressure sensors, and temperature sensors.

The installation of strain gages on the snubbers and sway struts was done to determine the exact load induced on each support when a water slug traveled down the pipe. The position on the extension pieces where the strain gages were to be placed had to be narrowed. This narrowing would amplify the strain induced on the extension piece thus allowing better accuracy and more sensitive readings. Narrowing of the carbon steel supports would have made the supports inadequate relative to the design requirements. Thus, to instrument the snubbers would require a complete replacement of the existing support extension pieces. It was decided to use 17-4PH stainless steel as the material of which the new extension pieces were to be made to take advantage of its higher strength.

The position detectors used to detect pipe movement were string potentiometers. There would be a potentiometer to detect pipe movement in the direction which the water slug would induce a force and also a potentiometer to detect pipe growth due to thermal expansion if this was different than the water slug direction.

The pressure detectors were installed to be able to determine if there was significant water hammer involved with the pipe movement. The water hammer effects could cause a pressure surge in the frequency range of approximately 2000Hz. To be able to detect such a pressure pulse, high temperature strain gage type pressure transducers were needed. The high temperature requirements were necessary due to the proximity of the sensors to the steam environment. It was desired to have the sensors as close to the steam as possible to be able to detect an undamped signal from the piping.

The temperature sensor requirements were similar to the pressure sensors in respect to the frequency response requirements. The temperature sensors were to be a RTD type of sensor which would give the required response. The difficulty with the temperature sensors would be the range requirements. There were no temperature sensors that were able to

35

withstand the temperature requirements for a given period of time and still be able to give the required frequency response. The frequency was determined to be more significant than the range of the temperature sensor, thus the RTD was used. The RTDs would last for only one run of the turbines. Multiple RTDs were calibrated to be able to have the multiple runs required by the test program.

The next obstacle was the recording medium to be used. Strip chart recorders presently in the possession of the company did not have the frequency response that would be usable with the high frequency transducers. FM tape recorders were the solution. The tape recorders had a frequency response of up to 40 KHz if used at the highest tape speed. The inputs to the recorders were also flexible enough to accept all of the transducers planned to be used.

Also to be monitored to determine total system performance were position indications for the motor operated valves, turbine speed, pump flow rate, pump discharge pressure, main steam pressure at the discharge of the steam generators, and main steam temperature at the discharge of the steam generators.

All of the above information was available from computer points, thus the only instrument requirements were to record those points on the FM recorders and synchronize that information with the other parameters.

The final problem of the test program was the synchronization of the test signals. The FM recorders were to be set up in various parts of the plant in close proximity to the transducers from which they were to get signals. Two signal generators were to be used. One would generate a 1KHz signal and the other would generate a 1Hz signal. These signal generators would have inputs to the FM recorders and would be used as reference points on all tape recorders.

Prior to conducting this test program, the June 9, 1985 event occurred which indicated that the quantity of water generated in the crossover lines was apparently sufficient to result in overspeed trips of the AFPT's.

To support this theory, several calculations were performed to quantify the amounts of water that could be formed as the steam condensed due to the heat transfer from the steam to the pipes.

The assumptions used to develop the calculations were:

1. The pipe lengths are initially at room temperature (70°F).

2. The pipe would heat up to the nominal steam generator temperature (535°F).

3. All heat given up by the condensing steam would be absorbed by the pipe and used to raise its temperature to 535°F.

The results of the calculations are shown in Table 1 and indicate the potential exists for excessive amounts of condensate to form in the steam supply piping to the auxiliary feed pump turbines. This amount of condensate can form very large slugs of water when it is able to migrate together.

The potential for the condensate to form large slugs resides in the configuration of the steam supply piping and the configuration of the steam
traps. The crossover piping has a very arduous path from the steam generators to the turbines. This path consists of many elbows and vertical
rises. This configuration and the inadequacy of steam traps (both in
quantity and location) to collect and discharge the condensate makes it
possible for the condensate to collect as slugs rather than be carried
along entrained in the steam flow.

A transient flow analysis was also performed to simulate the effects of
the June 9, 1985 event when both AFPT's were supplied with steam from
only the alternate steam generator. The computer model used solves the
mass, momentum, and energy conservation for the piping system including
the effects of the heat transfer to the pipe wall. The results of the
transient flow analysis indicated that within approximately 15 seconds
after the opening of the steam admission valves, both AFPT's are being
supplied with significant amounts of water. In addition, the analysis
showed that the water would not have cleared the lines prior to the
observed overspeed trips (25-30 seconds after the valve opening). To
further confirm the theory that the AFPT overspeed trips were caused by
water injection, a review was conducted of industry and vendor experience
with turbine overspeed trips and water induction. Several instances of
turbine overspeed trips caused by water induction were identified and
reviewed. The turbine vendor had performed limited testing on the
effects of water induction to both a running turbine and to a turbine
during startup. During these tests, the turbine speed would oscillate
as the water slugs passed through the turbine. These oscillations were
consistent with Toledo Edison test data obtained for turbine starts
with cold piping. Test data obtained for turbine starts with hot steam
inlet piping showed relatively smooth acceleration to speed.

CONCLUSIONS

Significant quantities of water were generated due to condensation
following admission of high temperature steam into the ambient temperature AFPT steam inlet piping. Steam traps have limited effectiveness
during the initial heatup of the piping due to the amount of water
generated and the formation of water slugs in the piping. These water
slugs were responsible for the damaged pipe hangers and the AFPT overspeed trips which occurred on June 9, 1985. The proposed test program
would have identified the magnitude of the forces caused by these water
slugs and could have identified the potential for AFPT overspeed trips
to occur. The installation of the new normally closed AFPT steam inlet
valve located approximately 20 feet from the turbines has resolved the
water slug formation problem.

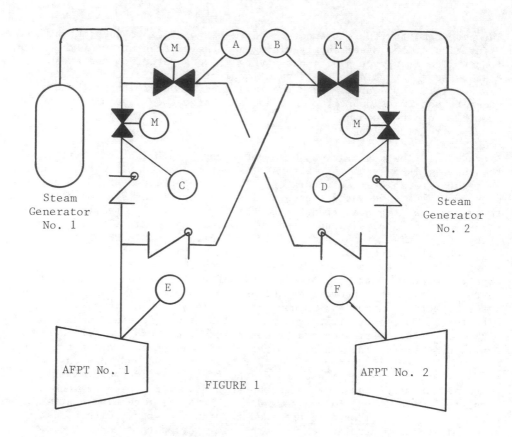

FIGURE 1

TABLE 1

Pipe Section (between nodes as shown in Figure 1)	Length of Pipe Run (feet)	Mass of Condensate (LBM)	Length of Slug in 6" pipe (feet)
Steam Generator 1 to AFPT 1 (nodes C to E)	368.8	785.6	70.2
Steam Generator 2 to AFPT 1 (nodes B to E)	686.2	1484.1	132.1
Steam Generator 2 to AFPT 2 (nodes D to F)	124.4	265.8	23.5
Steam Generator 1 to AFPT 2 (nodes A to F)	415.4	906.2	80.2

Probability of failure in BWR reactor coolant piping

G.S.Holman, T.Lo & C.K.Chou
Lawrence Livermore National Laboratory, Calif., USA

1 INTRODUCTION

The Lawrence Livermore National Laboratory (LLNL), through its Nuclear
Systems Safety Program, has performed probabilistic reliability
analyses of PWR and BWR reactor coolant piping for the NRC Office of
Nuclear Regulatory Research. Specifically, LLNL has estimated the
probability of a double-ended guillotine break (DEGB) in the reactor
coolant loop piping of PWR plants, and in the main steam, feedwater,
and recirculation piping of BWR plants. For these piping systems, the
results of these investigations have provided NRC with one technical
basis on which to address regulatory issues such as the need for pipe
whip restraints on reactor coolant piping or to make licensing
decisions concerning the replacement, upgrading, or redesign of piping
systems.

2 GENERAL APPROACH

Generic evaluations of reactor coolant loop piping were completed for
PWR nuclear steam supply systems manufactured by Westinghouse, Combustion Engineering, and Babcock & Wilcox (Holman 1984a, 1984b, 1985). In
these evaluations, LLNL estimated the probability of "direct" DEGB
(i.e., due to crack growth at welded joints) using a probabilistic
fracture mechanics model implemented in the PRAISE (Piping Reliability
Analysis Including Seismic Events) computer code. LLNL also estimated
the probability of "indirect" DEGB caused by the seismically-induced
failure of supports for the reactor pressure vessel, steam generators,
and reactor coolant pumps. For both causes of DEGB, extensive sensitivity studies identified key parameters affecting the probability of
pipe break while uncertainty studies quantified how uncertainties in
input data affected the final estimated probability of pipe break.
The results of these evaluations consistently indicated that the
probability of a DEGB in PWR reactor coolant loop piping is extremely
small (about 10^{-7} events per reactor-year from indirect causes, and
less than 10^{-10} events per reactor-year from direct causes) and that
thermal stresses, rather than stresses resulting from postulated
earthquakes, dominated the probability of direct DEGB. These results
suggested that the DEGB design requirement -- and with it related
design issues such as coupling of DEGB and SSE loads, asymmetric

39

blowdown, and the need to install pipe whip restraints -- warranted reevaluation for PWR reactor coolant loop piping.

Our BWR study has been essentially the same except that different dominant failure mechanisms were added. LLNL has so far limited its investigation to General Electric plants using the Mark I containment design, older plants which have recirculation piping particularly susceptible to the effects of stress corrosion cracking (SCC).

3 DOUBLE-ENDED GUILLOTINE BREAK CAUSED BY CRACK GROWTH

3.1 General approach

The probability of "direct" DEGB in reactor coolant piping is estimated using a probabilistic fracture mechanics model implemented in the PRAISE computer code and associated pre- and post-processing routines. Details of the model are documented elsewhere (Harris 1982) and will not be repeated here, but can be summarized as follows.

For a given weld joint in a piping system, the probability of failure (i.e., leak or break) is estimated using a Monte Carlo simulation technique. As diagramed in Fig. 1, each replication of the simulation -- of which a typical simulation may include many thousand -- begins with a pre-existing flaw having initial length and depth randomly selected from appropriate distributions. These distributions in turn relate the probability of crack existence. Fatigue crack growth is then calculated using a Paris growth model, to which are applied stresses associated with normal operating conditions and postulated seismic events. The influence of such factors as non-destructive examination (NDE) and leak detection is also considered through the inclusion of appropriate statistical distributions (e.g., probability of crack non-detection as a function of crack size). Leak occurs when a crack grows through the pipe wall, break when failure criteria based on net section stress (for austenitic materials) or tearing modulus (for carbon steels) are exceeded.

Completing all replications for a given weld joint and tabulating those cracks that cause failure yields the cumulative probability of failure as a function of time at that weld. If only pre-existing cracks are considered, then "stratified sampling" can be applied to assure that initial crack samples are selected only from those sizes that can potentially cause pipe break. Through this technique, very low failure probabilities (less than one in a million) can be reliably estimated from only a few thousand replications of the Monte Carlo simulation.

After the failure probabilities at all weld joints in a piping system have been estimated, a "systems analysis" combines these results with the non-conditional crack existence probability (a function of total volume of weld material) and seismic hazard (which relates the occurrence rates of earthquakes as a function of peak ground acceleration) to obtain the non-conditional probabilities of leak and DEGB.

3.2 Stress corrosion cracking

One significant factor complicating the evaluation of BWR piping, however, was the need to include effects of intergranular stress corrosion cracking (IGSCC). When present, IGSCC not only accelerates the growth rate of existing flaws, but also causes new cracks to initiate after plant operation has begun. The effect of these "initiated" cracks on the probability of DEGB must be therefore be considered in addition to that of pre-existing flaws.

Recirculation piping in older BWR plants, particularly those characterized by the General Electric Mark I containment design, has been found in recent years to be susceptible to intergranular stress corrosion cracking. Stress corrosion cracking occurs in stainless steel piping (in the Mark I plants, Type 304) when the "appropriate" (in an adverse sense) conditions of "sensitization" -- material properties conducive to IGSCC that result from prolonged exposure to high temperatures during welding -- environment, and stress are met. Earlier versions of PRAISE treated the effect of IGSCC on pre-existing cracks through a simple relationship between growth rate and the stress intensity factor at the crack front; crack initiation was not modeled at all. It is important to note that this model was not applied in our PWR evaluations because operating experience has indicated that IGSCC is not a problem in PWR reactor coolant loop piping.

As part of our BWR study we developed an advanced IGSCC model for the PRAISE code (Harris 1986). This model is semi-empirical in nature, and is based on experimental and field data compiled from several sources. Using probabilistic techniques, the model addresses the following IGSCC phenomena:

1. Crack initiation, including the effects of environment, applied loads, and material type (i.e., sensitization). Crack location, time of initiation, and velocity upon initiation are all defined by appropriate distributions based on experimental data. "Initiated" cracks are considered separately from pre-existing cracks until one of the following two criteria are satisfied: (1) the crack attains a depth of 0.1 inch, or (2) the velocity of the crack estimated according to the Paris growth law exceed the initiation velocity. Beyond this point, "initiated" and "fracture mechanics" cracks are treated identically.

2. Crack growth rate, including effects of environment, applied loads, and material type.

3. Multiple cracks. Because our earlier evaluations were based on pre-existing flaws only, each Monte Carlo replication included one crack only. Inclusion of crack initiation requires that multiple cracks be considered during each replication.

4. Crack linking. Treating multiple cracks requires that their potential linkage into larger cracks be considered. This is done using linkage criteria specified in Section XI of the ASME Boiler and Pressure Vessel Code.

The model covers not only the Type 304 stainless steel found in most Mark I recirculation piping, but Type 316NG ("nuclear grade") steel as well, a low-carbon alloy widely regarded as an IGSCC-resistant replacement for Type 304. Crack growth rates and times-to-initiation for each material are correlated against "damage parameters" which consolidate the separate influences of several individual parameters. The damage parameters are multiplicative relationships among exponential terms which individually describe the effects of the various phenomena on IGSCC behavior, including:

1. Environment, specifically coolant temperature, dissolved oxygen content, and level of impurities.

41

2. Applied loads, including both constant and variable loads to account for steady-state operation and plant loading or unloading, respectively.

3. Residual stresses. Steady-state pipe loads due to welding residual stresses are considered in addition to fatigue loads.

4. Material sensitization.

The damage parameters in the 304SS model were based on the results of both constant-load (CL) and constant extension rate (CERT) IGSCC laboratory tests. Many other factors were considered during initial model development, but were later excluded from consideration either because they were judged to be of secondary influence for 304SS, or because suitable operating data was not available to exercise them in a plant-specific evaluation. Although the present model was developed for 304SS, adapting the correlation scheme for Type 316NG was a relatively straightforward matter of defining new damage parameters based on appropriate laboratory data; the basic functional form of the model was otherwise left unchanged. Two features unique to the 316NG model are, however, noteworthy:

1. Where both CERT and CL data were available for 304SS, only CERT data was available for 316NG. These data were used to define constant-load growth rates and times-to-initiation in 316NG under the assumption that the creep behavior of both alloys is similar.

2. As noted earlier, three conditions are necessary for IGSCC in austenitic steels: stress, environment, and sensitization. In 304SS, whenever stress corrosion cracking occurs in laboratory tests intended to simulate operating BWR conditions, it is most often intergranular. In 316NG, however, CERT specimens fail by transgranular stress corrosion cracking (TGSCC), whereas IGSCC is observed in fracture mechanics specimens. Since the relative influence of environment and loading on TGSCC in 316NG appears similar to that of IGSCC in 304SS, the available TGSCC data were used to predict cracking in 316NG.

Residual stresses are treated as a random variable in the Monte Carlo simulation. Distributions of residual stress as a function of distance from the inner pipe wall were developed from experimental data for three categories of nominal pipe diameter. For large lines (20 to 26 inches), residual stresses took the form of a damped cosine through the wall as based on data collected by General Electric and Argonne National Laboratory. The nominal tensile stress at the inner pipe wall is about 40 ksi. For intermediate-diameter (10 to 20 inches) and small-diameter (less than 10 inches) lines, a linear distribution was assumed through the pipe wall with respective inside wall stresses of 9.3 ksi and 24.4 ksi.

The 304SS model was benchmarked by comparing predicted leak rates under nominal BWR applied load conditions against actual leak and crack indication data made available to us by the NRC Office of Nuclear Reactor Regulation (NRR). During benchmarking we quickly ascertained that residual stress was the parameter most influencing the predicted leak rates, and we therefore opted to adjust the model on this basis. A variety of schemes were considered before we settled on adjusting the stress magnitude (using a multiplication factor) to bring the model into agreement with the field data. In the course of this benchmarking, surprisingly severe factors (on the order of 0.2, i.e., reductions of 80 percent in stress magnitude) had to be applied to bring the model in line with the field data, suggesting that factors other than residual stress may be more influential than we first concluded.

3.3 Probability of direct DEGB

Applying this model in our Brunswick pilot study, we estimated the
leak and DEGB probabilities in the original recirculation piping (both
loops combined) to be about 0.68/ry and 9.0E-4/ry, respectively, when
IGSCC is present. During development of the IGSCC model, we found
that its complexity greatly increased computer time requirements for
its execution (up to three CPU hours per weld). We therefore grouped
the welds in the Brunswick recirculation loops by pipe diameter,
(conservatively) assuming that the welds in each group with the
highest applied loads were representative of the entire group. We
then estimated the leak and DEGB probabilities at each of these
representative welds and performed a systems analysis assuming that
these leak and DEGB probabilities applied to all welds in the
respective group. These weld-by-weld probabilities and the resultant
system probabilities are presented in Table 1 for both leak and
DEGB. These results apply to the existing recirculation piping at
Brunswick. A replacement configuration, fabricated from Type 316NG
and including fewer welds (eliminating, for example, the recirculation
pump bypass piping), has been proposed for the Brunswick plant.

  These analyses further indicated that if IGSCC is not a factor,
thermal fatigue is the primary cause of direct DEGB, and the
probability of break is similar to that in PWR reactor coolant loop
piping. As for PWR reactor coolant loop piping, earthquakes
contribute only negligibly to the probability of direct DEGB in the
BWR recirculation loops.

  Note that these results imply about one leak for each reactor-year
of operation. Field observations, however, indicate that the actual
leak frequency is closer to one in every ten reactor-years of
operation, which led us to conclude that our predicted leak
probabilities -- and by implication, DEGB probabilities as well --
were between one and two orders of magnitude high. Inclusion of an
impurities term in the 304SS model (which had not been done
originally) resulted in a slight reduction in the estimated failure
probabilities.

  Figure 2 summarizes cumulative leak probabilities for Types 304 and
316NG weldments over 20 years of operation. Results for 316NG are
presented both for initiated cracks and for pre-existing cracks, the
latter case reflecting only the effect of stress corrosion cracking on
crack growth and not only the addition of new "initiated" cracks to
the overall population. Two observations are significant here:

  1. At any given time, the estimated failure probability in 304SS is
some two to three orders of magnitude higher than in 316NG.

  2. The time required to reach a given leak probability is about six
times as long in 316NG as it is in 304SS.

  These results also show that where failure is 304SS is always
dominated by initiated cracks (i.e., resulting from stress corrosion),
while in 316NG the initiated cracks dominate the probability of leak
only after about 12 years. Once cracks are present, growth rates are
nominally the same in either material. Consequently, the predicted
difference in behavior between the two materials is due to differences
in the times-to-initiation.

4 DOUBLE-ENDED GUILLOTINE BREAK INDIRECTLY INDUCED BY EARTHQUAKES

If earthquakes and pipe breaks are considered as purely random events, the probability of their simultaneous occurrence is negligibly low. However, if an earthquake could cause DEGB, then the probability of simultaneous occurrence would be significantly higher. Our study of direct DEGB concluded that earthquakes were not a significant contributor to this failure mode. However, another way in which DEGB could occur would be for an earthquake to cause the failure of component supports or other equipment whose failure in turn would cause a reactor coolant pipe to break.

Two additional causes of pipe break were therefore considered in our evaluation of BWR reactor coolant piping: failure of "intermediate" pipe supports and supports for light loop components (e.g., recirculation pumps), and failure of heavy component supports.

4.1 Failure of heavy component supports

Evaluating the probability of indirect DEGB involves the following four steps:

1. Identify "critical" components whose failure could induce a DEGB. For each component, estimate the conservatism and the uncertainty in the calculated structural responses for various loading conditions, such as dead weight, thermal expansion, pressure, and seismic loads. In our PWR evaluations, we identified as critical components the supports for "heavy components" in the primary system: reactor pressure vessel, steam generators, and reactor coolant pumps. A BWR, of course, has no steam generators and the failure of coolant pump supports was considered as part of the direct DEGB evaluation. Therefore, the only critical components that we considered in our Brunswick indirect DEGB evaluation were those comprising the reactor support structure.

2. For each critical component, develop a fragility description for each failure mode. The fragility of a component may be based on several factors such as specific design allowables, or types of analysis used, or how such factors as soil-structure interaction or equipment damping were considered. Each fragility description relates the probability of structural failure conditioned on the occurrence of an earthquake of given peak ground acceleration.

3. Calculate the overall "plant level" fragility to account for all significant failure modes and the associated fragility descriptions.

4. Calculate the non-conditional probability of indirect DEGB by convolving the plant level fragility with an appropriate description of seismic hazard. "Seismic hazard" relates the probability of occurrence of an earthquake exceeding a given level of peak ground acceleration. In our Brunswick evaluation (Hardy 1986), as in our earlier PWR work, we used generic seismic hazard curves that we developed for plant sites east of the Rocky Mountains. For Brunswick, we found that the probability of indirect DEGB was about 2E-8 events per reactor-year, with a 90th-percentile value (confidence limit) of 5E-7 per reactor-year. We found that the "star" stabilizer at the top of the reactor pressure vessel, which restrains the RPV against lateral motion in the event of an earthquake, was the primary contributor to failure.

4.2 Failure of "intermediate" supports

Reactor coolant loops in PWR plants typically have small length-to-diameter ratios and, because of their stiffness, are supported solely by the major loop components (reactor pressure vessel, reactor coolant pumps, and steam generators); therefore, no additional supports are needed. However, recirculation loop piping in BWR plants is longer and smaller-diameter (typically 12 to 26 inches), and requires additional support from spring- or constant-load hangers. This piping may also have numerous snubbers to reduce stresses in the event that an earthquake occurs. Each recirculation loop at Brunswick, for example, has a snubber pair each on the inlet and outlet lines, as well as a snubber triplet at the top and at the bottom of the recirculation pump.

The potential effect of intermediate support failure on estimating the probability of direct DEGB is two-fold:

1. Support failure would redistribute applied stresses at weld joints, in turn affecting crack growth rates as well as the failure criteria used to define when pipe break occurs.

2. Accounting for stress redistribution would require an individual PRAISE evaluation for each support failure scenario, dramatically increasing the computational effort involved. For example, even if only four supports were addressed, sixteen separate PRAISE runs would be required to cover all possible combinations and permutations of support failure.

Our evaluations of indirect DEGB (i.e., support reliability) in PWR reactor coolant loop piping were based on the assumption that failure of a heavy component (e.g., steam generator) support unconditionally led to pipe break. This assumption was regarded as conservative (in reality a pipe would likely experience severe inelastic deformation before actually breaking) but nevertheless resulted in very low DEGB probabilities. To have assumed that failure of a snubber or a constant-load support would similarly cause a DEGB in BWR recirculation piping would have been unreasonably conservative. We therefore developed a more detailed approach to incorporate the effect of support fragility into the probabilistic fracture mechanics evaluation, which we used to investigate the effect of support failure on the probability of direct DEGB for our reference "pilot" plant (Lo 1986). After identifying support failure scenarios, we calculated the pipe stresses for each. We then performed sensitivity analyses -- again using the generic seismic hazard curves that we had developed for plant sites east of the Rocky Mountains -- which indicated that the probability of DEGB due to hanger and snubber failure was about the same (about 10^{-8} events per reactor year) as that of indirect DEGB due to the failure of the RPV support structure. We concluded that we could neglect support failure in our subsequent direct DEGB evaluations.

5 SUMMARY AND DISCUSSION

Not surprisingly, the results of our evaluations to date show that IGSCC dominates the probability of direct DEGB in recirculation piping fabricated from Type 304 stainless steel. In general, we have observed that:

1. Residual stresses appear to be the dominant factor influencing the number of cracks initiated and, to a lesser degree, their growth

rates. In the absence of IGSCC, thermal stresses dominate the probability of direct DEGB; earthquakes contribute only negligibly.

2. Initiated cracks always dominate the probability of leak (and by inference, break probabilities as well) in Type 304 stainless steel piping. By contrast, the influence of initiated cracks in Type 316NG piping first becomes apparent after about 12 years of operation.

3. The per-weld probability of leak in Type 316NG piping is between two and three orders of magnitude lower than that in piping fabricated from Type 304 stainless steel. Based on our past experience, this trend should hold true for break probabilities as well.

Type 316NG piping owes its improved SCC performance to crack initiation times being much longer than in 304SS piping. Crack velocities upon initiation are also significantly lower than in 304SS. Although the growth rate of "fracture mechanics" cracks in 316NG is somewhat lower than in 304SS, in general once cracks do initiate, their growth rates are nominally the same in either material. This result points to the number of cracks, rather than their growth behavior, as being the dominant influence on relative leak probabilities in Types 304 and 316NG stainless steel piping.

Although we are now in the process of explicitly estimating DEGB probabilities in Type 316NG piping (using the Brunswick pilot plant as our case study), we expect break probabilities to behave accordingly. Our past experience has shown that break probabilities are typically three to four orders of magnitude lower than leak probabilities. On the basis of the DEGB probabilities estimated for Type 304 recirculation loop piping (Table 1), and taking into account the fact that proposed 316NG recirculation loop configurations have significantly fewer welds than current 304SS configurations, we anticipate that system DEGB probabilities should be on the order of 10^{-6} per reactor-year.

REFERENCES

Hardy, G.S., et al., 1986. Probability of Failure in BWR Reactor Coolant Piping, Volume 4: Guillotine Break Indirectly Induced by Earthquakes. Lawrence Livermore National Laboratory, Report NUREG/CR-4792.

Harris, D.O., et al., 1982. Probabilistic Fracture Mechanics Models Developed for Piping Reliability Assessment in Light Water Reactors, Lawrence Livermore National Laboratory, Report NUREG/CR-2301.

Harris, D.O., et al., 1986. Probability of Failure in BWR Reactor Coolant Piping, Volume 3: Probabilistic Treatment of Stress Corrosion Cracking in 304 and 316NG BWR Piping Weldments. Lawrence Livermore National Laboratory, Report NUREG/CR-4792.

Holman, G.S., et al., 1984a. Probability of Pipe Failure in the Reactor Coolant Loops of Westinghouse PWR Plants. Lawrence Livermore National Laboratory, Report NUREG/CR-3660, Vols. 1-4.

Holman, G.S., et al., 1984b. Probability of Pipe Failure in the Reactor Coolant Loops of Combustion Engineering PWR Plants, Lawrence Livermore National Laboratory, Report NUREG/CR-3663, Vols. 1-3.

Holman, G.S., et al., 1985. Probability of Pipe Failure in the Reactor Coolant Loops of Babcock & Wilcox PWR Plants, Lawrence Livermore National Laboratory, Report NUREG/CR-4290, Vols. 1-2.

Lo, Ting-Yu, et al., 1986. Probability of Failure in BWR Reactor Coolant Piping, Volume 2: Pipe Failure Caused by Crack Growth and Failure of Intermediate Supports. Lawrence Livermore National Laboratory, Report NUREG/CR-4792.

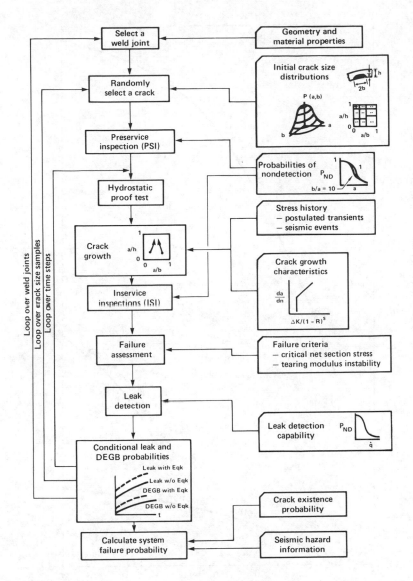

Figure 1. Flowchart of the probabilistic fracture mechanics model implemented in the PRAISE computer code.

47

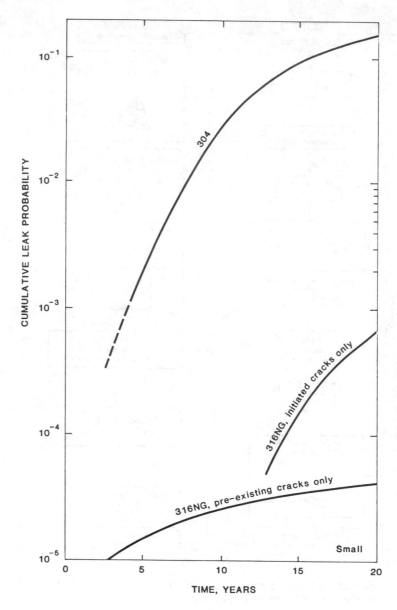

Figure 2. Cumulative leak probability as a function of time for small-diameter weldments fabricated from Types 304 and 316NG stainless steel.

Estimate of locations for repeated non destructive test procedures in pressurized components of LWR's

J.Jansky & L.Heger
Büro für Technische Beratung (BTB), Leonberg, FR Germany
G.Katzenmeier
Kernforschungszentrum Karlsruhe GmbH, FR Germany

Summary

A release of radiation into the environment through leakages in the pressure retaining components of light water reactor plants affects the public safety.
This danger is not perceptible through our human warning device.
Therefore intensified precautions must be made for nuclear installations to exclude these possible defects.
The structural integrity of the pressure retaining components of LWR is the precondition which has to be carried out before the commencement to operation of the nuclear installations.
If these requirements are not met and a leak in the system occures, the affected installations must be shut-down and have to be repaire to remain the structural integrity of the PRC. In installations of the size of Biblis B (appr. 4000 MW) such a shut-down means an impairment of the availability and this almost astronomical financial losses for the utilities.
Therefore the management of LWR-installations carries out extensive preventive correctional measures in times of planned shut-downs (change of fuel elements) which should result in discovering and repairing possible structural weaknesses in scheduled time.
One component of these efforts are the periodical nondestructive tests of the welding seams which are carried out in different periods of operational time. During this time again a base effort is to minimize the irradiation of the service and maintanance personnel.
A local estimation of testing cross sections in the PRC follows from the experience of failures (own and others) and own subjective criteria as f. ex. the accessibility of testing regions.
A calculated theoretically supported estimation of testing locations to be proved which results f. ex. from the maximum total degree of strain, chemical conditions of the pressurized medium and the quality of the material didn't exist up to now. Such a method yields in decrease of testing cost and total personnel exposure time.
In the article at issue fundamentals of such a conception

are being outlines for the theoretical estimation of the
locations in pipes of LWR from the
- process engineering assumptions
- geometrical boundary conditions and position of the cros-
  section in the pressure retaining components
- induced strains under operating and faulty conditions
- and quality of the material.
The procedure is documented at a case of corrosion induced
failure in pipes of HDR-research plant in Großwelzheim/FRG.
A comparison between the theoretical precalculation and the
fractographical findings is attached.

1. Introduction

Energy conversion requires facilities able to convert la-
tent chemical or atomic energies into high-grade forms of
energy easy to transport and use at the point of application.
The pressure retaining components of such facilities must
remain structurally and physically intact relative to the
environment because of the required functional status and
also because of the hazard potential they represent.
This consideration gives rise to three important aspects on
the basis of which pressure barriers should be designed
/1/:
(A) Protection of the environment from the destructive
    effects of the energy released.
(B) Leaktightness function, to maintain the energy poten-
    tial relative to the environment.
(C) Leaktightness function as far as the containment of
    hazardous substances is concerned.
The leaktightness function of pressure barriers is important especially for radioactively contaminated coolants, as
these may have an empact on their immediate vicinity within
and outside plants (installations, personal).
Preventing this hazard is exclusively in the domain of
plant operators, as man has no sensors allowing him to
detect radioactivity hazards and, consequently, is unable
to protect himself.
For this reason, nuclear power plants make provisions for
extensive in-service inspections to prevent potential incidents and accidents likely to result in radioactivity
releases. During refueling, most of the welds are being
inspected, as they are regarded as the weakest links in
pressure barriers. The testing personnel and auxiliary
personnel (e.g., for erecting scaffolds) are exposed to
radiation in this process. Even if the personnel does carry
monitoring devices so as not to exceed the legal limits,
the aggregate exposure data of the personnel depend on the
scope, i.e. the number, of inservice inspections in the
plant. Under the law, the operator of a nuclear facility is
obliged to minimize the total dose. For this reason, every
operator must try to prepare in-service inspections within
a plant in such a way that
* all "weak" cross sections respective components within
  the system are discovered in advance and
* work during operation is reduced to a minimum.
These requirements can be met in advance only in a detailed

preparatory phase. Boundary conditions such as
* process technology
* design geometry
* materials
* local stresses and loads
are used to pinpoint crucial weak spots within a pressure
retaining barrier. The methods applied in this instance are
described in this paper. The approach used, and its correctness, will be explained on the basis of a few examples
taken from the HDR facility.

2. Description of HDR Failures and Codes Used for Analysis

Pressure retaining barriers react to operating stresses and
loads resulting from internal pressures or temperatures in
accordance with the surface conditions of the components
wetted by the coolant.
If there are geometrical discontinuities in a wall sustai-
ning loads, they will grow after some initial scarfing, as
shown in Figure 1. If the wall of the component is smooth,
discontinuities first must form under operating conditions,
which will prolong the process of component damaging.
Whether the discontinuities forming and growing are detected in time depends on the analytical methods used to pinpoint the cross sections of weak spots and, ultimately,
also on the quality of the in-service inspections performed.
Failure modes are a function of crack forms, overall geometry, loads and stresses and, finally, the interactions
between the material and the medium.
If crack development is controlled mainly by corrosion,
long incipient cracks will form on the wetted inner surface, in which case the ligament relative to the outer surface cannot be seve-red in a corrosion-driven mode. Only if
the load limit is reached in the weakened cross section,
that cross section can be severed, which mostly leads to
major crack extension (katastrophical failure). Similar
types of failure can be expected to arise in corrosion/
erosion processes on the inside of a pressure retaining
barrier /2/, by high flow velocities of medium.
Similar damage in pipes carrying coolant has been registered in the Superheated Steam Reactor (HDR)Test Facility run
by the Karlsruhe Nuclear Research Center at Karlstein near
Frankfurt, which is being used for safety research at light
water reactors.
In some phases of experimental operation, in which the reactor pressure vessel had been subjected to an artificial
thermal shock load, the coolant circuit (p=110 bar,
T=310°C) was filled with pressurized water enriched with
oxygen (8 ppm O_2). The oxygen in the demineralized water
causes the electrochemical potential to shift by 200 mV. As
a consequence, the pressure barrier will be more susceptible to corrosion in regions subjected to higher loads/
stresses.
The following cases of damage encountered in the HDR, which
were traced back to corrosion, will be used as a basis for
the analyses outlined below:
* Circumferential break in the region of an underdimensio-

ned reducer made of 15Mo3, experienced in 1983 /3/.
* Leakage in the region of a reducing pipe made of 15Mo3 in
  the penetration through the steel shell /6/, encountered
  in 1986.
* Several discontinuities in dismantled pieces made of
  15Mo3 removed from the cooling circuit /4/, encountered
  in 1986.

To determine theoretically the critical piping cross sections, the Stuttgart version of the Pipe Stress program
package supplied by Control Data /5/, together with the BTB
follower program, P-STRAIN, was used.

2.1. HDR Circumferential Break

The geometry of the reducing pipe, which is underdimensio-
ned for the test conditions (p = 110 bar/T = 310°C), is
shown in Figure 2. This reducing pipe was severed by means
of a large circumferential failure under the operating
conditions mentioned above.

The main stresses resulting from the internal pressure
load, as calculated for the actual geometry of the reducing
pipe by MPA /7/, are attached. That load was superimposed
the bending stresses of approx. 30 N/mm² resulting from the
weight of the closing valves (approx. 450 kg on a lever arm
of approx. 1000 mm) acting excentrically to the axis of the
pipe. For the longitudinal stress in the critical cross
section this resulted in a maximum contribution of
380 N/mm².

The circumferential break was some 2 mm downstream of the
large diameter of the reducer. This would mean that a theoretical stress intensity factor of approx. 30 MPa√m would
have to be assumed.

2.2. Leak in the Penetration through the Containment shell.

Figure 3 shows a reducing pipe of 15Mo3 with ND 100, which
had been installed for a coolant bypass line in the region
of the penetration through the containment shell. The pipe
isometry indicated in Figure 4 shows that the movement of
the pipe was impeded by additional steel walls installed in
the facility.

The longitudinal strains calculated in the maximum cross
section ranged up to the 1 ‰ limit in the critical cross
section (node No. 3). On this basis an stress intensity
factor in the order of 20 MPa√m can be estimated.

3. Metallographic and Fractographic Finding
3.1. Circumferential Break

In the schematic representation of the area of the break, a
number of semielliptical cracks facing the inside of the
pipe were find.

The depth of the cracks are ranging up to through-cracks
(point of the first leakage). The through-crack ist dis-
placed by some 180 degrees relative to the position of the
closing valves in the vertical pipe section.

From the point of leakage on the circumference, shear lips
of a tough, but fast break can be seen extending in the two
circumferential directions.

Detailed fractographic evaluation, shows traces of stable

crack growth under constant load and corrosive conditions in the flat areas of the crack perpendicular to the longitudinal direction of the pipe (6, 7).

3.2 Leackage in Penetration sleeve through the containment shell.

As before several semi-elliptical cracks open in the direction of the wetted surface, which extend right to the outer suface. The notch required for stable crack extension development in fracture mechanics (singularity) was produced by weld root sagging, Figure 5.
The magnitude of this root sagging was theoretically estimated with a = 3 mm relatively well in the theoretical analysis /6/.

4. Remedial Measures for NDT Inspections

As can be seen from the three examples discussed above, it is possible to determine cross sections in pressurized water systems in which cracking must be expected to occur preferably.
Those spots must meet conditions listed in Figure 6.
The data enumerated under Items 1 to 6 in the table first of all indicate whether the inside of the pressurized enclosure can be wetted at all by corrosive media (enriched by O_2 or CO_2).
This indication requires sound knowledge of the technical process aspects of the facility.
Items 7 to 11 in the table refer to existing geometric notches in the pressurized enclosure and the materials examined. This often requires extensive work in plant archives an on the spot. Finally, Items 12 to 14 furnish information on loading/overloading of the regions investigated. Also this work need a long study of controller readings as far as they installed.
The strains composed of internal pressure, impeded strains and, perhaps, dynamic loadings are subsequently determined by means of the Pipe Stress and the P-STRAIN follower programs.
If the strain calcul. for the actual geometry then is found to be in the region of notches due to fabrication, such as
* the joint between a reinforced pipe elbow and the pipe,
* a sag in the root of the circumferential weld,
* a step in geometry,
it is possible to expect stable crack opening development starting from $\varepsilon_L > 0.15$ %o ($K_I > 10$ MP$\sqrt{m}$). The smaller the notches, the higher the strain must be, in order to furnish crack extension development in the mode observed.
Theoretical analysis of the systems allows specific and also more frequent checks to be made of cross sections potentially impairing plant safety and, hence, also availability.

5. References

/1/ J. Jansky: Bildung und Erweiterung von Rissen bei wiederholten thermischen Transienten in druckführenden Bauteilen. Techn. Fachbericht PHDR 47-85, Oktober 85
/2/ NRC-Bericht zur Surry II - Bruch am 09.12.1986

/3/ MPA-Dokumentation und Interpretation des HDR-Rohrab-
     risses, Stuttgart 23.1.1984
/4/ Ergebnisse der Festigkeitsanalyse unter Berücksichti-
     gung der Korrosionsanfälligkeit der HDR-Versuchslei-
     tung aus 15 Mo3; BTB-Bericht 38, 19.5.1986
/5/ Schadensuntersuchung am HDR-Thermoschock-Versuchs-
     kreislauf, MPA-Bericht - 8152212 A/Dr.Bl./Sl/Ng. von
     12/86
/6/ Pipe stress - Manuell, Control Data, 1977
/7/ C. Kußmaul, D.Blind und J. Jansky: Rißbildungen in
     Speisewasserleitungen von Leichtwasserreaktoren. Ur-
     sachen und Abhilfemaßnahmen; Sonderdruck aus VGB
     Kraftwerkstechnik 64. Jahrgang, Heft 12, Dezember 84,
     Seite 1115 bis 1129
/8/ J. Jansky: Studie zum Korrosionseinfluß bei thermo-
     schock belasteten Druckbehältern und Rohrleitungen,
     BTB-Leonberg, PHDR-Arbeitsbericht 2.254/85, Dezember
     1986

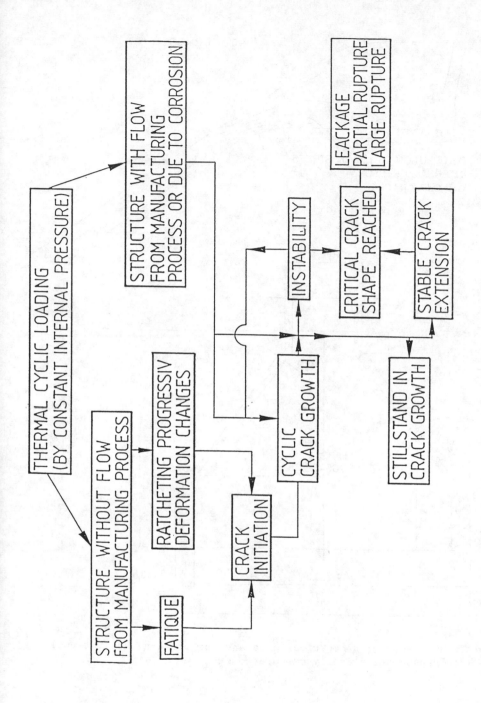

Figure 1: Distortion path for pressure retaining components under cyclic loading

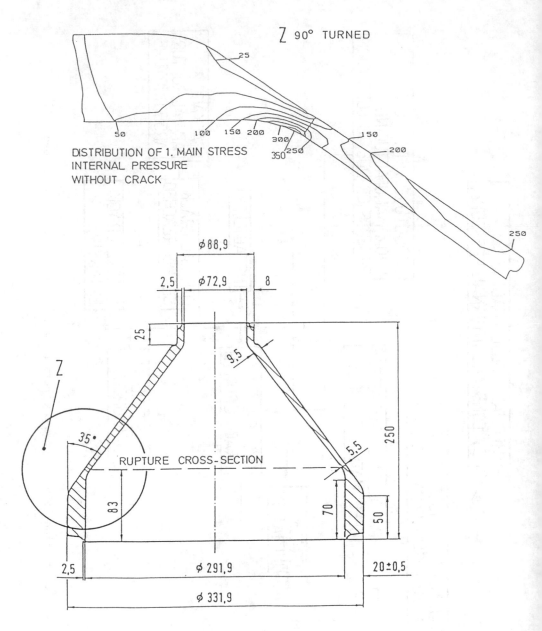

Z 90° TURNED

DISTRIBUTION OF 1. MAIN STRESS
INTERNAL PRESSURE
WITHOUT CRACK

Figure 2: HDR-reducer; geometrical data and for loading case internal
pressure elastically calculated stresses

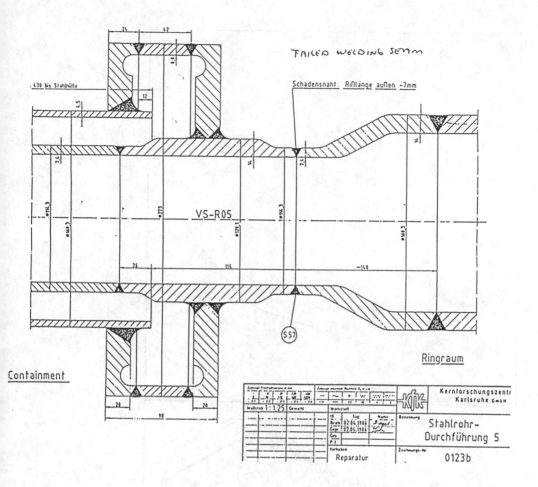

Figure 3: Geometrical shape of pipe reducer in the containment

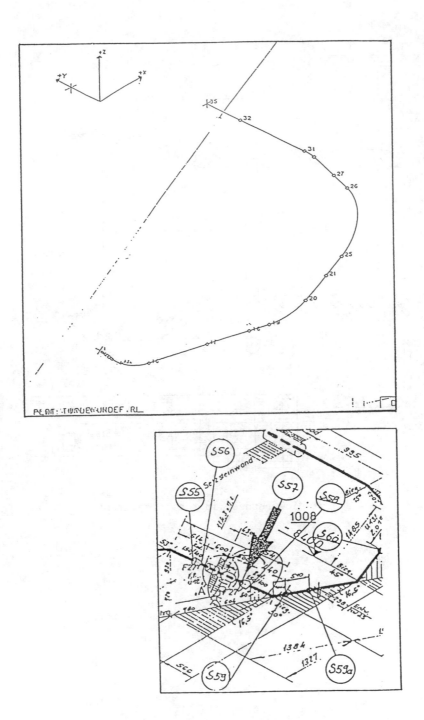

Figure 4: Detail of piping isometry, between the containment and steel wall

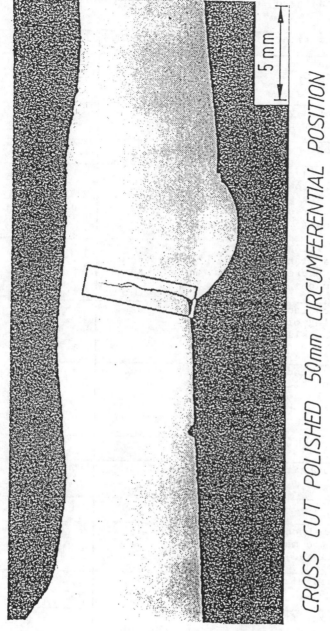

TUBE

REDUCER

5 mm

CROSS CUT POLISHED 50mm CIRCUMFERENTIAL POSITION

Figure 5: Metallographical-cut of the leackage area in the reducer/pipe welding root

Weak Spots to be Analyzed in a Pressurized Enclosure
Plant:
System:

Pos.	Boundary conditions	Test result	Source/comment
1	Is a condensation area in the vicinity (can O_2 or CO_2 be suct-in)?		
2	Downstream of a vacuum jet or a vacuum pump?		
3	Wet steam region?		
4	Area without any flow or with only minimal flow (branchings leading to safety systems, measuring systems, startup lines)?		
5	Closed to a flanged joint, steam-sealed glands, vented atmosphere, drainage, suction from open vessels?		
6	Is the region located in a horizontal leg or has it even been installed at a negative slope?		
7	Are any pipe bends, whipping protections, supports or hangers, nozzles or compensators nearby?		
8	Heat exchanger, vertical?		
9	Pipes with longitudinal welds?		
10	Circumferential welds?		
11	Connection between ferritic and austenitic materials (welded, flanged)?		
12	Are any fluid mixing areas nearby?		
13	Possible stratification region?		
14	Are impact type loads possible?		

Figure 6: Testing matrix for estimation of system/cross section affected by strain induced corrosion

The evolution of NRC regulatory requirements for standby diesel generators based on recent operating experience

C.H.Berlinger

Nuclear Regulatory Commission, Washington, D.C., USA

1 INTRODUCTION

The Nuclear Regulatory Commission requires that each licensed reactor
facility provide both an onsite and an offsite electric power system to
assure the functioning of all structures, systems and components important to safety. The safety function to be served by both the onsite and
offsite power systems is to provide sufficient capacity and the capability to assure that:

1. The specified acceptable fuel design limits and the design
conditions/criteria of the reactor coolant pressure boundary are not
exceeded as a consequence of any anticipated operational occurrence, and

2. the reactor core is adequately cooled and the containment integrity
and other vital functions are maintained in the event of any postulated
accident.

In addition, both the onsite and offsite electric power systems are
required to have sufficient independence, redundancy and testability to
perform their required safety functions assuming an additional single
failure. These requirements are specified in General Design Criteria #17
in Appendix A to Part 50, Title 10 of the Code of Federal Regulations.
The purposes of onsite electric power systems are (1) to provide power
promptly to engineered safety features if a loss of offsite power and an
accident occur simultaneously and (2) to provide power to equipment
needed to maintain the plant in a safe condition given an extended loss
of offsite power.

Typically at U.S. nuclear power plants, onsite AC power systems rely on
the use of diesel generator sets to provide the required emergency AC
power. A high level of reliability must be designed into the diesel
generator units and it must be maintained throughout their service
lifetime by the establishment of appropriate testing, maintenance and
operating programs. The Nuclear Regulatory Commission has provided
guidance to the nuclear industry in Section 8.3.1 of the Standard Review
Plan for the Review of Safety Analysis Reports for Nuclear Power Plants
(NUREG-0800). Section 8.3.1 defines an acceptable basis for meeting
these design criteria and references several guidance documents
including:

1. Regulatory Guide 1.9, which addresses the proper selection, design
and qualification of emergency diesel generators as onsite AC power
systems. A diesel generator set selected for use as an emergency onsite
power supply should have the capability to start and accelerate a number

of large motor loads in rapid succession and be able to sustain the loss of all or any part of such loads while maintaining voltage and frequency within acceptable limits, and must supply power continuously to emergency equipment needed to maintain the plant in a safe condition during an extended loss of offsite power.

2. Regulatory Guide 1.108, which provides guidelines for the preoperational and periodic testing of emergency diesel generators to help ensure that they will meet their performance and reliability requirements.

3. IEEE-387 (1977), the IEEE Standard Criteria for Diesel Generator Units Applied as Standby Power Supplies for Nuclear Power Generating Stations, which defines the principal design criteria and qualification testing requirements for application of emergency diesel generators at nuclear facilities. If followed, this standard will help ensure that performance and reliability requirements will be met.

4. The Diesel Engine Manufacturers Association (DEMA) Standard, which describes industry standard practices for the design and operation of low and medium speed stationary diesel and gas engines.

In conjunction with these design requirements, the U.S. Nuclear Regulatory Commission requires all applicants and licensees to establish and maintain a quality assurance program to assure that structures, systems and components important to safety are designed, fabricated, erected and tested to quality standards commensurate with the safety functions to be performed, and, to assure that a quality product is provided which will perform its required safety function (General Design Criteria #1, Appendix A to 10 CFR 50). The detailed description of these quality standards and recordkeeping requirements are included in 10 CFR 50, Appendix B. The NRC staff may also specify additional requirements within individual plant Technical Specifications and special license conditions. These additional requirements include enhanced periodic maintenance and surveillance programs and limiting conditions for operation which have evolved based on past operating experience. These requirements provide a means of assuring their long term reliability and operability.

Since the late 1950s, the U.S. NRC's regulatory requirements regarding the use of diesel generators as emergency onsite AC power sources have continued to evolve. As a result of recent operating experience, reliance upon qualification testing and periodic maintenance and surveillance testing has assumed even greater importance. In the future, as additional operating experience is accumulated, these regulatory requirements will be modified to maintain high levels of operability and reliability. The significant improvements in regulatory requirements, which have developed recently, are described in this paper.

2 CHRONOLOGY OF REGULATORY REQUIREMENTS

During the period from the late 1950s to the early 1970s, the U.S. NRC required that emergency diesel generators (EDGs) meet IEEE standards and industry practices. Included among these requirements were, for example, routine periodic tests of EDGs on a monthly basis and during refueling outages. In 1975 the NRC issued a report which addressed the reliability of onsite EDG sets, which was based on over a decade of operating experience. That report identified failures to successfully start as the dominant failure mode. In 1977 Regulatory Guide 1.108, Revision 1 was issued. In that revision the NRC established a new surveillance testing requirement, which was based on the frequency of failures to start on

demand during the previous 100 engine tests. Table 1 reflects the
frequency at which every EDG at a nuclear plant was to be tested, as
required by Revision 1. As noted, once four failures to start had
occurred in one engine, all EDGs at the plant were required to be tested
every 3 days, whether or not they had experienced failures to start of
their own.

Table 1. Frequency of surveillance testing based on number of failures to
start in past 100 tests.

Number of failures to start	Required frequency of surveillance testing
0 to 1	31 days
2	14 days
3	8 days
4	3 days

With the implementation of this requirement for increased frequency of
testing, utilities which had been experiencing operability or reliability
problems with their EDGs took action to improve their maintenance programs. These efforts over the next several years appeared to be successful in reducing their susceptibility to failures to start on demand.

However, in 1982 the Brunswick nuclear plant experienced a diesel
engine failure which was the result of fatigue type failures. It was
believed that excessive engine start testing was a contributor to this
failure. The regulatory guides had specified that engine start tests
were to be conducted from ambient conditions to at or above full rated
load. The lack of clear definitions of ambient conditions and the proper
load for conducting these surveillance tests permitted many nuclear
facilities to conduct these tests as cold/fast starts, without the
benefit of existing keepwarm systems, and at loads in excess of their
rated load.

In conjunction with a study conducted by the NRC regarding the total
loss of onsite and offsite AC power (station blackout), an analysis of
EDG failure rates was undertaken to quantify the reliability of EDGs. In
1983 NRC published its report (NUREG/CR-2989) which concluded, in part,
that the required increased frequency of surveillance testing can contribute to a reduction in overall EDG reliability. In 1983 the NRC staff
issued Generic Letter 83-41 which requested that applicants and licensees
provide information relative to their individual surveillance programs
and to identify their current approach to conducting fast/cold engine
start tests. In 1984 the NRC staff issued Generic Letter 84-15 which
advised licensees to take appropriate actions to reduce the number of
cold/fast start tests. This recommendation was based on the fact that
diesel engine keepwarm systems, which were in general use at nuclear
facilities, were available to maintain lube oil and jacket water temperatures at levels commensurate with those recommended for reducing thermal
stress levels during fast start and fast load surveillance tests, and
therefore, it would not be necessary to test engines at conditions not
normally experienced.

In 1985, following a large number of repeated failures of several EDGs
at the North Anna nuclear power plant, a significant number of changes
were made to their EDG related Technical Specifications. These modifications were implemented for the following purposes:

1). To establish a 95% reliability goal for each engine, independent of all other engines at the plant.

2). To change the monthly surveillance test from fast/cold start and fast/full load tests to slow starts with positive engine prelubrication from ambient (keepwarm) conditions, followed by gradual loading up to a lower indicated "full-load" level; not to exceed its full rated load.

3). To reduce the frequency at which accelerated periodic surveillance testing would be conducted, based on the number of failures in the past 100 and/or 20 tests. This was applicable to those individual engines which had experienced repeated failures to start, as a means of maintaining the 95% reliability goal, and it removed the requirement that an EDG which was not having problems be tested unnecessarily. Tables 2 and 3 reflect those changes made regarding the increased frequency of required surveillance testing.

Table 2. North Anna technical specification for frequency of surveillance testing based on number of failures to start in past 100 tests

Number of failures to start	Required frequency of surveillance testing
0 to 4	31 days
5 or more	14 days

Table 3. North Anna technical specification for frequency of surveillance testing based on number of failures to start in past 20 tests

Number of failures to start	Required frequency of surveillance testing
0 to 1	31 days
2	7 days

4). To modify Limiting Conditions for Operation (LCOs) associated with special surveillance testing requirements when plant safety equipment was in a degraded condition. Previously, when an offsite AC power circuit or an onsite EDG was either lost or out of service, all operable EDGs were required to be tested within one (1) hour and at eight (8) hour intervals thereafter until repairs were completed and the equipment was returned to service. The new LCOs were changed such that:

1). With loss of an offsite power circuit, test only the EDG on that circuit within 24 hours, if not tested during the previous 24 hour period, and do not repeat.

2). With loss of an EDG, test the redundant EDG within 24 hours, and do not repeat.

3). With loss of both an offsite power circuit and an EDG, test the other redundant EDG within 8 hours, do not repeat.

Between 1977 and 1985 the start reliability of EDGs did improve but, as was mentioned above and as described more fully below, by as early as 1982 the wear and tear on EDGs, perhaps due in part to the increased testing requirements, was becoming more evident from the number of engine operability failures. With the approval in April 1985 of the North Anna Technical Specifications, a basis was established for modifying other plant technical specifications and the Standard Technical Specifications.

The intention was to reduce the number of cold/fast starts and to increase the reliability of EDGs by providing an incentive to perform more
frequent major overhauls and to implement enhanced maintenance and
surveillance programs. Therefore, in the future with wider application
of the North Anna Technical Specifications, the U.S. NRC will affect a
significant reduction in EDG testing requirements and an overall improvement in EDG operability and start reliability in nuclear plant
applications.

3 SIGNIFICANT RECENT OPERATING EXPERIENCE

3.1 North Anna Units 1 and 2

The EDGs at North Anna are Fairbanks Morse diesels rated at 2750 KW, at
900 RPM. They are 12 cylinder, opposed piston, turbocharged engines
without cylinder heads, intake or exhaust valves. These engines utilize
a lube oil keepwarm system which continuously prelubricates the lower
crankline. In addition, they have an upper crankline prelubrication
system which is manually operated for 2 minutes prior to any planned
engine start and a 2 gallon capacity lube oil booster system which
operates on engine starting air.
   Prior to December 7, 1984 one of the Unit 2 EDGs (2H) experienced
several spurious trips. On December 7, 1984 it was taken out of service
for repairs. The plant Technical Specifications in force at that time
required that the other EDG (2J) be tested within 1 hour and every 8
hours thereafter. When EDG-2J experienced a trip, it was taken out of
service and EDG-2H was placed back in service, tested and then retested
every 8 hours. On December 9, 1984 EDG-2H tripped. With EDG-2J inopera-
ble the Technical Specifications required the nuclear plant to be shut-
down. On December 16, 1984 all repairs were completed and the plant was
restarted. On January 13, 1985 EDG-2J experienced a failure and tripped
on high crankcase pressure. Subsequent inspection revealed excessive
ring wear which required replacement of 1 piston and its rings. Based on
the number of failures in the past 100 tests, the Technical Specifications required that both EDGs be placed on a 3 day testing cycle.
   On March 13, 1985 EDG-2J experienced another trip on high crankcase
pressure. Inspections revealed a large number of cylinders, pistons,
insert bushings and lower rod bearings which had failed and required
replacement. In accordance with Limiting Conditions of Operation contained in the plant Technical Specifications, after 72 hours, Unit 2 was
shutdown. A preliminary failure analysis by the licensee and manufacturer failed to clearly identify the root cause of the failures and in
part attributed the increased frequency of failures to NRC requirements
for accelerated testing at excessive engine loads. The NRC disagreed
with the licensee's determination. The required surveillance testing had
succeeded in only surfacing these problems and attributed the recent
failures to either inadequate or degraded lubrication. As a result of
these preliminary findings, the NRC did relax the accelerated surveillance testing requirements but, the licensee was required to perform more
frequent inspections and maintenance.
   In October 1985 EDG-2H experienced another engine failure. Inspections
revealed that the floating piston pin bushings on the upper pistons had
suffered some elongation. At about the same time, during a surveillance
test, EDG-2J experienced an inadvertent engine overload. Inspections of
EDG-2J revealed several elongated floating bushings with some of the

65

bushings having expanded to contact with the insert bushing. At least one of the insert bushings had been extruded.

In March 1986 the licensee completed its failure analysis, identifying the following failure sequence:

1). Elongation of the wrist pin floating bushing.
2). Loss of the 0.035 inch gap between the floating bushing and the insert bushings.
3). Extrusion of the insert bushings.
4). Contact between the piston skirt and cylinder liner.
5). Cylinder liner cracks.
6). Engine trip on high crankcase pressure.

The root cause for these failures was identified as inadequate lubrication, which was corrected by a change to a higher grade Mobil oil.

3.2 Fermi Unit 2

Fermi utilizes 4 Fairbanks Morse EDGs which are very similar to those used at North Anna. One relevant difference is in the lube oil booster system, which has a smaller capacity of approximately 1.25 gallons. Prior to January 1984 the Fermi EDGs relied on a manual prelube system for prelubrication of the upper crankline before all planned starts. In January 1984 the lube oil booster system was installed and for the next year only this booster system was used for prelubrication (manual system inactive). In January 1985 several upper crankline main bearings failed. The cause was identified as inadequate lubrication during engine starts. Thereafter, both manual and booster systems were used. The booster system capacity was also found inadequate but it was not modified.

In November 1985 EDG-13 experienced a failure. Inspections found severe damage to the upper connecting rod bearing, the upper piston skirt and rings and the cylinder liner in cylinder number 3. The licensee's and manufacturer's failure analyses concluded that the upper connecting rod had failed in fatigue, that the bearing cap had been inadequately clamped (as evidenced by fretting), but that adequate lubrication had been provided (no overheating). Since no root cause had been identified, the NRC required that the other 3 engines at the plant be inspected. EDG-11 inspections revealed minor scoring and all upper crankline main bearings were replaced. The engine was retested, during which several main bearings failed due to misalignment of the bearing caps during the preceding reassembly. During engine retesting, a lube oil foaming problem was observed. EDG-12 was found to have a broken oil ring and a fuel oil leak into the lube oil system. EDG-14 was also found to have fuel oil in the lube oil system.

The inspections and failure analyses concluded that there had been some gritty foreign material in the lube oil system, that the lube oil was inadequate for the required engine service and should be changed to reduce its foaming tendencies, and that the bearing preconditioning and run-in procedure, which used a mixture of gritty polish and lube oil to precondition the bearings and journals, was inadequate. Although the lube oil problems were suspected as a contributor to the failures, the root cause of all these problems could not be identified with certainty. Therefore, the NRC established a requalification program for these engines which included a demonstration test and frequent periodic inspections in conjunction with an enhanced maintenance program. The 2 engines which had experienced most of the problems (EDG-11 and -13) were selected for requalification. The lube oil system was thoroughly flushed to remove contaminants. The lube oil was changed to improve its

anti-foaming characteristics. A 100 hour bearing break-in run was
conducted to season the new bearings. A series of 10 prelubed fast
starts and a 7 day continuous run under load was conducted, and each of
these tests was followed by gap checks and a limited visual inspection of
the upper crankline main bearings.

The predominant failure contributors were maintenance related and all
operations and maintenance procedures and practices were reviewed and
improved. In addition, gap checks of both the upper and lower crankline
main bearings were required every 6 months or after any non-prelubed
engine start. To assure that the operability and reliability of these
EDGs was maintained, only fully qualified and well trained personnel were
to be permitted to work on these engines in the future.

3.3 Transamerica Delaval diesel generators

During a preoperational surveillance test on August 12, 1983, the main
crankshaft failed on 1 of the 3 EDGs at the Shoreham Nuclear Power
Station. The EDGs at Shoreham were manufactured by Transamerica Delaval,
Inc. (TDI), which has supplied 54 EDGs to 14 other nuclear power plants
in the USA. While evaluating the Shoreham failure, the NRC and the owner
utilities became aware of a broad pattern of deficiencies involving
critical engine components in TDI engines at Shoreham and other nuclear
and non-nuclear facilities. These deficiencies appeared to stem from
inadequacies in design, manufacture, and quality assurance/quality
control on the part of TDI. In addition to the crankshaft failures,
other problem areas which were identified included an engine block
failure, piston failures, cracked and leaking cylinder heads, excessively
worn turbocharger thrust bearings, and the rupture of a defective fuel
line and fire. Thirteen of the affected nuclear utilities formed an
Owners Group to establish a program for upgrading and reaffirming the
adequacy of the TDI diesels for nuclear service.

Phases I and II of the Owners Group program were completed in 1984 and
1986, respectively. Under Phase I, the Owners Group developed technical
solutions to the most significant known problems. This served as the
basis for proceeding with plant licensing during the period prior to the
completion of the Phase II program. Phase II of the Owners Group program
proceeded beyond known problem areas to systematically consider approximately 150 to 170 component types per engine, which were important to the
operability and reliability of the engines. Phase II was intended
primarily to ensure that significant new problem areas did not develop in
the future because of deficiencies in design or quality of manufacture.
The Owners Group performed design reviews and recommended needed component upgrades, modifications and inspections to validate the quality of
manufacture and assembly. The preparation of a comprehensive engine
maintenance and surveillance program to be implemented by the individual
owners was a major element of the Phase II program.

During 1986 the NRC completed its final evaluation of the Owners Group
program and issued its final report, NUREG-1216 (August 1986). The NRC
concluded that implementation of the Owners Group and NRC recommendations
concerning quality revalidation inspections, component modifications and
replacements, load restrictions and operating restrictions and precautions would establish the adequacy of the TDI EDGs for nuclear service,
as required by General Design Criteria #17 of Appendix A to 10 CFR 50.
The NRC further concluded that these actions had ensured that the design
and manufacturing quality of the TDI engines were within the range
normally assumed for diesel engines designed and manufactured in

accordance with 10 CFR 50, Appendix B. Continued reliability and operability of the TDI engines for the life of the facilities would be ensured by implementation of an enhanced maintenance and surveillance program.

The NRC concluded that several Phase I components merited special emphasis in the areas of load restrictions and/or maintenance and surveillance. These included the DSRV connecting rods, DSR-48 crankshafts, DSRV-20 crankshafts, cylinder blocks, cylinder heads, type AF piston skirts and turbochargers. Engine load restrictions were incorporated in plant Technical Specifications, license conditions, engine operating procedures and operator training, as appropriate. The most critical periodic maintenance and surveillance actions were incorporated as license conditions.

During 1985, several events occurred involving TDI diesel generators at operating nuclear facilities which are worthy of note. The assessment and resolution of these failures have enhanced the reliability of TDI engines in nuclear standby applications.

3.4 Grand Gulf

During an engine test on November 6, 1985, immediately following an extensive maintenance outage, a TDI diesel underwent an overspeed event, which caused extensive damage to bearings and other critical components. The engine startup test being conducted at the time was intended to verify that the preventive maintenance operations just completed had been performed correctly. Such a test is performed routinely after maintenance of this type, prior to placing the EDG into an operable status. During the preceding outage, considerable work had been done on the Woodward governor in which the lube oil/hydraulic fluid was totally drained from the speed control unit. During refill of the governor lube oil system, it was not properly vented.

During the startup test, the governor, which controls engine speed, did not function and the engine accelerated to a speed well in excess of its design/operating speed. As a consequence, bearings, bushings and other critical components were damaged. In addition, the overspeed trip mechanism, which is designed to rapidly shut down the engine during an overspeed event, failed to trip the engine quickly enough to prevent damage.

Based on the inspections, the cause of the governor malfunction, which led to the overspeed and subsequent damage, was clearly identified as inadequate lube oil level. New procedures were developed to properly refill the governor after maintenance operations. This information was promptly forwarded to owners of similar engines. In addition, a potential design deficiency in the overspeed protection trip system was identified and a design modification was developed by Woodward and TDI to reduce the trip response time and enhance the required engine overspeed protection system.

3.5 Catawba

Two separate failures of the number 7 main bearing on a Catawba EDG occurred on November 20 and December 5, 1985. The first failure occurred after 180 hours of operation. The second failure occurred after less than 90 seconds. The inspections identified 2 factors which may have contributed to the failures:

1). Misalignment of bearing shells during assembly
2). Excessive wear and scoring of the bearings and crankshaft journals from operation of the engine with contaminated lube oil

The NRC concluded that improper installation was not a common causal factor between the 2 events since the first failed bearing was installed at the TDI factory while the second was installed by licensee staff at the reactor site. In addition, no other failures of this type had been reported for TDI engines in nuclear service. Bearing failures due to improper installation or misalignment characteristically occur, as did the second failure, within minutes. Although the second failure is believed to have been caused by installation problems, the first failure could not be attributed to a similar cause. No physical or dimensional anomalies were found during the inspections which could cause repeated alignment problems.

Inspections before and after the failures identified an unusual degree of wear and scoring of bearings and journals. In addition, solid metallic particles found imbedded in the aluminum bearing shells and in the lube oil system were analyzed and found to be shot blasting material used during field piping modifications and repairs to the lube oil system. It is believed that this gritty material contributed to the scoring and wear leading up to the first failure.

The NRC had considerable uncertainty as to the cause(s) of the bearing failures, especially the first failure. The recommendation was made to carefully install and confirm proper alignment of the lower bearing shells by lifting/jacking the crankshaft during assembly. In addition, extensive flushing and cleaning of the lube oil system was performed prior to placing the engine back into service. The effectiveness of these measures was verified by a 100 hour confirmatory endurance test, during which the number 7 main bearing was disassembled and carefully inspected after 1 hour and again at the completion of the 100 hours of testing.

4 CONCLUSIONS

The NRC's regulatory requirements pertaining to emergency diesel generators will continue to evolve. The need for enhanced maintenance and surveillance programs has clearly been identified and reaffirmed through recent operating experiences. The assessment and resolution of these recent events has further contributed to the increased level of reliability of diesel engines used for nuclear standby applications. These events have also served to emphasize the importance of training and the development of proper maintenance procedures. The reliance upon qualification testing and periodic surveillance testing has been retained, but the test frequencies and conditions have been relaxed to minimize the potential for creating unnecessary problems. An analysis of the requirements imposed by the Japanese and the resulting reliability levels achieved by them indicates that the NRC requirements appear to be headed in a similar direction. The Japanese require EDGs to be routinely tested every 2 weeks, but they do not require these to be fast/cold starts. A fast start is only performed at each refueling outage. Between refuelings, they do not change the frequency of surveillance tests based on the number of past failures and they do not require additional special tests when either an EDG or an offsite power circuit is lost. In addition, each EDG is completely overhauled and inspected every 2 years. These requirements have resulted in almost failure free diesel operation at their nuclear power plants.

Development of the nuclear maintenance and surveillance program for the TDI diesel engines

R.J.Deese
Duke Power/TDI Owners Group, Charlotte, N.C., USA
C.H.Berlinger
Nuclear Regulatory Commission, Washington, D.C., USA

1 INTRODUCTION AND BACKGROUND

During the 1970s, many utilities ordered diesel generators from Transamerica Delaval, Inc. (TDI) for installation at nuclear plants in the USA to provide standby emergency power. The first of these engines to become operational at an operating plant was at San Onofre Unit 1 in 1977. However, nuclear plant operating experience with TDI Emergency Diesel Generators (EDGs) remained limited until preoperational test programs were commenced at Shoreham Nuclear Station and Grand Gulf Nuclear Station Unit 1 in the early 1980's.

Concerns regarding the reliability of large bore, medium speed diesel generators manufactured by TDI for application at domestic nuclear plants were first prompted by a crankshaft failure at Shoreham in August 1983. However, a pattern of other concerns in critical engine components subsequently became evident at Shoreham and at other nuclear facilities using TDI diesel generators.

In response to these concerns, 13 U.S. nuclear utilities formed a TDI Diesel Generator Owners Group to address operational and regulatory issues relative to TDI diesel generator units. The Owners Group submitted a proposed program to the NRC which was intended to provide an in-depth assessment of the adequacy of the respective utilities' TDI engines to perform their safety related function (Reference: TDI Owners Group, "TDI Owners Group Program Plan, March 2, 1984").

The Owners Group Program Plan involved three major elements:

1. Phase I: Resolution of known generic problem areas to serve as a basis for the licensing of plants during the period prior to completion of Phase II of the Owners Group Program.

2. Phase II: A design review/quality revalidation (DR/QR) of a large set of important engine components to assure their adequacy.

3. Expanded engine tests, inspections, and maintenance as needed to support Phases I and II.

The program was managed and performed under the principle directorship of Duke Power Management and Technical Services, Inc. (Now Duke Engineering & Services, Inc.) During the 14 month program over 2,000 components were reviewed, producing 12 site specific reports (one for each member utility) representing a total of 58 volumes of information. The Technical Review Program was completed on schedule in February, 1985.

A significant activity of the Owners Group effort involved the assembly of operational experience data pertinent to the TDI engines. Using input from various nuclear industry sources (including Institute of Nuclear Power Operations, Significant Event Report - Others, Licensee Event Reports, 10CFR 50.55E Reports and 10CFR21 notices) as well as non-nuclear industry sources, a substantial data base of engine/component operational experience was accumulated. This data base contributed significantly in the development of enhanced maintenance and surveillance requirements, as well as in the development of design review and quality revalidation recommendations.

As the Owners Group Technical Staff began its Design Review and Quality Revalidation efforts, the wisdom of an enhanced maintenance/surveillance program was very evident. The design reviews confirmed that component design was or would be (after completing recommended modifications) adequate for the anticipated service requirements, and quality revalidations provided an "as built" check to assure the components were of the correct materials and manufacturing tolerances and processes. While these items verified the current engine condition and served to increase reliability, maintenance/surveillance was identified as the key item to assure continued reliability. The surveillance portion of the program was developed based on available engine operating data to establish a formal trending analysis program to help anticipate component problems before they could result in an unplanned forced outage of the diesel generator unit.

2 NUCLEAR REGULATORY COMMISSION (NRC) REVIEW AND INVOLVEMENT

A key factor that contributed to the timely completion of such a comprehensive program was that NRC personnel were regularly and actively involved in the review process such that their comments were readily factored into the technical investigations. This cooperation between the regulatory agency and industry was likely the most significant contributor to the timely resolution of this issue. The NRC staff and its consultant, Pacific Northwest Laboratories (PNL), completed their review of the Owner's Group documentation and on June 24, 1986 issued a Safety Evaluation Report (SER). The memorandum transmitting the SER noted "The staff has concluded the implementation of the Owner's Group and PNL recommendations concerning quality revalidation inspections, component modifications and replacment, local restrictions, operating precautions, etc. will establish the adequacy of the TDI diesel generators for nuclear standby service......" (Source: NRC Memorandum, June 24, 1986, C Berlinger to H Thompson, et al.)

3 TDI REVIEW AND INVOLVEMENT

TDI participated actively with the Owners Group from its formulation. The supplier, in addition to their own efforts, put forth a considerable amount of effort and technical resources to assist the Owners in obtaining a resolution to NRC concerns which included making available personnel, manufacturing drawings and other technical data.

4   MAINTENANCE CRITERIA AND FREQUENCIES REVISITED

During the initial development of the maintenance and surveillance
program, member utilities urged that maintenance frequencies correspond
with scheduled plant outages to the extent possible to permit prudent
economical performance of the maintenance plan. This was generally
followed, however little attention was given to the length of downtime
required to perform the overall maintenance and surveillance program
during an outage. It became evident that, in order to comply with
plant Technical Specifications, one engine had to remain operational
while the other engine was down for maintenance. However, this
required sequencing in many cases made the diesel generator maintenance
the critical path item for plant outages. In addition, the NRC SER
required that several Owners Group inspections that were intended as
one time inspections be performed on a routine basis. (Ref: PNL-5600,
December, 1985.)

The NRC staff met with Owners Group representatives in October, 1985,
and agreed that the Owner's Group would undertake a technical review to
address member utility concerns and review the NRC required items to
develop a generic maintenance program that would be more easily
accomplished during the limited outage time and to keep diesel
maintenance from being a critical path scheduling item.

A seven member technical committee was formed to study the overall
maintenance/surveillance (M/S) program to identify areas for
improvement. The specific activities to accomplish this were:

1. Collect and organize M/S program exceptions submitted by all
Owners Group utilities.

2. Evaluate technical basis for each exception proposed and its
corresponding Owners Group recommendation.

3. Develop a draft list of potentially acceptable/unacceptable M/S
program exceptions with corresponding justification data.

4. Develop draft generic M/S documents for R-48, V-16 and V-20
engines based on potentially approved exceptions.

5. Study overall M/S program, clarify any open issues and revise as
necessary to account for potentially adopted exceptions.

6. Synchronize M/S activities where possible to streamline the
program.

7. Standardize M/S interval measurement where possible throughout the
document with emphasis on outage events.

8. Review TDI standard maintenance recommendations and assure that
all pertinent issues are included. Identify rationale for any not
included.

Once completed, the revised draft generic program was forwarded to
member utilities and the NRC for review. A committee was then formed
composed of one qualified representative from each participating
utility, one representative from TDI, members of the above technical
review committee, and NRC representatives. This committee finalized
the generic M/S documents and forwarded the documents to member
utilities for submission to NRC for final approval.

5   COMPONENT MAINTENANCE CHANGES AND JUSTIFICATIONS

As noted previously, the Phase I portion of the Owners Group program
focused on generic problem areas. These areas involved sixteen
components which were: air start valve capscrews, connecting rods,

connecting rod bearing shells, crankshaft, cylinder block, cylinder
heads, cylinder head studs, cylinder liners, engine base and bearing
caps, engine mounted electrical cable, high pressure fuel injection
tubing, jacket water pump, piston skirts, push rods, rocker arm
capscrews, and turbochargers. The balance of the program addressed
some 170 components. The primary focus of Owners Group and NRC reviews
involved the sixteen generic components because functioning of these
components is required to start and operate the engine and these
components, as noted, exhibited generic operating problems. This
discussion addresses significant maintenance requirements for some of
these components.

It should be noted that between the time that the Owners Group
completed its initial maintenance recommendations (November, 1984) and
Owners began making site specific requests for changes (October, 1985),
significant operating hours and experience in nuclear service had
accumulated to supplement the original data accumulated. This
experience included:

1. Shoreham EDG103 endurance run completed in November, 1984.

2. Approximately 1,631 hours and 748 starts on the Division I engine
and 1,050 hours and 350 starts on the Division II engine at Grand Gulf
Nuclear Station.

3. Approximately 994 hours on the Train A engine and 966 hours on the
Train B engine at Catawba Nuclear Station, Unit 1.

4. Significant testing during plant start-up and hot functional
testing at other nuclear stations.

For many components, no changes were made to the maintenance/
surveillance frequencies established for the original maintenance/
surveillance plan, but a frame-work was added, where appropriate, to
provide for extending the frequency in the future based on additional
positive operating experience. Thus the maintenance/surveillance plan
was developed as a "living document" to which changes could be made
without requiring a formal NRC licensing review under a pre-approved
arrangement.

The most significant change to the overall program was the
redefinition of the overhaul frequency. Originally the overhaul
frequency was established at a 5 year interval based on the Shoreham
experience. Prior to the crankshaft failure, Shoreham had established
a very strict maintenance/surveillance plan to improve engine
reliability. Following the quality revalidation inspections, it was
obvious that this frequency was very conservative especially in light
of the number of operating hours expected over a five year period.
(TDI experience has been to overhaul engines in the range of 10,000 to
15,000 hours and the five year intervals for nuclear standby operation
is estimated to log only 500 hours). Thus an extension of the overhaul
period to ten years was considered prudent, allowing flexibility and
performing overhauls in parallel with the reactor 10 year inservice
inspections. However, several of the Owners Group and NRC consultant
technical reviews had assumed that certain components would be
available for inspection at five year intervals. To satisfy this
requirement and hold valid these technical reviews, it was agreed that
25% of each of those components which were assumed to be inspected at
five year intervals would be inspected via a partial disassembly. If
these inspections found the component within the wear limits
established by the manufacturer, this would be justification for
reassembly and continuation to the ten year interval. Inspections from
that point would continue at 10 years. Should an inspection identify

an unreviewed problem, subsequent disassembly would be undertaken to determine the cause. Additionally, maintenance/surveillance frequencies were established with proper justification to permit engine disassembly on an alternate cycle basis, thus precluding disassembly of more than one engine during each outage.

The original maintenance/surveillance program relied heavily on inspections to foresee developing problems. The revised plan relies more on trending of recorded data that can readily be taken from engine instrumentation. This trending permits a more orderly and responsive assessment of future maintenance/surveillance needs.

For example, the lube oil and jacket water heat exchangers are monitored as part of this trending program. Originally the method of monitoring was to be via disassembly and visual inspection for scaling and corrosion. Disassembly of these components is time consuming and generally difficult. Disassembly prior to acknowledging a need to do so was not considered warranted. The trending program set forth for these components incorporates frequent sampling of the fluids they cool. This sampling and subsequent analysis are intended to recognize the presence of corrosion products. In addition, a heat balance is performed on the exchangers to monitor their ability to perform their design function. Loss of cooling over a sustained period would indicate problems and this information coupled with the cooling medium sampling provide sufficient timely information to permit teardown and cleaning at more appropriate times. Additionally, a daily walkdown of the engine is performed to detect any significant leakage that may be occurring.

Similarly the jacket water and lube oil heaters were originally scheduled to be disassembled, cleaned and checked every outage. The trending program established requires daily monitoring of the temperatures of the jacket water and lube oil. Any deterioration of the heating elements will appear in the trend data and permit orderly maintenance for these components.

The maintenance/surveillance program has been established to be a "living document." This approach was adopted to take credit for modifications, enhancements and positive operating experience to prudently reduce inspection frequencies. The procedures to justify these reductions are already set forth in the maintenance plan and preapproved by NRC so that these specified changes can be made without licensing action. Take, for example, the cylinder liners. Originally this component was to receive a boroscopic visual inspection every outage to check for abnormal wear. Typically, abnormal wear which might occur, will show up quickly. Therefore, the maintenance procedures allow the suspension of the boroscopic inspection after completion of two such inspections following piston-liner reassembly. These inspections would demonstrate proper installation and predict normal wear until the next disassembly when the procedure is repeated.

The surveillance documentation also provides a basis to suspend other monitoring. For example, some nuclear related engines have developed what have been determined to be benign cracks in the engine base. The surveillance program permits suspension of this crack monitoring, should they occur, once the cracks have been monitored for a sufficient period to determine that they are indeed benign.

6   RESULTS AND CONCLUSIONS

Preliminary feedback from Owners Group members and NRC is that
licensing and implementation of the new maintenance/surveillance
program is proceeding well. Over half of the Owners have received
operating licenses and all licenses have successfully implemented the
new maintenance/surveillance program.

Proper maintenance is important in ensuring long, reliable and
satisfactory service of the emergency diesel generators. Maintenance
work, in order to be effective, must be carried out thoroughly and
regularly. It is the belief of NRC and the Owners that this program
has and will assure the continued reliability of the TDI Diesel
Generators required for application in nuclear service.

REFERENCES

US NRC Safety Evaluation Report, "Safety Evaluation Report of the
    Operability/Reliability of Emergency Diesel Generators Manufactured
    by Transamerica, Delaval, Inc." Washington, D.C., June 24, 1986.
TDI Owners Group, "TDI Owners Group Program Plan,
    March, 1984.
TDI Owners Group, "TDI Diesel Generator Design Review and Quality
    Revalidation Report prepared for Texas Utilities Generating Company
    Comanche Peak Steam Electric Station - Unit 1, Revision 2, April
    1986.
Pacific Northwest Laboratories, PNL-5600, "Review of Resolution of
    Known Problems in Engine Components for Transamerica Delaval Inc.
    Emergency Diesel Generators," December 1985.
Pacific Northwest Laboratories, PNL-5336, "Review of Design Review and
    Quality Revalidation Report for the Transamerica Delaval Diesel
    Generators at Shoreham Nuclear Power Station Unit 1," October 1985.
Pacific Northwest Laboratories, PNL-5444, "Review of Design Review and
    Quality Revalidation Report for the Transamerica Delaval Diesel
    Generators at Comanche Peak Steam Electric Station Unit 1," October
    1985.
Pacific Northwest Laboratories, PNL-5718, "Review of Transamerica
    Delaval Inc. Diesel Generator Owners Group Engine Requalification
    Program - Final Report," December 1985.
NRC Memorandum, C H Berlinger to Hugh L Thompson, et al, June 24, 1986.

Service life prediction of components and systems

H.Eisele

Technischer Überwachungs-Verein Baden e.V., Mannheim, FR Germany

BACKGROUND

In many countries of the OECD member states the operating licences of
nuclear power plants will expire about 40 years after their date
of issue. The licenced operating period is derived from settings in
the ASME Code, which might have been fixed according to an empirical
basis and experiences with fossil plants taking into account some safety
margin for uncertainties in nuclear applications (1).

It is obvious that there may be an economic impact if nuclear plant
life extension allows nuclear electric generating units to continue to
operate beyond their original 40 years licenced period (2), (3), (4).
This question nowadays is strongly influenced by
- a serious acceptance crisis of nuclear energy and the necessary
efforts for restoring lost confidence among the public
- the difficulty to achieve the licence for a new site for a nuclear
power station caused by this acceptance crisis in the public.

Therefore, appropriate life-extending measures can be a way for the
utilities to demonstrate safety and reliability of nuclear power stations
in future, too.

In the Federal Republic of Germany an operating licence for a nuclear
power station is not directly restricted in its exploitation by a certain
number of years. But from the licencing point of view it must be
kept in mind that the construction permits of all plants are based
on documents, records and data in which an operating time of a plant
of 32 years is assumed; in combination with a load factor of 0.8
German power plants require relicencing activities after an operation
period of 40 years, too. However, the validity of the licence to
operate a nuclear power plant depends essentially on the success
to meet with the main demand of the German Atomic Law (5):

It is to provide by all technical and administrational efforts and
measures during planning, construction as well as operation of the
plant to bring and to keep all installations and procedures in a
condition without exertion of risk to environment and public, judged
on the basis of present state of science and technology.

Deviations from the licenced status of safety or new findings and
settings in regulations and guidelines may lead to backfitting activities
for the plant. If essential demands for the safety of components, systems
or procedures are missed or if out of this intolerable risks arise for
the health of public, the plant must be shutdown for corrective actions.

Therefore, if these general demands for the safe operation of nuclear power plants may be warranted, out of the German legal basis reasons for the restriction of the licence may not be derived. But our present experience shows that a new licence in combination with a legal public hearing will be necessary. The success of this procedure will be dubious if in the coming years no new nuclear power plant will be built and the state of science and technology is not obviously established.

EXPERIENCES

In a nuclear power plant a great variety of components must be kept in safe and reliable conditions. In order to predict the service life of the plant's components and systems all modes for the degradation of reliability must be considered. Table 1 shows typical influences due to aging and wear on components and systems. These influences are not of the same importance for all plant areas, but many degradation mechanisms are caused by designers, construction planning and the fabrication procedures in an early stage of the plant life.

The components which dominately control the plant life are those whose replacement or repair is difficult or impossible. A categorization on the basis of such aspects (1) is summarized in table 2.

However, the result of such a categorization strongly depends on conceptional items of a plant, which will be pointed out in this contribution.

The importance of such plant properties may be demonstrated by the replacement of the steam generators in the nuclear power plant Obrigheim which was realized within a quarter of a year approximately, Fig. 1.It can be shown that the best measures to extend service life of components and systems are based on lessons learned from events caused by failures.

In Table 3 some degradation mechanisms are attributed to the components and systems and their operational reliability. Further on, Table 3 gives reference to the main reasons by which deterioration phenomena are operative. Most of the degradation mechanism shown in Table 3 are important for the long term integrity of pressurized components. Each aspect in this concern must be considered in order to predict service life of components and systems.

The experiences with operating plants and the assessment methods to evaluate the integrity of the pressurized components of watercooled reactors have influenced guidelines and regulations in the Federal Republic of Germany in recent years to a great extent (6).

One of the results of these efforts, essential for service life prediction of components and systems, is the cumulative usage factor resulting from specified loadings due to transients for a modern plant design as shown in Fig. 2. It can be show that the results influence the activities for in-service-inspections an their frequencies.

CONCLUSIONS

1. The knowledge about the history of technology and the experiences with operating nuclear power plants show that the service life of a plant is basically determined by economic considersations.

2. The basis for life extension of nuclear power plants will derived from experiences with operating plants within the next ten years. It must be demonstrated that the basic design (software) of a complex plant will be kept in mind within two generations of operating staff.

3. There is no chance to operate a plant-Methuselah without planning and building up new plants.

REFERENCES

(1) Nakajima, H. et al, Life prediction study of reactor pressure vessel as essential technical foundation for plant life extension, OECD Nuclear Energy Agency Symposium on plant life extension, Paris, 24th-27th February, 1987

(2) Makovick, L. et al, U.S. National and regional impaets of nuclear plant life extension, OECD-NEA Symposium, (see Ref. (1))

(3) Hostetler, D.R., The Surry I plant life extension pilot study, OECD-NEA Symposium, (see Ref. (1))

(4) Neils, G.H. et al, The Monticello nuclear generating plant life extension pilot study, OECD-NEA Symposium (see Ref. (1))

(5) Federal Republic of Germany, Law on the peaceful use of nuclear energy and the protection against its risks (Atomic Law, 1980)

(6) RSK Guidelines for Pressurized water Reactors, 2nd Edition (4/80), January 24, 1979.
Köln: Gesellschaft für Reaktorsicherheit (GRS) mbH

Table 1   Influences of aging and wear

	DEFINITIONS	TYPICAL INFLUENCES
AGING	CHANGE OF SPECIFIC PROPERTIES OF METALLIC OR NONMETALLIC MATERIALS DEPENDENT ON BOTH - ENVIRONMENTAL INFLUENCES - TIME	LOADING COLLECTIVE, RESULTING LOCAL STRESSES FATIGUE AND CRACK FORMATION CHEMICAL COMPOSITION EFFECTS IN CONJUNCTION WITH NEUTRON IRRADIATION AND FRACTURE TOUGHNESS THERMAL AGING CORROSION AND CRACK GROWTH RATE IRRADIATION-ASSISTED STRESS CORROSION CRACKING (IASCC)
WEAR	PROGRESSIVE LOSS OF MATERIAL FROM THE SURFACE OF A SOLID BODY CAUSED BY THE MOVEMENT OF ANOTHER BODY (SOLID, LIQUID, GASEOUS)	FRICTION EROSION, EROSION-CORROSION (FRETTING), PITTING, WASTAGE, CREVICE FORMATION

Table 2   Categorization of components for service life prediction

	CATEGORY 1	CATEGORY 2	CATEGORY 3
ATTRIBUTE REPLACEMENT REPAIR CONTROLING QUALITY	DIFFICULT, IMPOSSIBLE POSSIBLE ENGINEERING LIFE (SERVICE LIFE)	POSSIBLE POSSIBLE ECONOMIC LIFE	PLANNED -- NO DIRECT RELATION TO SERVICE LIFE
EXAMPLES	REACTOR PRESSURE VESSEL CONCRETE STRUCTURES ELECTRIC CABLES	STEAM GENERATORS HEAT EXCHANGERS PIPING ETC	PUMP-, VALVE- INTERNALS FUEL ASSEMBLIES CONTROL RODS, DETECTORS ETC.
TYPE OF DEGRADATION	TIME (a) DEGRADATION	TIME (a) or (b) DEGRADATION	TIME (b) DEGRADATION

a): repair
b): replacement

—·—·—Limitation (safety)

Table 3   Origin of degradation mechanism

	DEGRADATION MECHANISM	INFLUENCED COMPONENTS AND SYSTEMS	CAUSED BY
(1)	IRRADIATION EMBRITTLEMENT	REACTOR PRESSURE VESSEL (RPV) INTERNALS OF RPV	DESIGN, MATERIAL, IMPURITIES IN STEEL (COPPER, PHOSPHORUS)
(2)	FATIGUE, CORROSION FATIGUE	PIPING AND ITS FITTINGS AND SUPPORTS, NOZZLES, VALVES, MIXING REGIONS OF FLUIDS WITH DIFFERENT TEMPERATURES	RESULTING LOCAL STRESSES, OPERATIONAL LOADING, SYSTEMS ENGINEERING
(3)	GENERAL CORROSION, PITTING, WASTAGE	SYSTEMS WITH LOW VELOCITIES OF FLUIDS, CONDENSATE IN STEAM LINES, SERVICE WATER SYSTEMS, SAFETY INJECTION SYSTEMS, INTERNALS OF PUMPS AND VALVES	SYSTEMS ENGINEERING, OPERATIONAL INFLUENCES, CREVICES (DESIGN), MATERIALS
(4)	STRESS CORROSION CRACKING (SCC)	WELD VICINITY IN COMPONENTS (STAINLESS STEELS), STEAM INDUCED FERRITIC PIPING/HIGH TEMPERATURE AND OXYGEN-CONTENT OF FLUID)	MATERIAL, OPERATIONAL CONDITIONS, OF CHEMISTRY CONDITIONS, INSULATION AND CASKET MATERIAL
(5)	WELD RELATED CRACKING, HYDROGEN EMBRITTLEMENT	ALL KIND OF WELDS, INTERFACE BETWEEN STAINLESS STEEL CLADDING AND VESSEL MATERIAL	STEEL-COMPOSITION AND MANUFACTURING PROCESS
(6)	EROSION-CORROSION	STEAM AND FEEDWATER PIPING, STEAM SEPARATORS, HEAT EXCHANGERS, TURBINE BLADES	CHEMISTRY CONDITIONS (PH-VALUE) MATERIALS, SYSTEMS ENGINEERING

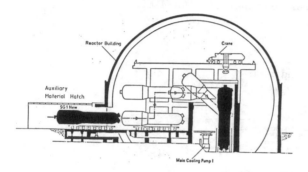

<u>Fig. 1</u>   Nuclear power plant Obrigheim, Installation of SG 1 (New)

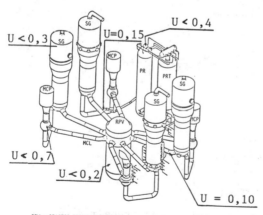

RPV = REACTOR PRESSURE VESSEL
MCP = MAIN COOLANT PUMP
MCL = MAIN COOLANT LINE
SG = STEAM GENERATOR
PR = PRESSURIZER
PRT = PRESSURE RELIEF TANK

U = USAGE FACTOR

Fig. 2

PWR-fatigue analysis, usage factor for main components

80

E Fast reactor core and coolant circuit structures

The European strategy for Fast Reactor core static mechanics studies

C.H.Jones

United Kingdom Atomic Energy Authority, Preston

A.Bernard

Centre d'Etudes Nucleaires de Cadarache, France

R.Di Francesca

ENEA-DRV, Bologna, Italy

D.Coors

INTERATOM, FR Germany

T.van Steenberghe

BELGONUCLEAIRE, Brussels, Belgium

1 INTRODUCTION

The European strategy for Fast Reactor core static mechanics studies has been formulated and agreed. It identifies the outstanding work in addition to current national programmes, which is needed in support of the three reactors to be built collaboratively under the inter-governmental agreement. It also identifies the new facilities which are needed.

A fast reactor core has a large number of sub-assemblies, with hex-agonal wrappers, nested together. Each wrapper is separated from neighbours by spacer pads located above and/or below the active core. Clearances are needed for thermal expansion, void growth, dilation and bow of the wrappers. Restraint is provided "naturally", as in SPX2, or by a surrounding barrel as in CDFR or by any intermediate style. Restraint causes interactive loads, which are offset by creep.

The magnitude of the loads and displacements varies widely with the choice of wrapper material and the style of restraint. Uncertainties in predicting them can be separated into those caused by variance in the swelling and creep of the wrapper material and those arising from mechanical processes of large arrays.

The economics of fast reactors must be improved for them to compete with PWR's and fossil fuelled power generation. This requires higher burn-up and longer dwell. It calls for low swelling wrappers which tend to give a slack core with attendant problems of vibration or rattling of sub-assemblies. Flexible design is required so that a gradual change can take place from the use of swelling materials initially, to low swelling materials ultimately.

2 OBJECTIVES

The strategy is designed to meet the needs of participating European countries by achieving the following objectives:
- to provide generic information and data on the static mechanics of all styles of core of interest, validate codes and optimise new designs.
- to endorse specific features of SPX2 and SNR2 designs.
- to provide the new facilities which are needed.
- to present the programme of work required and timescale.

83

3 UNCERTAINTIES

In designs with light restraint, vibration may cause frictional restraints to become negligible. The importance of friction may vary
considerably, depending on the style and the tightness of restraint.
If frictional forces are zero, variance between calculations and
reality results only from approximations in input data and assumptions
which simplify the codes.

If friction is significant, then the calculation of core loads and
displacements is much more complicated and expensive. Friction permits excess elastic-strain energy, above the minimum, to be stored in
the core. This can be released, by any transient event or reduction
in the friction coefficient. There are no unique solutions. The restraint system must prevent excessive deviations from prediction.

4 BENEFITS

Benefits sought from the proposed work are that:
 - increased assurance will come from validation of codes, and
measurement of variances from predicted loads and displacements.
    - scope for less conservative design and improvements in safety
margins can be exploited.

5 NATIONAL PERSPECTIVES AND PROGRAMMES

5.1 France

The french development programme, initially based on SPX2 project,
will be extended to advanced reactors studies, which exploit core
design options.

The HARMONIE code system, together with modules such as FORCEZ for
core-handling machine interaction and HARMOREA for calculating
reactivity effects, now includes detailed models of core mechanics.

The French rig for large arrays studies, R106E, simulates the
naturally restrained core in SPX1 at full scale. It has been used to
support the SPX1 design and to validate HARMONIE for this core style.
The rig has been extended to meet SPX2 requirements by including a
central island of 7 short-spike S/A's. A new device to simulate
charge discharge operations completely, with a parametric representation of the real handling machines stiffnesses, has just been
installed on R106E. It will enable an extended validation of FORCEZ
to be made.

Some limitations in the capabilities of R106E, are too fundamental
for a complete validation of recent HARMONIE developments, such as
geometrical non linearities and the friction model, and from a more
general point of view, for reducing significantly the uncertainties
for different core styles. These features will be obtained in the new
CHARDIS IIC rig.

Code validation from in-air tests is complemented by measurements
obtained from reactor operation and compared with a very fruitful, but
difficult code simulation. The operation of Phenix, and now of SPX1,
has produced, and will produce, a large amount of data. Handling
forces are systematically recorded on the two reactors and most of the
S/A's distortions are measured in Phenix by PIE.

5.2 Italy

ENEA activities in core static mechanics are directed primarily to qualification of the PEC core design including assessment of all mechanical, structural components and thermal distorsion effects.
  The main Italian objectives comprise development and validation of HARMONIE code including SISCO (developed by ENEA) and a new code to predict forces throughout handling operations.

5.3 Germany & Belgium

The German experience is based on operation of KNK II and on the experimental programme performed in support of SNR-300 in co-operation with Belgium. For SNR2, the need for handling tests with arrays of at least 7 full scale sub-assemblies is emphasised. They will cover the range of possible geometrical variables, simulation of refuelling movements with bowed and dilated sub-assemblies and tests of initial core loading without dummy sub-assemblies.
  For the core mechanical design and for the treatment of problems such as fuel management, the 2d-code FIAT as well as the 3d code system DDT is available at Interatom. For detailed fuel assembly wrapper tube mechanical design and analysis the 3d-code STRAW developed by Belgonucleaire is used within the German-Belgian collaboration.

5.4 UK

The development programme described in succeeding papers in this session will remain based on CDFR until a new reference design is available. Continued study of core mechanics is needed to arrive at an optimum style of restraint. The code CRAMP is the basis of core design and of PFR core management. It will be extended to calculate interactions with the handling machine throughout refuelling operations and to include spike clearance.
  Experiments with the charge-discharge rig CHARDIS I showed reasonable agreement between the axial load to charge or discharge a sub-assembly, but not for individual loads. CHARDIS II provides more accurate and comprehensive data. It will soon have a generic and adaptable handling machine. Data produced will be used to study mechanical interactions of core operation and refuelling and to validate CRAMP for barrel restrained cores.
  Test sections have been designed to expose all guidance and locating features to the full range of reactor duties in the sodium components test facility at Risley and at reactor operating conditions. The effective friction coefficients will be measured throughout. In the "Monkey's Paw", a section of dilated wrapper will be extracted from neighbour sub-assemblies.

6 FACILITIES FOR TESTING LARGE ARRAYS

Existing facilities are listed in Table 1. None of them could yield the information with large arrays of sub-assemblies needed to validate the codes used for SPX2, SNR, CDFR or successors. The European view is that a single new facility is needed to meet both current and future requirements, which is easily adaptable to changes in styles of restraint or handling machine design. The best way to meet this need is to replace the existing rig by CHARDIS IIC. This uses modified dummy SPX1 sub-assemblies.

7 PRINCIPLE FEATURES

CHARDIS IIC retains the basic principle that the boundaries of the array of sub-assemblies are adjusted in a pre-determined manner to represent the displacement of the corresponding imaginary boundaries within a real core. These displacements and the resulting loads are measured at the boundaries and at all spacer pads surrounding the central sub-assembly. The principle applied to simulation of "natural" restraint is illustrated in Fig. 1. It requires an increase in lattice pitch (of 9%) to provide room for adjustment of sub-assembly bow and dilation, attaching load measuring pads and use of a tilting chandelle. All S/As have a joint at the core mid-plane which can be preset and clamped to any required bow. All have detachable spikes with adjustable clearance and stiffness.

The new features of CHARDIS IIC are:
- A diagrid replaces the bottom plate of CHARDIS II. It locates tubes (representing chandelles) which can be tilted to simulate bow.
- Two stiff frames are provided at restraint planes. They can be moved vertically to represent different styles of restraint.
- Rams with large displacements are provided at each restraint plane. Transducers measure the load and displacement of each ram.
- A charge machine guide tube of larger diameter, with a grab to suit, will be fitted.

To use the rig, the procedure is:
(a) Calculate loads, gaps and displacements for a reactor core through one or more runs and shut downs.
(b) Calculate a matching set of rig adjustments.
(c) Measure actual initial conditions.
(d) Calculate the expected consequences of applying (b).
(e) Measure them and compare.

To simulate refuelling, set up the shut-down conditions, as above, then with constant displacements convert each ram from its stiff mode to a flexible mode. Springs within the ram then carry the load and

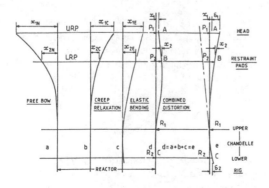
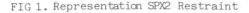
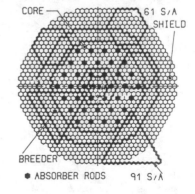

FIG 1. Representation SPX2 Restraint   FIG 2. Core Locations & Boundaries

represent the remainder of the core. The central (distorted) S/A may
then be discharged and an undeformed S/A may be inserted.

8 SIZE AND SHAPE OF ARRAY

A key question is how can we be sure that large arrays will behave as
predicted and without unpleasant surprises? The authors and their
colleagues judge that the best value for money and time will come from
using hexagonal arrays of 61 S/As. Fig. 2 shows how they may be used
to represent various regions of the core. After a few years work, any
doubt about the subtleties introduced by the shape of the core
boundary, could be resolved by tests with 91 S/As.

9 OTHER FACILITIES REQUIRED

The French R106E is being adapted to test 7 SPX2 S/As in an array of
54 SPX1 S/As. A representative charge machine is being added. This
will meet the French need for a rig with standard S/As to study handling problems and contribute to code validation.
     Each other country constructing one of the "European" reactors will
require a simple rig with 7 standard S/As and a handling machine.
     Italy has a facility for testing SPX2 spike/chandelle components in
air and water. Mulinox is used for tests in sodium on PEC spikes.
     The UK has constructed SOD 1 to test the spike/support tube/diagrid
bush interactions in CDFR. It can be adapted to test specimens similar to the SPX2 spike and chandelle. SOD 2 has been designed to study
the friction and wear of spacer pads. It can also perform Monkey's
Paw extraction tests.

10 PROGRAMME

   Handling tests in R106E with bowed S/As for SPX2 and with distorted
S/As and variable grab stiffness for SPX1 will be made in 1987. Tests
with 19 S/As in CHARDIS and of the PEC spike and socket in MULINOX
will be completed. Code development will include introducing friction
and clearance in HARMONIE and development of FORCEZ. DDT will be
verified and endorsed.
   By 1989, it is planned to complete CHARDIS tests with 61 S/A's,
CRUPER tests with 91 S/A's (to study loading mode, pad area, high
friction) and to complete construction of CHARDIS IIC. By this time,
CRAMP will be validated for barrel restrained cores and extended to
cover spike clearance and refuelling; DDT will include multi force pad
flexibility. Generic tests of CDFR spike/support tube/diagrid in sodium will be completed and tests to evaluate the durability and friction of contact faces will be in progress. It is envisaged that the
whole programme will be completed by 1993.

11 CONCLUSION

The European strategy recognises the valuable contribution of existing
national programmes, and the common need for improved validated codes
for core static mechanics and refuelling load calculations.

In the European view a single major adaptable rig is required. It will be used to study the behaviour of large arrays of sub-assemblies and of interactions with the handling machine. This rig should be a new version of CHARDIS II known as IIC. It will have an initial complement of 70 sub-assemblies made from SPX1 dummy core sub-assemblies used in arrays of 61.

France already has the R106E rig. Each other country constructing a new reactor under the inter-governmental agreement, will require a simple rig with 7 standard sub-assemblies and a handling machine replica.

The programme includes tests in sodium rigs to endorse the durability of guiding features and bearing surfaces and also to establish the best values for and variability of the friction coefficient.

The SPX2 design will be endorsed in 1987.

The programme shows that the new work identified in the strategy will provide soundly developed and validated codes with the complimentary mechanical design information needed by 1993.

TABLE I
Existing core static mechanics rigs

Medium	Rig	Country	Remarks
Air	CHARDIS II	UK (UKAEA)	Represents 'barrel' restrained cores with a generic charge machine.
	CRUPER	UK (NNC)	Single plane of 91 S/As.
	R106E	France (CEA)	7 SPX2 S/As and 54 SPX1 S/As with fully representative charge machine.
	CHARLIE	France (CEA)	Measurement of bending stiffness.
	PROMETHEE	France (CEA)	Measures pad elastic and thermal creep distortion at operating or fault temperatures.
	FOOT/ CHANDELLE	Italy (U of PISA)	Mechanical test of Foot to Chandelle hard facing.
Sodium	SOD 1	UK (UKAEA)	Tests Spike/Support Tube/Diagrid bushes.
	SOD 2	UK (UKAEA)	Tests spacer pad and head surfaces.
	SOD 3	UK (UKAEA)	SOD 2 for Monkey's Paw Tests.
	EXPRESSO	Italy (ENEA)	Thermal Shock + dwell.
	CEDI	Italy (ENEA)	Thermal Shock + dwell.
	RSB	Germany (INTERATOM)	Tests reactor handling mechanisms.
	AKB	Germany (INTERATOM)	Tests core components.

A review of the UK fast reactor core mechanics programme

P.J.Bramah, C.H.Jones & J.C.Duthie
United Kingtom Atomic Energy Authority, Springfields Nuclear Power Development Laboratories, Salwick, Preston

K.F.Allbeson
National Nuclear Corporation, Risley, Warrington, Cheshire, UK

1. INTRODUCTION

Fast reactor cores contain large numbers of hexagonal sub-assemblies, S/As in a closely packed array. Irradiation causes the S/As to deform and to contact each other, generating inter S/A loads. The deformations and loads vary during a reactor run, during power changes and when the reactor is shut down. They are dependent on the core 'style' or design, the S/A materials and on the magnitude of the friction coefficient. Although a core relaxes as power is reduced, refuelling loads are governed by the S/A deformation, the friction levels and also by the way in which the charge machine interacts with the core.

The ability to calculate the behaviour of large arrays of S/As is fundamental to the development of fast reactor cores, for the study of alternative core designs and for predicting the behaviour of operating reactors. It is also vital to be able to demonstrate that reactors can be refuelled reliably. These needs require the development of appropriate analytical techniques and the construction of experimental facilities, firstly, to validate the codes and to assess their confidence limits, and, secondly, to generate data on friction under appropriate conditions. A co-ordinated programme to satisfy these needs, funded jointly by UKAEA and NNC, has been carried out in the UK for over ten years.

This has resulted in the development of the computer code CRAMP (Duthie 1987). This code is the cornerstone for all UK core mechanics studies and a large part of the development strategy is concerned with its validation. In particular, the CHARDIS and CRUPER rigs are operated to model the S/A interaction process and to study refuelling operations. In addition, the series of rigs SOD RIG 1, 2 and 3 have been constructed or are in design to determine the behaviour of the rubbing surfaces during charge/discharge operations.

During the past two years, considerable debate about the development of a European strategy for Fast Reactor core mechanics studies has been undertaken by the European collaborative partners. This has resulted in the development of a unifed strategy which will require the construction of a new rig CHARDIS IIC. This is described in a separate paper to this conference (Jones 1987). However, these discussions have also confirmed the soundness of the basic UK programme. The objectives and strategy underlying this large experimental and theoretical programme are described in this paper.

89

2. BACKGROUND

Design of the UK Prototype Fast Reactor, PFR, was frozen before the
significance of neutron induced voidage, NIV, swelling and irradiation
creep was fully appreciated. As a result, the reactor was constructed
with a 'free standing' core style in which the S/As are mounted in
groups of six around a stiff leaning post. Some inter-action between
groups occurs but only modest loads are permitted. The development of
significant wrapper bow is prevented by rotating or shuffling the
S/As.

Initially the UK commercial demonstration fast reactor, CDFR, had
the same core style. However, in the mid nineteen seventies it was
realised that more positive restraint was required, partially to limit
S/A distortion and head displacements but also for reasons associated
with the dynamic behaviour of the core. As a result, the 'barrel'
restrained style was adopted in which the S/A has a flexible spike and
the wrappers are padded at two levels called restraint planes. These
are at the top of the S/A and just above the core. The core is
surrounded by a stiff barrel which supports restraint rings at the two
planes. In this style, the S/A bow displacement and spike loads are
minimised, easing refuelling, but the on-power inter S/A loads are
moderate.

On the continent of Europe, the 'natural' restraint style has been
adopted. In this design, the S/As have a stiff spike and one main
restraint plane above or below the core at which the majority of the
interactions occur. However, the bow displacements are larger in this
style and, in the outer regions of the core, some interactions are
generated at the tops. On power loads are generally less than with
barrel restraint but the refuelling loads are higher. The S/As also
have lower natural frequencies closer to earthquake excitation values.

Currently, barrel restraint is the reference style in the UK.
However, this is being reviewed with the objective of optimising the
design to generate one which has the very best mix of dynamic, on-
power and refuelling characteristics. It is expected that the optimum
will have many of the properties of the barrel style and so work is
continuing vigorously on the core mechanics development programme
which is based on the CDFR design.

3. CRAMP DEVELOPMENT

In parallel with the decision to develop the barrel restrained core,
it was realised that a fully comprehensive code was needed to analyse
its mechanical behaviour. Two codes SABOW and CRAMP were in existence
but it soon became obvious that CRAMP was the better and had the
greater potential for development. Since then CRAMP has been improved
continuously. Some of the details of its solution technique and its
capabilities are described by Duthie (1987). The code calculates the
progressive change in the shape of S/As caused by NIV swelling and
creep and computes the changes in the mechanical interactions within
the array. It contains an extremely efficient solution algorithm and
is capable of solving 30° and 60° symmetric solutions and full 360°
arrays.

Having developed the basic non-friction versions of the code, effort
over the recent past has been concentrated on the incorporation of the
mechanical effects of absorber rods and of inter S/A friction. The

rods connect the core to the above core structure and introduce complexity because of the variety and number of contacts which are possible. The friction algorithm in CRAMP is an exact solution which follows each event as the core deformation develops. The development of the solution routine was troublesome because after each event many possible solutions must be tested to find the correct one and because true instabilities can occur within the array. Nevertheless, these developments have been successful. In addition, a special version of CRAMP called RIG CRAMP has been developed to model the operation of the core model rigs CHARDIS and CRUPER.

Substantial effort has also been expended to enable CRAMP to be used as a routine tool for managing the PFR core. The code has been incorporated within the COSMOS suite of programs which are used to perform the neutronic calculations for PFR. Flux and temperature data are transferred automatically to CRAMP and so the code can be used to determine when fuel needs to be rotated, shuffled or replaced. In addition, speculative calculations can be performed to assess future refuelling strategies.

In the near future it is planned to introduce additions to represent, firstly, the clearances around S/A spike bearings and, secondly, charge/discharge operations. The clearances are important for core designs which include stiff S/A spikes. The refuelling additions are by far the more complex and the techniques which will be used have not yet been defined.

However, at present the majority of the development effort is concerned with the verification and validation of the code. CRAMP is being verified through internal procedures and through the IWGFR exercise in which specific examples are solved using a variety of codes and the results compared.

Validation is being carried out in two separate exercises. Firstly, the parts of the code which compute the mechanical behaviour of an interacting array of S/As are tested using the CHARDIS and CRUPER rigs. This is described below. The second exercise is to assess the accuracy of the prediction of the neutron induced S/A deformations. This can only be done by the comparison of CRAMP and PIE results obtained either from specific PFR experiments or from measurements on standard PFR S/As. The planned experiments are described below. Comparison of the CRAMP results and measurements of S/A bow, length and dilation changes are fraught with difficulties, firstly, because of the general uncertainty about the accuracy of the material rules, secondly, because PFR has operated at a range of temperatures which complicate the use of the rules and, thirdly, because the as-made S/A dimensions and bows are not known to the high levels of accuracy needed for this purpose. As a result, useful comparative data can only be obtained from a limited number of S/As. These must have had a simple operational history and have generated large displacements which are significantly greater than the possible initial dimensional uncertainties or have had a number of intermediate PIE inspections. Suitable S/As from PFR have been identified and currently calculations are in-progress to provide comparative data.

4.   CHARDIS/CRUPER

As described above, the CHARDIS and CRUPER rigs are the basic tools for validating the mechanical portions of the CRAMP code. They were designed to reproduce, as well as can be done outside a reactor, the bowing and dilation deformation processes which occur within an array; bowing causes multi point bending of the wrapper and spike. In the barrel restrained core the principal loads are at the foot of the wrapper and at the two restraint planes; the bending moments in the spike are small. Dilation generates squashing loads throughout the array.

CRUPER (Barnes 1984) was designed to verify the operation of CRAMP within a single restraint plane. The rig incorporates an array of 91 short (600 mm) 'S/As' of known stiffness and dimensions and which have a single plane of contact pads. Servo controlled rams impose known displacements at the periphery of the array. Ram loads and inter pad gaps and loads are measured and are compared with values calculated from the ram movement using CRAMP.

The CHARDIS 1 and 2 rigs are described in separate papers presented to this Conference, Edwards and Jones (1987) and Darke et al (1987). The rigs model a barrel restrained core and contain an array of S/As within a stiff framework which supports a base plate/diagrid and two restraint planes. The dimensions of the S/As are known very accurately. The array can represent any area within the core, for example at the core centre or near the core breeder boundary. The rig central S/A can therefore be any S/A within the core. S/A displacements at the restraint planes are small in a barrel restrained core and so bow loads are generated by moving the bottom of each individual S/A by an amount equal to its free bow deflection. Dilation loads are generated by rams at the restraint planes which act separately on the external faces of the peripheral S/As in the array. The rams can be operated in either a 'stiff' or a 'flexible' mode. When stiff, the rams position the boundary of the array; when operated flexibly, they simulate the effect of the remainder of the core. The S/A pitch is increased, firstly, to allow the bottom end movements and, secondly, so that pads can be attached to the wrappers at the restraint planes. These pads can have a variety of surfaces so that friction coefficients may be changed. Some pads are also straingauged to permit the measurement of inter S/A loads.

The CHARDIS 1 Rig contained an array of 19 PFR length S/As. Idealised experiments were carried out and the measured loads around the central S/As were compared to CRAMP calculated values derived from the known ram and S/A movements. Edwards and Jones (1987). These tests indicated that CRAMP was working well but also showed the need for improvements to the rig.

CHARDIS 1 was reconstructed as CHARDIS 2. The improvements are described by Darke (1987). This rig is initially being used with 19 S/As but will soon be converted to a 61 S/A array. To operate the rig, data on S/A shapes and positions for the area and core state of interest are extracted from a CRAMP calculation of the operation of CDFR. The corresponding off power data are also known. The sizes of the rig S/As are then modified to the off power dimensions by the use of shims under the pads. The ram head and S/A bottom end displacements needed to generate the load pattern occurring at the particular core state are then calculated using a separate program. This is not an exact process and only produces an approximation to the core condition. The displacements are then applied in systematic increments in the rig and the resulting loads and inter S/A gaps are measured. These are then compared with RIG CRAMP results.

The first indications are that excellent validation data are being generated (Darke et al 1987). The change in the size of the rig array from 19 to 61 S/A will show whether array size has any effect on the level of experimental and calculational agreement. This is important because the results must be extrapolated to real arrays which may well contain over 1000 S/As.

In addition to the tests to study the behaviour of the core alone, CHARDIS 2 will also be used as the basic tool for studying the mechanics of charge machine/core interactions during refuelling operations. A representative charge machine which includes typical freedoms, clearances and stiffnesses has been designed and is to be constructed above the rig. The rig will be set up to the appropriate off-power condition for the core area of interest. When carrying out charge/discharge tests the rams must be operated in their two modes. Firstly, in their stiff mode they are used to fix the boundary of the array. They are then converted to flexible mode without changing the displacement, thus giving the boundary the flexibility of the whole core. The S/As surrounding the central one will be bowed and straked to make an appropriately shaped hole and then the highly bowed or dilated central S/A will be discharged. The effects of errors in the positioning of the machine will also be determined. The axial load and the lateral loads at the head of the S/A and at the restraint planes will be measured. Equivalent charging tests will also be carried out. Checks will be carried out to ensure that no snags develop and that rotational alignment occurs automatically. The data will be compared with results from Charge/Discharge CRAMP.

5. IN-REACTOR/PIE TESTS

The CRAMP validation exercise described above will generate the major part of the data needed to provide a full understanding of the behaviour of a barrel restrained core. However, the free standing core in PFR, which allows almost unrestrained S/A bowing, does not provide representative conditions to test the combined effects of NIV swelling and creep which are generated by S/A bowing and pad deformation in more firmly restrained cores. Therefore two special PFR experiments, the restrained bowing experiment, RBE, and the loaded pad test, LPT, have been devised to test the UK ability to calculate deformations in these key areas.

In the RBE a specimen wrapper of reduced across flats dimensions will be mounted inside a thickened PFR wrapper and will be irradiated in a region of high flux gradient which will cause both wrappers to bow. Periodically, the experiment will be removed from the reactor, inspected, the outer wrapper will be rotated through 180° and then returned to the reactor with the inner wrapper in its original orientation. Thus the outer wrapper will remain approximately straight while the inner one bows. The measured distortions and contact loads will be compared with CRAMP results.

The loads on pads impose high local stresses and strains. The creep deformation will be difficult to calculate. In the LPT, CDFR pad specimens will be mounted in special PFR assemblies and will be irradiated under representative loads, temperatures and flux levels. The load will be exerted using bellows pressurised by sodium at S/A inlet pressure. PIE distortion measurements will be compared with results drived using the defined material rules.

6. EXPERIMENTS IN SODIUM

As mentioned above, a series of tests will be carried out in SODIUM RIGS 1, 2 and 3 to assess the endurance and friction levels which will be developed at the inter S/A and spike contact points in sodium with very low oxygen levels. Adhesion and/or wear may be possible at these points even though they will be hardfaced.

In SODIUM RIG 1, a S/A spike support tube and its diagrid bushes will be mounted in a representative manner. These will be subjected to a simulated duty cycle representing the most arduous combination of applied loads including the effects of variations in S/A distortion during life, flow induced vibration and charge/discharge. Vibration and friction will be monitored during the tests and all the bearing surfaces will be inspected for damage after the test.

The endurance and friction of the pads at the upper and lower restraint planes will be tested in SODIUM RIG 2. Pads representative of the designs at the two planes will be mounted on a short central wrapper. Appropriate loads, which can be changed during operation, will be applied through mating pads on six surrounding wrappers. On power loads will be applied and the rig will be soaked in sodium at 600°C. The loads and temperatures will then be reduced to refuelling values and the pads will be drawn past each other firstly to full disengagement and then to represent reinsertion. Axial loads will be measured and, after the tests, the pad surfaces will be examined for wear.

Rig 2 will be modified to RIG 3 to represent the withdrawal of a dilated wrapper section past the pads at the two restraint planes.

CONCLUSIONS

In this paper, a strategy has been presented for demonstrating the satisfactory static mechanical behaviour of the UK fast reactor core. The programme is designed to provide a fully validated code for predicting static core mechanics and refuelling behaviour. The appropriate parts of the code will be validated using out of pile rigs. Specific in-reactor experiments, PIE inspection or in sodium tests will be used to validate other parts of the code or to provide the necessary basic data.

REFERENCES

Barnes, W.D. 1984. A review of the UK core mechanics experimental programme. Paper 3/1, IWGFR specialist meeting on prediction and experience of core distortion behaviour (IWGFR-54) Wilmslow UK.

Darke, A.C. Ridehalgh, F, and Jones C.J. 1987 Commissioning and initial operation of CHARDIS 2 and comparison with CRAMP. Paper E1/5 9th SMiRT Lausanne

Duthie, J.C. 1987. Development and Application of the CRAMP code for Fast Reactor Core Assessment. Paper E1/3 9th SMiRT Lausanne.

Edwards, J.A. and Jones, C.J. 1987. Analysis of data from the core mechanics rig CHARDIS 1 using the program CRAMP. Paper E1/4 9th SMIRT Lausanne

Jones, C.H. 1987. The European Strategy for the Static Mechanics of Fast Reactor Cores. Paper E1/1, 9th SMiRT, Lausanne.

Analysis of data from the core mechanics rig CHARDIS I using the program CRAMP

J.A.Edwards & C.J.Jones
United Kingdom Atomic Energy Authority, Preston

1 INTRODUCTION

Fast Reactor Cores consist of a large number of closely packed hexagonal sub-assemblies which contain the fuel, the breeder and con-trol rods. Each of the sub-assemblies becomes distorted during its life in the core and exerts forces on the neighbouring assemblies.

In the case of CDFR these forces are eventually reacted against two stiff restraint rings which are carried on a barrel surrounding the core. The sub-assemblies are fitted with pads at the appropriate heights.

It is necessary to ensure that the forces and displacements of the sub-assemblies do not either impair the life of the assembly or pre-vent removal of the assembly at the end of its' life. The computer code CRAMP[1] was written to predict these loads and displacements. A large scale experimental rig (CHARDIS I) was constructed to help with the validation of CRAMP and also to investigate the mechanics of the discharge process. The rig contained virtually full size assemblies having the correct geometry under representative loading conditions.

In this paper the rig together with details of one particular sequence of experiments are described and the application of the CRAMP program to the analysis of the results is discussed.

2 CONCEPT OF THE RIG

The loads developing on a sub-assembly in a fast reactor core may be divided into two broad categories, firstly, those caused by temp-erature or neutron dose and, secondly, those caused by radial grad-ients in these parameters.

The major difference between the two is that the average temperature and the cross sectional averaged effects of neutron induced swelling and creep at any level along a S/A causes it to expand or contract at that level. However, the temperature and dose gradients across the S/A causes it to bow and so generates a moment on the assembly.

It is possible to simulate the bow loads by bending the sub-assemblies in an array and to simulate the others either by applying loads at the periphery of the array or by introducing deliberately over size sub-assemblies.

3 DESCRIPTION OF CHARDIS I

The CHARDIS rig, which is illustrated in Fig. 1 consisted of a hexagonal array of 19 sub-assemblies supported on a baseplate. The S/As
had the correct mechanical properties and were provided with removable
pads at the two restraint plane heights. The pads were designed to
retain the correct local wrapper flexibility. At each of these
levels, the array was surrounded by a stiff framework which carried
adjustable rams. The baseplate and the frames were supported by three
legs.
   The line of action of the rams was between two S/As, see Fig. 2, and
equal loads were applied to both. This was achieved because the rams
were pivoted at the head and at their attachment to the frame. The
rams could be operated in either a stiff or a flexible mode. When
flexible, they were designed to represent the remainder of the S/As in
the core.
   The rig S/As, which were modified from PFR dummies, were 3720 mm
long and 142 mm across flats. Each S/A had a spigot attached to the
lower end and a handling head attached to the upper end. The spigots
were carried within slotted cups which were mounted below the rig base
plate. The orientation of the slots could be adjusted relative to the
base plate and the spigots could be moved within the slots by set
screws causing elastic bending of the S/As. The 19 cups were arranged
on a triangular lattice of 165 mm giving a nominal gap between assemblies of 23 mm. The whole array was surmounted by a screw driven
charge machine which could be positioned over any of the sub-
assemblies.
   The centre sub-assembly of the hexagonal array was slightly
different in that it was fitted with a flexible spike typical of those
used at the bottom of sub-assemblies used in PFR. Its lower end was
also contained in a slotted cup.
   The pads which surrounded the centre sub-assembly at the lower restraint plane were of a slightly different design to the remainder and
were also fitted with strain gauges. The configuration had been
specially developed to measure the loads on these pads.

4 OPERATION OF THE RIG

In one set of experiments a series of movements were imposed on the
rig to generate a simplified, idealised representation of the loads
and displacements which would be generated in the centre of a
restrained core. The test started with the rig in its 'at rest' condition. That is, the spigots and the sub-assemblies were at their
nominal pitch positions and the rams were in their stiff mode just
touching the pads at the upper and lower restraint planes. The slots
in the support cups were oriented in a radial direction pointing
toward the central S/A. In the first stage, the spigots at the
bottoms of the outer ring of sub-assemblies were moved inwards
sequentially in increments of 1.25 mm to a maximum of 11.4 mm. The
spigots on the inner ring of sub-assemblies were moved in a similar
manner during the second stage. These bending movements generated a
squashing load on the array of S/As at the lower restraint plane and
an outwards load onto the framework at the upper plane. During this
sequence the centre assembly was removed and replaced at each of the
measurement stages. The discharge loads were measured and the strain
gauge amplifiers which measured the pad loads were re-zeroed.

In the final stage an additional squashing load was applied at the lower restraint plane from the rams. Their adjusting screws were advanced sequentially in steps of 1.2 mm to a maximum of 9.0 mm which gave a mean load of 4500 N per ram. Finally the centre assembly was discharged and charged twice. The rig was then returned to its 'at rest' condition and the gaps between the pads at the upper and lower restraint planes were measured.

5 CRAMP CALCULATIONS

CRAMP[1] was written to calculate the S/A deformations and consequent mechanical interactions which are caused in a fast reactor core due to the combined effect of swelling and irradiation creep. A modified version of the code was developed to allow modelling of the operations in the CHARDIS I rig. The modifications to CRAMP allowed the movements of the rams and the sub-assembly bottom ends to be represented. The freely pivoting ram heads were modelled thus imposing zero torque on the ram.

Prior to the CHARDIS tests the flexibility of the S/As in the rig and the cross-sectional flexibility of the wrapper had been measured. The latter was defined using a "Miller Matrix", the terms of which gave the normal or tangential deflections of a face due to unit normal or tangential load on any of the faces[1]. In the wrapper flexibility tests, the pads were parallel and contacted along their full 75 mm length. Different values of the local flexibility were obtained for standard rig S/As and those fitted with instrumented pads. These were allowed for in the calculations.

The initial gaps between S/As at the restraint planes must be specified as input data to CRAMP. The gaps measured after the test with the rig in its 'at rest' condition were used in the calculations.

During the rig operation in which the S/A bottom ends were moved, the discharge and charge operations caused distortions which allowed the rig to 'relax' towards the non-friction condition. Therefore, that part of the experiment could be treated as a non-friction case. The remaining part of the experiment, in which the rams were moved, was also initially treated as having zero friction.

The results obtained, for the mean load on the six faces of the centre S/A are compared with the measured loads in Fig. 3. This figure shows excellent agreement between the calculated and measured loads during the stages in which the S/A bottom ends were moved. However, once the rams were moved in on the array the calculated loads increased more rapidly than those measured. The calculated increase was 1646 N (370 lbs) compared to the measured increase of 1152 N (259 lbs). At the completion of the rig movements the calculated and measured mean centre S/A loads were 2927 N (658 lbs) and 2393 N (538 lbs) respectively.

The calculated loads on each face of the centre S/A were compared with the measured loads. At the completion of rig operations, the range of the calculated loads was 414 N (93 lbs). This was less than the measured range of 725 N (163 lbs). There was little correlation between the calculated and measured individual face loads.

5.1 Sensitivity studies

To establish the cause of the discrepancy in mean load between calculation and measurement during the ram movement phase, several other
calculations were carried out. Firstly, the calculation representing
the ram movement phase was repeated assuming finite constant, (that is
equal static and dynamic) friction coefficients. For these calculations, a slightly different ram model was assumed in which the ram
heads did not pivot. With a coefficient of 0.2 the calculated additional mean centre S/A load due to the ram movement was only 22 N
(5 lbs) higher than in the non-friction calculation. With a coefficient of 0.5 the additional load was 40 N (9 lbs) higher. These results showed that friction had virtually no effect in this experiment.
　It was realised that there could be small errors in the measured
gaps. Therefore the sensitivity of the results to the assumed initial
inter-S/A gaps was then examined. This was done using another set of
gaps which were, on average, smaller than those used previously. In
particular, the LRP gaps between the S/As near the centre of the array
were all zero. However, the change had little effect. The calculated
increase in load during the ram movement phase was almost identical to
that obtained in the original calculation. Any error in the initial
gaps used in the original calculation was therefore unlikely to be the
source of the discrepancy between the calculated and measured mean
loads generated by ram movement. However, errors in the initial gaps
may have been responsible for the errors in the calculations of
individual face loads.

5.2 Face tilting

It was then realised that during the sequence of CHARDIS operations
described above, movement of the S/A bottom ends tilted the S/As causing misalignment and reducing the contact length between adjacent
pads. The misalignment may have been sufficient to cause contact to
occur only at one end of the pads. Point contacts probably occurred
in different areas of the array in different phases of the rig operations. As the outer ring was deflected, point contacts would be generated on the faces between S/As of the inner and outer rings. This
would tend to be eliminated as the inner ring was moved. However,
this movement would, in turn, generate point contact on the faces of
the vertical central S/A. The radial movement of S/As in a ring would
cause some tendency to point contact within the ring. However, the
movements would be small and the point contacts would probably be
eliminated by elastic deflections. The most severe mis-alignment was
therefore between the centre S/A and the first ring during the third
phase of rig operations.
　So that the effect of point contacts could be investigated, flexibility measurements were made for a wrapper under point loads at one
end of the LRP pads. These measurements showed that the coefficient
in the Miller Matrix which defined the deflection of a face due to
unit normal load on that face was 35% larger under the point load
condition than in the full pad contact condition.
　Calculations were then performed that allowed for S/A tilting and
consequent point contacts within the array. Two representations of
the array were investigated. These were:
　i) Using point contact throughout the array, through all three
stages of the rig operations.

ii) Using point contact around the central S/A only, through all three stages of the rig operations.

The results for case i) and ii) are shown in Fig. 4. During the first stages the loads were mainly caused by bending of the S/As and were insensitive to the details of the contact. Fig. 4 shows that the increased flexibilities slightly reduced the agreement with the measured values in these initial stages. During the ram movement stage the loads increased due to compression of the cross-section and were sensitive to the local flexibility. The calculated rate of increase in load with ram movement reduced. When point loads were assumed to occur around only the centre S/A, the rate of increase in load with ram movement was in very close agreement with the measured rate. The final calculated average centre S/A face load was 2509 N (564 lbs). It was obvious that the main cause of the original discrepancy was because of poor modelling of the real nature of the pad to pad contact. Comparison of the calculated and measured individual centre S/A face loads showed the maximum error to be 38%, with the calculated load the higher.

The full significance of this finding is not yet clear. Similar misalignment of the pads will tend to occur in the reactor. As the reactor shuts down significant changes in S/A bow occur. This will certainly lead to point rather than line contacts at the pads and point contact flexibilities must be used to define off power loads. However, in on-power conditions any tilts would tend to be removed by creep. Further analytical work is planned to determine the exact nature of on-power contacts.

6 CONCLUSIONS

1. The CRAMP simulation of the sequence of CHARDIS 1 operations has contributed significantly to understanding the interaction of S/As within the rig. It was possible to calculate the mean face loads occurring at the centre S/A accurately. However, the calculated individual centre S/A face loads showed poor agreement with measurement.

2. CRAMP was shown to give good agreement with measurements for those stages of the rig operations in which S/A bending was important. However, when there was significant compression of the S/As, the calculations showed that the different local flexibilities due to point and full contacts between pads were important.

3. For the sequence of rig operations studies, friction effects were shown to be insignificant.

REFERENCE

Duthie, J.C. Development and Application of the CRAMP code for fast reactor core assessment. Paper E 1/3. 9th International Conference on Structural Mechanics in Reactor Technology, Lausanne.

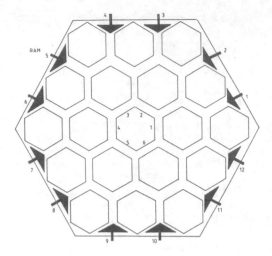

Figure 1.   CHARDIS 1   Figure 2.   RAM LOCATIONS

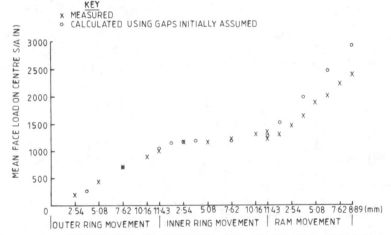

Figure 3.   COMPARISON OF MEASURED AND CALCULATED RESULTS

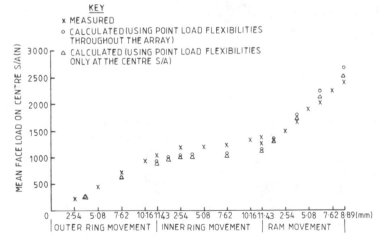

Figure 4.   RESULTS ASSUMING POINT CONTACTS

100

Fluidelastic instability on Super Phénix LMFBR weir system: A survey

S.Aita, Y.Tigeot, R.J.Gibert

Commissariat à l'Energie Atomique, IRDI/DEMT/SMTS, CEN Saclay, Gif-sur-Yvette, France

1 INTRODUCTION

In Super Phénix LMFBR, a cold flow derived from core bottom insures keeping the main vessel at a suitable temperature. The cold Sodium involved in this "Main Vessel Cooling Circuit" (figure 1), is run between the main vessel and a weir vessel, and is then discharged to an internal restitution collector.

During reactor hot tests, this weir system shows to be subject to a fluidelastic instability, never observed before. It is due to the coupling between the weir fluid collectors (feeding and restitution) because of the flexibility of the weir shell. The energy source is that of gravity during fluid fall down [1,2,3].

This paper makes the point concerning the theoretical, computational and experimental work, already done on this problem. A theoretical model was established, and an instability thresholds analysis procedure was elaborated [1]. This procedure was found relevant for retrieving Super Phénix 1 situation, and was effectively correlated to the parametric study realized with several mock-ups [3]. The theoretical model was also used to predict vibratory levels in steady non-linear regimes [2]. Calculated vibratory levels were in accordance with those observed experimentally.

A convenient solution was then found and validated for Super Phénix 1 reactor, and a design procedure was recommended to avoid such problem in future reactors.

2 THE EXPERIMENTAL WORK

Soon after the discovery of the instability on the reactor, a large experimental program was decided [3]. A plane mock-up was quickly realized at the "Commissariat à l'Energie Atomique". It reproduced the instability, and showed that fall height and feeding flow rate, are leading parameters of the instability. It showed also that, depending on these parameters, the instability can occur with two different typical mechanisms.

For the first mechanism, large free surface motions were noticed, especially for the restitution collector, with spatial shapes corresponding to typical sloshing modes. The periodicity of these motions correspond to sloshing modes resonance frequencies. For the second

101

mechanism, the weir shell itself sustained large amplitudes of displacement, while restitution collector free surface motions were of lower amplitudes and of irregular shapes. Here, classical fluid-structure interaction resonances, with negligeable gravity effects, were involved. In both cases, discharge over the weir shell was pulsating.

Other mock-ups more representative of the axisymmetrical geometry of the reactor were constructed. In these mock-ups of different scales (see table 1), the thickness and material of the weir shell were varied in order to obtain a parametric exploration of the influence of the weir shell flexibility.

In fact, the similitude on a mock-up is essentially governed by a fluidelastic Froude number represented by the ratio Ω_n of resonance frequencies between the sloshing mode of the collectors and a classical fluidelastic mode [3]:

$$\Omega_n = \frac{fg_n}{fs_n} \tag{1}$$

This ratio depends on the shape of the mode, represented here by the azimuthal order n for axisymmetrical geometry. The ratio is maximal for n values corresponding to the modes giving the lowest fluidelastic resonance frequencies.

Experimental parametric exploration showed that no instability can occur, if the maximal Ω_n is lower than 0.2. More detailed analysis, were conducted, comparing instability thresholds on the mock-ups, for the two typical mechanisms described here above, with those retreived from the theoretical model [3].

3 THE THEORETICAL MODEL

A theoretical model of the instability has been elaborated and discussed by authors [1,2]. In this model, the system formed by the flexible weir shell and adjacent Sodium collectors (restitution and feeding, figure 2) is described by its modal basis, which can be calculated using a finite element fluid-structure interaction computer code. The calculated modes take into account the gravity effects on the collectors free surfaces. These modes may also include the effects of other shells and collectors which are present in the reactor [4].

The main feature of the model is the description of the flow and force sources, introduced by the discharge at collectors free-surfaces, as non-linear source relations between the different modes of the weir system. These relations are written around the steady-state regime of a permanent discharge without instability. Only the fluctuating terms need to be considered.

3.1 Discharge sources

For the feeding collector, the source written is a flow rate time derivative, expressed as:

$$\Delta \dot{q}_1 (t, \Theta) = - \dot{q}_d (t, \Theta) \tag{2}$$

102

Here q_d is the discharge flow rate per unit length. It is assumed to be critical, so that it is related to free surface elevation Z_1 by:

$$
\begin{array}{ll}
q_d\ (t,\Theta) = K\ \sqrt{g}\ Z_1^{3/2}\ (t,\Theta) & Z_1\ (t,\Theta) > 0 \\
q_d\ (t,\Theta) = 0 & Z_1\ (t,\Theta) \leqslant 0
\end{array}
\tag{3}
$$

K is a shell crest form coefficient, adjusted using permanent flow experiment* [1].

For the restitution collector, a similar flow source is written expressing that the fluid mass extracted from the feeding, is given to the restitution, with a delay τ corresponding to the fall time.

$$
\dot{\Delta q}_2\ (t,\Theta) = +\ \dot{q}_d\ (t-\tau,\Theta)
\tag{4}
$$

A third source term is derived, for the restitution collector, from momentum equilibrium, between the falling fluid film and the fluid volume of the collector. This source is a force which can be expressed as the fluctuating part of:

$$
F(t,\Theta) = -\ \rho_f\ q_d\ (t-\tau,\Theta)\ V_f
\tag{5}
$$

Here V_f is the velocity at which the falling fluid film arrives at top of restitution collector.
The delay time τ in the model takes into account the complex behaviour the falling fluid film, where the friction along the weir shell wall gives velocities V_f far less than those valid for free fall. The friction model was correlated with specific experimental results [1].

3.2 Conservation relations

Mass conservations equations are also written for the two fluid collectors. They describe that the total flow injected into a collector (or rejected from it) as being divided into a discharge flow, and a mean free surface elevation. This gives for the feeding collector:

$$
\int_0^{2\pi} (e_1\dot{Z}_1 + q_d)d\Theta = 2\pi\ q_o
\tag{6}
$$

and for the restitution collector:

* In fact, the weir shell crest form was designed in order to have such a critical discharge. This is to avoid other hydraulic instabilities and gaz dragging in the core. The falling fluid is, in this design, always adherent to the weir shell wall.

$$\boxed{\int_0^{2\pi} (-e_2\dot{z}_2 + q_d)d\Theta = 2\pi\, q_o} \tag{6}$$

Here q_o is the mean flow rate injected to the feeding collector, or rejected from the restitution collector. It can be assmued constant, when one may use the model for instability thresholds analysis. On the contrary, when estimating vibratory levels for non-linear steady regimes, it has to be considered fluctuating. This means that here, the hydraulic behaviour of the whole cooling circuit plays a significant role, which can be modelized as q_o verifying a momentum equation of the form:

$$\underbrace{\frac{g}{L} \int_0^{2\pi} (e_1 z_1 - e_2 z_2)d\Theta}_{\substack{\text{restoring force due to}\\ \text{gravity}}} + \underbrace{2\pi\dot{q}_o}_{\substack{\text{fluid}\\ \text{inertia}}} + \underbrace{f(q_o)}_{\substack{\text{term due}\\ \text{to pump}\\ \text{head}}} = 0 \tag{8}$$

The term due to pump head can be taken as a viscous damping ; i;e.:
$f(q_o) = A\, q_o$ [2].

4 INSTABILITY THRESHOLDS ANALYSIS

An instability thresholds analysis procedure was derived from the theoretical model. This is by taking a first order development when writing the discharge relations on typical eigenmodes involved in the instability mechanisms.

The first typical mechanism is a two modes mechanism. These are the two sloshing modes of feeding and restitution collectors, close in resonance frequencies, and coupled by the flexibility of the weir shell. A simplified stability criteria was derived for this 2 degree of freedom mechanism [1,3]. It shows that the stability depends here on the ratio of modal damping* to the relative difference in natural frequencies between the two modes. Otherwise the stability domain can be completely expressed in a feeding flow rate versus fall height diagram, as shown in figure 3. Calculated stability domains correspond to those identified on SPX 1 reactor.

The second mechanism is a unique eigenmode one, which is a classical fluidelastic mode of weir shell and collectors, with low effect of gravity. Again a simplified criteria was developed for this case, showing that the stability depends on the ratio of free surface modal displacements between restitution and feeding collectors. Stability domains in fall height – flow rate diagrams were also calculated and compared with those experimentally observed [3].

* Modal damping ratios "in air" were measured during full scale tests on the reactor [6]. Fluid damping effects due to viscosity and permanent flow were estimated to be low. This fact was verified measuring responses to core and heat exchangers outlet flow on the reactor, with no instability.

The validation of this thresholds analysis procedure allowed it to be used for the design of the 1500 MWe LMFBR project weir system [5]. Let's note here that the objective of a design procedure is to insure being in the stable domain for all the fall heights and flow rates involved in nominal and transient operation.

5 STEADY VIBRATORY LEVELS

The non-linear theoretical model was implemented in the OSCAR module of CASTEM 2000 computer code. This module is a general purpose modal synthesis code using, for time behaviour calculation, an explicit time integration scheme [7]. Several sets of calculation were performed for various configurations of fall heights and flow rates. These calculations considered all the modes of weir shell and collectors.

The aim of the calculations was first to identify the predominant mode in the steady vibratory regimes, among those predicted instable by the linear instability thresholds analysis. Secondly, it was of interest to interpret several aspects of the time histories of the signals measured on the reactor (figure 4). The last and most significant aim was to retrieve the vibratory levels experimentally observed.

Specifically these calculations showed how a modulated vibratory motion can occur when two modes, belonging to the same mechanism, are instable with similar instability rates. They also showed that a mode instable in the first mechanism (sloshing) can be predominant over the one, instable in the second mechanism, even having a lower instability rate. Also, good agreement was found comparing estimated and measured vibratory levels.

One must note here that the non-linearity responsible of level limitation was identified as being the overall mass conservation relations, discussed in 3.2.

6 DISCUSSION AND CONCLUSIONS

The above described work helped to find and validate a solution for the instability for SPX 1 reactor situation. This was achieved by changing the nominal hydraulic conditions of the reactor, in order to be in a stable zone in the weir system fall height - flow rate diagram.

The effective confrontation of the stability thresholds analysis procedure derived from the theoretical model with the experimental observations, had a second important consequence. This is for the design of similar weir systems, where the help of scale mock-ups is limited. In fact, limit conditions on weir shell bottom are difficult to represent on a scaled mock-up as well as dissipative-damping effects. These two parameters are of importance respectively, in the definition of the resonance frequencies involved in the instability, and in the possibility to get it. The instability thresholds analysis procedure can overcome these difficulties for design purposes.

Finally, the steady regime vibratory analysis shows the ability of a non-linear theoretical model to understand and quantify the behaviour of a complex fluid-structure interaction problem including flow.

REFERENCES

[1] S. AITA, R.J. GIBERT: "Fluidelastic Instability in a Flexible Weir: A Theoretical Model", ASME PVP Conf., Vol. 104 – Chicago (USA) 1986.
[2] S. AITA, T. BENASSIS, R.J. GIBERT and D. GUILBAUD: "Fluidelastic Instability of a Flexible Weir: Non-Linear Analysis", Int. Conf. on Flow Induced Vibrations, Bowness-on-Windermere (U.K.), 1987.
[3] S. AITA, Y. TIGEOT, C. BERTAUT and J.P. SERPANTIE: "Fluidelastic Instability of a Flexible Weir: Experimental Observations", ASME PVP Conf., Vol 104 – Chicago (USA) 1986.
[4] S. AITA, C. BERTAUT, R. HAMON: "Vibration Analysis of a Pool Type LMFBR: Fluid-Structures Coupled Modes and Response to Flow Excitations", 8th SMIRT Conf., paper E7/8 – Bruxelles (B) 1985.
[5] A. CROS, S. AITA, R. DEL BECCARO, M.P. BARREAU: "Thermal Protection of the Rapide 1500 Main Vessel by Means of a Spillway", 9th SMIRT Conf., paper E5/2 – Lausanne (CH) 1987.
[6] S. AITA, F. GANTENBEIN, Y. TIGEOT, C. BERTAUT, J.P. SERPANTIE: "Vibration Analysis of a Pool Type LMFBR: Comparison between Calculation and Full Scale Test Results", 7th SMIRT Conf., paper E4/3 – Chicago (USA) 1983.
[7] M. FARVAQUE, D. GUILBAUD, F. GANTENBEIN, R.J. GIBERT: "OSCAR, a New Computer Code for the Dynamic Analysis by Substruction", 8th SMIRT Conf., paper 88/3 – Bruxelles (B.) 1985.

TABLE 1

FLUIDELASTIC SIMILITUDE FACTOR Ω_s FOR REACTOR AND MOCK-UPS

	Scale factor	Width of weir shell e_c (mm)	Material	$\frac{\Omega \ max}{n}$	n_{max}	first mechanism	second mechanism
SPX1	1	e_c SPX	Steel	0.48	n1	largely	possible
LDP	~ 1/10	2	Steel	0.54	–	possible	possible
		2	Aluminium	0.91	–	largely	largely
LDC	1/12	2	Aluminium	0.19	11	no	no
		1.5	Aluminium	0.25	10	limit	limit
		1	Aluminium	0.36	9	possible	possible
EPOC	1/4	8	PVC	⩾ 1 !	6-11	largely	not applic.
		8	Aluminium	0.28	8-9	possible	possible

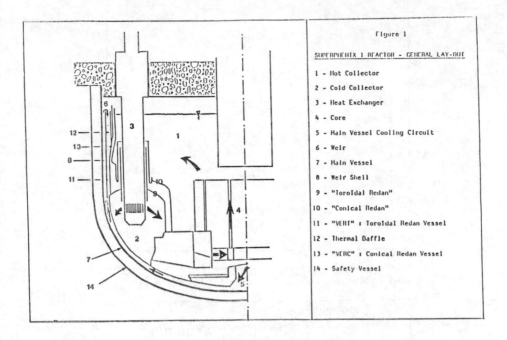

Figure 1

SUPERPHENIX 1 REACTOR - GENERAL LAY-OUT

1 - Hot Collector
2 - Cold Collector
3 - Heat Exchanger
4 - Core
5 - Main Vessel Cooling Circuit
6 - Weir
7 - Main Vessel
8 - Weir Shell
9 - "Toroidal Redan"
10 - "Conical Redan"
11 - "VERT" : Toroidal Redan Vessel
12 - Thermal Baffle
13 - "VERC" : Conical Redan Vessel
14 - Safety Vessel

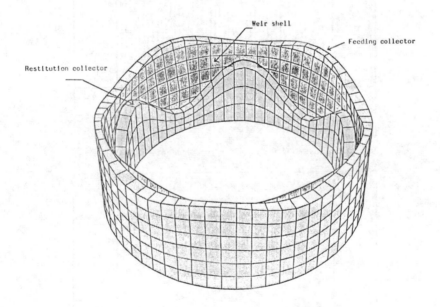

Figure 2 - Typical shape of a sloshing mode of the weir system

107

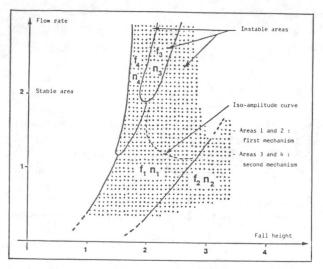

Experimentally observed instability domain on SUPERPHENIX Reactor

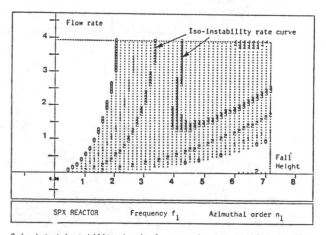

Calculated instability domain for type 1 mechanism (sloshing)

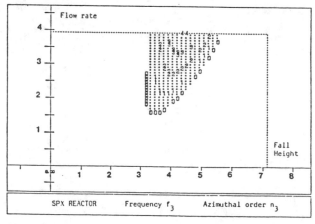

Calculated instability domain for type 2 mechanism (fluidelastic)

Figure 3

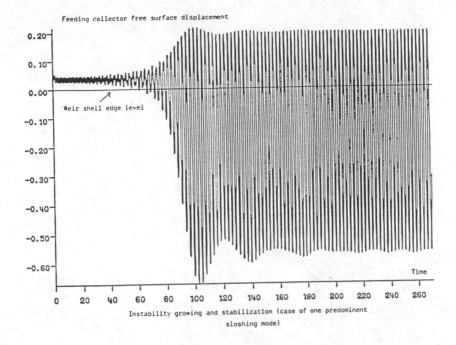

Feeding collector free surface displacement

Weir shell edge level

Time

Instability growing and stabilization (case of one predominent
sloshing mode)

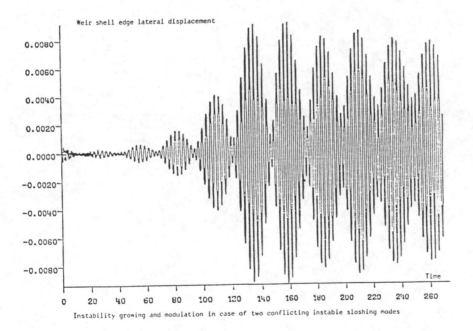

Weir shell edge lateral displacement

Time

Instability growing and modulation in case of two conflicting instable sloshing modes

Figure 4

109

F LWR pressure components

Nonlinear dynamic response of piping subjected to pressure waves: Calculation and test results

L.M.Habip

Kraftwerk Union AG, Offenbach, FR Germany

H.Hunger

Kernforschungszentrum Karlsruhe GmbH, Projekt HDR-Sicherheitsprogramm, FR Germany

1 EXPERIMENTAL DYNAMIC LOADS

In the course of a recent series of full-scale piping blowdown tests conducted at the decommissioned HDR-plant, originally housing an experimental superheated-steam reactor system, the nonlinear dynamic response of a feedwater line without snubbers has been investigated.

The blowdown was initiated by suddenly opening the cross-section of the pipe at one end, with the help of a rupture disc assembly located in a nozzle near a fixed point. Intense hydrodynamic loads were generated by the closing of an undamped check valve following the simulated pipe rupture. The measured valve lift time history is compared in Figure 1 with the results of a calculation which was performed before the test. A corresponding pressure time history is shown in Figure 2. The agreement between measured and calculated results is satisfactory during the initial loading phase which is essential when considering maximum nonlinear structural response.

Hydrodynamic forcing functions of low amplitude are typical of damped valves that are commonly used in nuclear power plants. The high-amplitude loads due to the undamped valve were used to assess quantitatively the dynamic load-carrying ability of a piping system without snubbers.

2 PIPING CONFIGURATION

The routing of the feedwater line with check valve utilized in the tests is shown in Figure 3. As piping material, mainly the ferritic steels 15 NiCuMoNb 5 and 15 Mo 3 of high and low yield strength, respectively, were used. In particular, all reducers were made of 15 Mo 3.

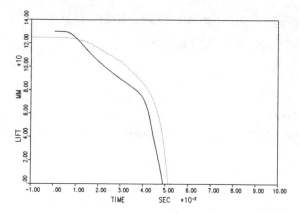

Figure 1. Valve lift time history: —— calculation, --- test 21.3.

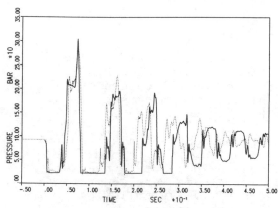

Figure 2. Pressure time history: —— calculation, --- test 21.3.

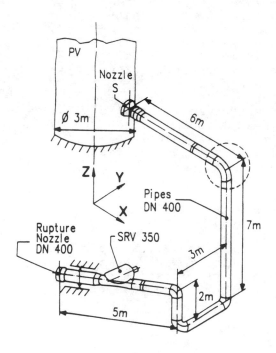

Figure 3. Routing of feedwater line with check valve (SRV 350).

Table 1. Natural frequencies, Hz.

Analysis	Tests (snapback)
3.9	4.5
5.5	6.2
7.2	7.5
11.5	12.7
15.1	16.8

There were no intermediate pipe supports of any kind between the pressure vessel nozzle and the fixed point.

3 CALCULATION AND TEST RESULTS

Some of the calculated and measured natural frequencies of the piping system are given in Table 1.

A dynamic, linear-elastic structural analysis performed on the basis of preliminary forcing functions during the planning stage of the tests gave a first indication of highly stressed areas. These were confined to parts of the piping system close to the fixed ends.

A nonlinear analysis using calculated hydrodynamic forcing functions and the elastic-plastic stress-strain diagrams of the two piping materials with different yield strengths led to a maximum strain amplitude of about 4 % at the upper reducer near the nozzle of the pressure vessel. Such geometrical discontinuities had been modelled locally with a lower wall thickness than actually measured in order to simulate roughly the influence of a possible stress intensification.

Piping strains were measured at several locations and have been evaluated in considerable detail elsewhere (Kussmaul et al. 1987). Plastic deformation of the piping system accompanied by permanent strains is clearly exhibited by the measured longitudinal strain time histories, such as shown in Figure 4, where the corresponding results of the calculation are also presented.

Values of strain recorded on the outer surface of the pipe close to the fixed ends indicate that, in this particular case, the overall result of the present nonlinear calculation may indeed have been a conservative and useful estimate of maximum strain. Obvious discrepancies concerning the actual location of the highest strain and details of the time history of the response remain, however.

In comparison with previous tests using a somewhat different piping system in which maximum strains of about 0.6 % were measured (Habip 1983, Habip et al. 1985), the new series allowed reaching a considerably

115

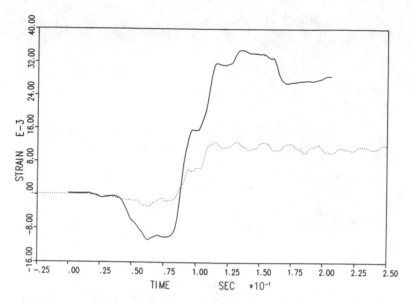

Q01: reducer at pressure vessel nozzle.

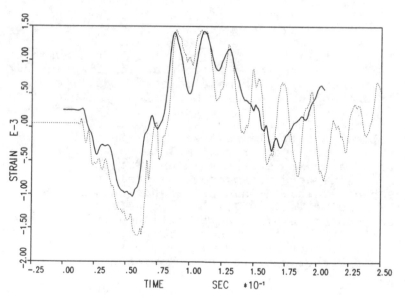

Q2 : pipe near pressure vessel nozzle.

Figure 4. Longitudinal strain time histories:
—— calculation, --- test 21.3.

116

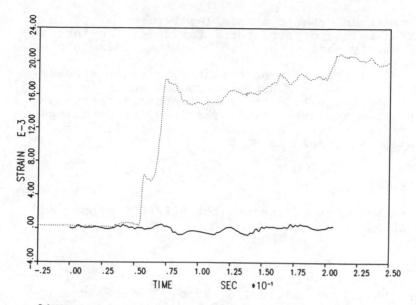

Q9 : reducer near valve.

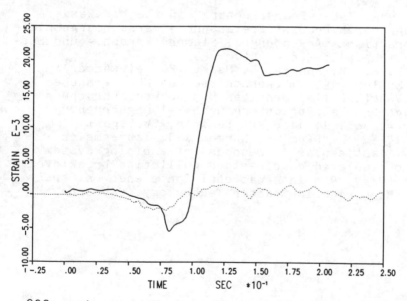

Q92: reducer between valve and blowdown nozzle.

Figure 4. Longitudinal strain time histories:
—— calculation, --- test 21.3.

117

higher measured dynamic strain level (about 2 %) in the pipe. The absence of structural failure is an indication of safety margins available in the piping system considered.

These findings are in agreement with other observations of nonlinear structural response under dynamic loads of short duration and lend further support to earlier attempts (Habedank et al. 1981) at defining strain criteria for piping steels, especially for application at Stress Level D.

ACKNOWLEDGMENT

This work was part of the project HDR/RORB sponsored by the BMFT.

REFERENCES

Habedank, G., L. M. Habip & J. Hess. 1981. Spannungs- und Dehnungsabsicherung bei Rohrleitungen für dynamische Lastfälle mit extrem niedrigen Eintrittswahrscheinlichkeiten. Atomkernenergie-Kerntechnik 39: 137 - 140.

Habip, L. M. 1983. Beanspruchung einer Speisewasserleitung durch Druckstoßbelastung - Versuchsergebnisse und praktische Anwendungen. Siemens Forsch.- und Entwickl.-Ber. 12: 386 - 394.

Habip, L. M., H. Hunger, D. Meier, F. Beißwänger & K.U. Müller. 1985. Erkenntnisse aus HDR-Versuchen über Ventilschließverhalten und Rohrbeanspruchung infolge Druckstoßbelastung nach Leitungsbruch. Paper 4, Vol. 1, 11th MPA-Seminar, Stuttgart.

Kussmaul, K., H. Diem, H. Hunger & G. Katzenmeier. 1987. Elastic-plastic response of a piping system due to simulated double-ended guillotine break events. Paper F2/4*. 9th International Conference on Structural Mechanics in Reactor Technology, Lausanne.

Critical cracking potential for stress corrosion cracking of nuclear pressure vessel steels in pressurised high temperature waters

T.Shoji & H.Takahashi

Research Institute for Strength and Fracture of Materials, Tohoku University, Sendai, Japan

1. Introduction

Quantitative assessment of subcritical crack growth caused by stress corrosion cracking (SCC) and corrosion fatigue (CF) of nuclear pressure boundary materials in pressurized high temperature water has been receving a world wide concern from a view point of structural integrity assurance of components. Metallurgical, mechanical and environmental aspects in subcritical crack growth have been examined and, consequently, electrochemical processes at crack tip is considered as a primary important factor in crack growth enhancement. Therefore, effects of electrochemical potential upon evnironmentally assisted cracking have been examined in connection with effects of dissolved oxygen content and temperature upon these cracking as a international cooperative research (International Cyclic Crack Growth Rate Group, ICCGR)[1].

Mechanistic works on environmentally assisted cracking have also been carried out in relation to electrochemical potential in BWR and PWR environments and several cracking mechanisms are suggested. Especially, effects of sulphur in steels upon crack growth evhancement is emphasized in relation to a revision of evaluation curve for wet condition of ASME Boiler and Pressure Vessel Code, Sec.XI, Appendix A[2]. On the other hand, slow strain rate tests has been used as a simple techique to evaluate a cracking susceptibility of a given material/ environment system.

In this work, the cracking susceptibility of nuclear pressure vessel steels with different sulphur contents in simulated BWR and PWR water is examined in terms of critical crack potential by means of SSRT under the controlled potential chemically by disolued oxygen or potentiostatically.

2. Experimentals

The materials used in this work are nuclear pressure vessel steels of SA508 Cl. III with the low sulfur content and SA533B Cl I containing low and medium amount of sulphur in steels. The chemical composition and mechanical properties of the materials are shown in Table 1 and Table 2, respectively. Plain round specimen with gauge length of 40 mm and diameter of 3 mm is used as shown in Fig. 1. Tensile direction is coincided with T-direction. Slow strain rate tests are carried out in pressurized high temperature water at the temperature range of 150 C to 288 C.   Fig. 2 shows a autoclave used in this work. One end of the specimen is covered with thermal shrinkable Teflon tube and fixed to autoclave by use of swage lock fittings and the other end which is kept low temperature can slide easily during the extension.

The use of teflon tube to cover specimen grip is for electrical insulation of specimen from a autoclave and also to minimise the frictional forces. Fig. 3 shows a water loop for SSRT where flow rate is 4.2 - 6.0 l/hr.   Water condition used in this work are summarized in Table 3. Temperature and dissolved oxygen effects are examined by SSRT on A508 in deionized water and sulphur effects contained in steels as well as in the environment are investigated in the simulated PWR water under potential control conditions at 288 C. Tests are performed by Instron type Machine at the strain rate of 2.1 x 10^{-6} and 4.2 x10^{-6}1/sec. After the tests, fracture surfaces are fractographically examined by SEM.

3.   Critical cracking potential in simulated BWR and PWR waters

3.1 SSRT results in a simulated BWR water environment

As has beenn pointed out previously[3], dissolved oxygen content in water has drastic effects upon corrosion potential of steels in high temperature water.   However, dissolved oxygen has no exactly unique relationship with potential since the electrode potential influenced by the history of the sample after immersion into water.   Therefore, direct measurement of potential by use of high temperature reference electrode would be recommend to see what the potential is.

Fig. 4 and 5 show the SSRT results obtained at 150 C, 200 C, 250 C and 288 C where susceptibility of environmentally assisted cracking is evaluated in term of the maximum SCC crack (quasi-cleavage type) length observed on a fracture surface as a function of dissolved oxygen contents and corrosion potential, respectively. SCC suscepti- bility is strongly affected not only by dissolued oxygen content, in other words, specimen potential but also by testing temperature as has been pointed out by J. Congleton et al.[4] on SA533B steels with sulphur content of 0.006%.

As can be seen clearly in Fig. 5, there is a threshold potential for cause of EAC depending upon temperature.   This critical potential above which EAC is observed on a fracture surface can be denoted as a critical cracking potential and can be regarded as a measure of SCC susceptibility under agiven condition.   The critical cracking poten- tial showed the minimum value of about −200 mV (Ag/Agcl) at a tempera- ture range of 200 C to 250 C, above and below which the potential shifted to more noble side, especially at 150 C.   This interesting trend can be seen in Fig. 6. There has been several experimental data on a temperature effects of this kind of material in high temperature water and some conflict results were observed on a temperature range where the maximum susceptibility of SCC was observed. This temperature effects on cracking behavior should be considered in relation to electrochemical potential of specimens. Fig. 7 shows the temperature effects on crack growth rate as a function of potential. Above the critical cracking potentials, crack velocity increase with the in- crease of potentials and then tends to saturate at noble potential region. The crack growth rate in a saturated region was about 5 x 10^{-5}

mm/sec irrespective to testing temperature except 150 C where only one data is available.

Recently, the significance of number of crack initiated in SSRT tests along the specimen gauge length has been pointed out[4,5] from a view point of real crack tip strain rate. It seems reasonable to conclude that the effective crack tip strain rate will be some function of the number of cracks present and that for a given applied strain rate, the crack tip strain rate will decrease as the crack number increases. Fig. 8 shows the results on the number of cracks observed along the gauge length of specimens after the tests. The considerable difference in the number of cracks are clearly shown between the data at 250 C and 200 C, and those at 288 C. Particulary, the crack numbers at 250 C and 200 C show the maximum at the potential of 100 mV in spite of the fact that a single crack was developed at 288 C.

3.2 SSRT results in a simulated PWR water environment

The effects of sulphur contents in steel were examined under the potentiostatically controlled potentials in the simulated PWR water environment at 288 C. Typical load-elongation curves obtained at various potentials for the medium sulphur steel are shown in Fig. 9. Reduction of maximum load and fracture elongation became more clear when tested at more noble potentials than free corrosion potential, - 680 mV. This strong effects of potentials upon SCC susceptibility is similar to those for BWR water environment as described previously. The similarlity of the cracking potentials for low and medium sulphur steels can be demonstrated in Fig. 10 where the reduction in area is taken as a measure of SCC susceptibility and the critical cracking potential can be easily determined. The drastic change in reduction in area is observed at the potentials of -125 mV and -225 mV for low and medium sulphur steels, respectively. It would be worthwhile to note here that the clear decrease of reduction in area is also observed at the cathodic potential of -1000 mV where Fe must be stable. The SCC characteristic as a function of potentials can be delineated more directly by measuring the SCC crack length on fracture surface as shown in Fig. 11. Sulphur content in steel has a significant effects on SCC susceptibility. Sulphur, in general, exists in steels as a form of MnS and by SEM observation, it is clear that most of the SCC cracks were associated with MnS.

3.3 Dependence of the critical cracking potential upon sulphur content in steels and environments.

Combining the SSRT results obtained on A508 in simulated BWR water and on A533B in simulated PWR water, the critical cracking potential can be clearly related to the sulphur contents in steels. As summarised in Fig. 12, no influence of different environments and steels upon the critical cracking potential is observed. In the figure, data obtained by ICCGR as a round robin tests on A508 is also plotted with good agreement. Also, Fig. 13 shows the effects of sulphate contamination upon the critical cracking potential. As well as the effects of sulphur in steels, sulphate concentration in environments also has significant effects upon the critical cracking potential. Only a few ppm sulphate ion can shift the critical cracking potentials, 400 mV to the basic direction. Taking the corrosion potentials of RPV in plant during service operation as -220 mV for BWR and -

121

500 to -800mV in PWR into consideration, medium sulphur steels in BWR water or even low sulphur steels in PWR contaminated by sulphate ion may have susceptibility to environemntally assisted cracking when stressed above yield strength.

4. Conclusions

Following conclusive remarks can be drawn based upon the SSRT tests performed both under chemically controlled potential and under potentiostatically controlled potential on three kind of RPV steel with differnt sulphur contents.
(1) Temperature dependence of the critical creacking potentials was quantitatively evaluated. The critical cracking potential showed lowest value at the temperature range of 200 C to 250 C.
(2) The dependence of the critical cracking potential upon sulphur contents in steels and environments was also evaluated and at 288 C in PWR environment, the potential was -125 mV and -225 mV for the low and the medium sulphur steels, respectively. However, a addition of small amout of sulphate, SO_4^{2-} in the system shifts the critical cracking potential drastically towards the more basic potential.

Acknowledgements

This work was jointly supported by The Ministry of Education, Science and Culture, Grant-in Aid for General Scientific Research, 60550045 and The Kurata Foundation, The 18th Kurata Research Grant.

References

[1] Proc. 2nd IAEA Specialists' Meeting on Subcritical Crack Growth, May 15-17, 1985, Sendai, NUREG/CP-0067(MEA-2090) April 1986.
[2] ASME Boiler and Pressure Vessel Code, Division 1 1980, Ed., Appendix A, (1980) ASME.
[3] M. E. Indig et al., Review on coating and corrosion, Ed. R. N. Parkins, Freund Publication House Ltd., Israel, Vol. 5, 1982, p.173.
[4] J. Congleton et al., Corrosion Science., Vol. 25 No. 8/9, 1985, p. 633.
[5] R. N. Parkins, Corrosion, Vol. 43, No. 3 1987, p. 130.

Table 1 Chemical compositions (wt%).

Material		C	Si	Mn	P	S	Ni	Cr	Mo	Cu	N
SA508 CLIII	Low S	0.19	0.20	1.45	0.003	0.003	0.78	0.15	0.48	0.05	0.0088
SA533 B	Low S	0.17	0.26	1.37	0.003	0.004	0.60	0.07	0.46	0.02	—
SA533 B	Medium S	0.21	0.29	1.45	0.007	0.014	0.65	0.03	0.51	0.03	—

Fig. 1 Specimen geometry.

Table 2 Mechanical properties(300 C).

Material		Yeild stress (MPa)	Tensile strength (MPa)	Elongation (%)	Reduction of area (%)
SA508 CLIII	Low S	404	553	19.7	70.8
SA533 B	Low S	418	586	20.8	70.0
SA533 B	Medium S	422	608	18.3	57.2

Table 3 Water chemistries.

	BWR	PWR
Pressure (MPa)	8.3	13.3
Test temp. (°C)	150 ~288	288
Conductivity (μs/cm)	<0.2	<20
PH (at R.T.)	5.9 ± 0.1	6.9 ± 0.1
Flow rate (l/hr)	6.0	4.2
D. O. (ppb)		<10
Li⁺ (ppm)		2
B (ppm)		500

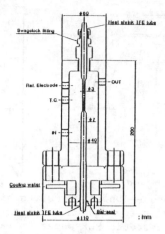

Fig. 2 Autoclave used.

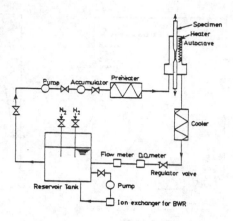

Fig. 3 Water loop for the simulated
BWR and PWR waters

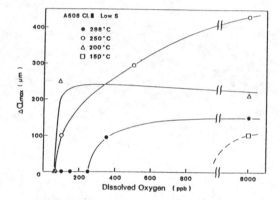

Fig. 4 Cracking Susceptibility to D. O.

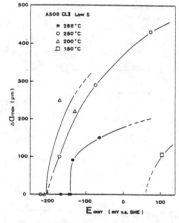

Fig. 5 Cracking susceptibility
to potentials of speci-
men.

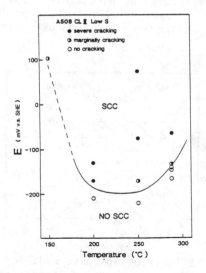

Fig. 6 Effects of Sulphur content on
critical cracking potentials.

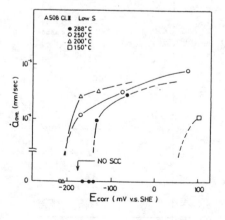

Fig. 7 Effects of potential on the
average crack growth rate

123

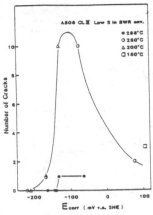

Fig. 8 Effects of potential on
the number of cracks.

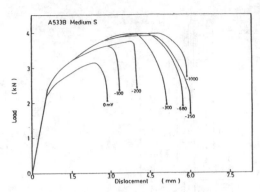

Fig. 9 Typical load-elongation curves of
A533B 1 with the medium sulphur
content tested in PWR water.

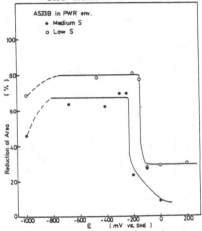

Fig. 10 Potential dependence of
reduction in area.

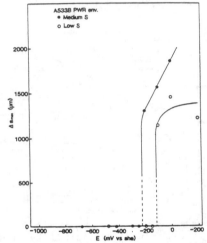

Fig. 11 Effects of potential on the
maximum crack growth length
measure on the fracture surface.

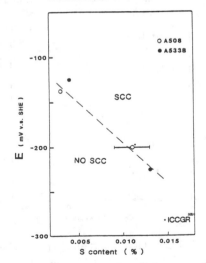

Fig. 12 Effects of sulphur on E_c.

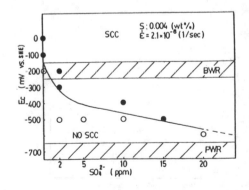

Fig. 13 Effects of sulphate addition to
PWR water on the critical cracking potential of low S steel.

G Fracture mechanics and NDE

Influence of mix ferritic-bainitic microstructure on the fracture performance of 2 1/4 Pct Cr – 1 Pct Mo steel

M.Holzmann, Z.Bílek, J.Man & B.Vlach
Institute of Physical Metallurgy CAS, Czechoslovakia

1 INTRODUCTION

The heat treatment of thick plates or forgings made from low alloy steels used in pressure vessels production may produce the mix ferritic-bainitic microstructure in dependence on cooling rate. Fracture performance of tempered mixed microstructure is substantialy different from fracture behaviour of tempered bainitic microstructure. Some results selected from the broadly based research program aimed at investigating the influence of mixed microstructure on the fracture behaviour are presented in this paper.

2 MATERIAL, EXPERIMENTAL TECHNIQUE

To investigate the effect of ferrite portion on fracture toughness we used the commercially available low alloy steel having the chemical composition 0.12 pct C, 0.57 pct Mn, 0.27 pct Si, 2.3 pct Cr, 1.04 pct Mo, 0.015 pct S, 0.015 pct P. The steel was supplied in the form of plate 4200 mm long, 1800 mm wide and 30 mm thick. This plate was cut into the blanks with dimensions 260 mm x 55 mm x 30 mm which were heat treated by austenitizing at the temperature 940 °C//one hour and by controlled cooling (from austenitizing temperature transferred to the salt bath 700 °C, holding 3 min., 8 min., 18 min., air cooling). The ferrite portion was determined by quantitative metallography analysis (Quantimet). The average values of ferrite fraction were 25 %, 38 % and 54 % according to the holding time at 700 °C. All the blanks were tempered, 640 °C/two hours and air cooled.

From the blanks the specimens were prepared for tensile tests, Charpy tests and for static (K_I = 2 MPam$^{1/2}$s^{-1}) fracture toughness measurements by three-point bend technique. The tests were performed in the temperature range from 20°C to –196°C. In this temperature interval, the transition occurs from upper shelf fracture toughness values with ductile fracture initiation to the region with cleavage unstable fracture initiation after different extent of plastic deformation. In the case of unstable cleavage fracture initiation the fracture toughness was described either

127

by K_{IC} or by K_{IJ} values determined from the critical values of J_I-integral.

3 RESULTS AND DISCUSSION

3.1 Tensile properties

The dependences of the yield stress $R_p0.2$, the tensile
strength R_m, the uniform elongation ε_{Rm}^p and the hardness
HV 10 measured at the ambient temperature, on the percent
volume fraction of ferrite (α-Fe) are given in Fig. 1.
The yield stress $R_p0.2$ depends linearly on the percent volume fraction of α-Fe

(1) $R_p0.2_{\%\alpha\text{-Fe}} = R_p0.2_{100\%B} - k\ (\%\ \alpha\text{-Fe})$

where k for the tempering treatment used equals 6.24. As
the ferrite component, due to its lower resistance against
yielding, is the controlling parameter for the onset of
plastic deformation in mixed bainitic-ferritic microstructure, the decrease of the yield stress with increasing ferrite content results from local yielding of the individual
ferrite grains in the matrix with lower volume fraction of
α-Fe, or from bulk yielding of ferrite-grain islands at
higher volume fraction of α-Fe.

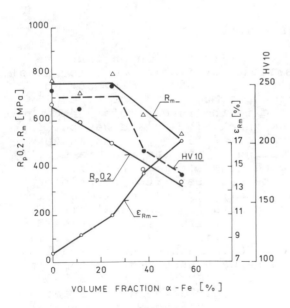

Figure 1. Mechanical properties and hardness as a function of percent α-Fe.

The pattern of the ul-
timate tensile strength
R_m is quite different.
Up to 25 % volume frac-
tion of α-Fe, R_m is constant and only with furt-
her increase of α-Fe begins to decrease. The
hardness HV 10 as a func-
tion of α-Fe content followed the similar trend
as R_m. Obviously up to
25 % α-Fe content, the
strength fall due to de-
crease of the harder bai-
nite component in micro-
structures may be still
compensated by work har-
dening of the ferrite
component, as it can be
judged from the increased
value of ε_{Rm} (Fig. 1).
At a higher α-Fe volume
fraction the work harde-
ning of ferrite component
is, even in spite of the
larger uniform elongation
ε_{Rm} (Fig. 1), insuficient to replace the strength fall caused by the decrease of the harder bainitic component.

128

3.2 Transition temperature behaviour of Charpy specimens

For estimating the ductile-brittle transition a concept similar to that described by Helms et al. (1982) has been applied. An example of the CVN-energy pattern and fracture appearance vs temperature for a microstructure with 38 % of ferrite is shown in Fig. 2. The following transition temperatures are defined:

t_p - propagation transition temperature (at and below this temperature, the fibrous propagation is changed into the cleavage one);

t_i - initiation transition temperature (at and below this temperature, the initiation of fracture is cleavage);

FATT - fracture appearance transition temperature (50 % of fracture surface is of ductile appearance). The region A represents the region of the upper shelf CVN-energy, the region B is a transition region, the region C is the region of the lower shelf CVN-energy.

The transition temperatures for the investigated microstructures determined from the above mentioned dependence are given in Tab. 1. From this table it follows that after the same tempering treatment the pure bainite gives the lowest transition temperatures. The occurence of ferrite leads to a significant transition temperature shift to higher temperatures. The largest shift occurs for the smallest proportion of ferrite. As it will be shown further, this fracture behaviour correlates with that of larger fracture toughness specimens.

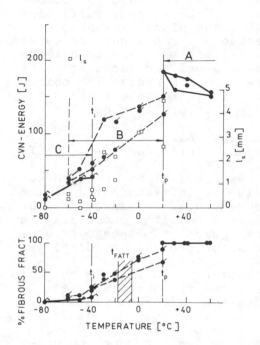

Figure 2. CVN-energy vs temperature, l_s- stable crack growth.

Table 1. Transition temperatures.

Microstructure	t_i °C	t_p °C	FATT °C
Bainite (Ba)	-60	-20	-35
Ba - 25 % Fe	-20	+100	+10 - +15
Ba - 38 % Fe	-40	+20	-5 - -15
Ba - 54 % Fe	-30	+30	±0 - -15

3.3 Fracture toughness

An example of dependence of the fracture toughness on the temperature for a microstructure with 25 % of ferrite proportion is presented Fig. 3. The points marked ⊥ do not represent actual values of fracture toughness but they only show that, at the given temperature, the fibrous initiation has taken place. The points marked K_{IC} represent the values fulfilling the ASTM conditions for the valid determination of fracture toughness. The points marked K_{IJ} represent the cleavage fracture toughness determined by using the J-integral concept. All the critical J_{IC}-values fulfilled the condition a, B, W−a = 50 J_{IC}/R_e+R_m. The temperature t_i^s is the cleavage-fibrous initiation transition temperature. The large scatter of test results, especially in the region of cleavage fracture after general yielding, was typical for the mixed microstructures. To be able to compare the fracture toughness values for different microstructures, the experimental data have to be treated by a statistical method. For the description of fracture toughness dependence on the temperature the equation was applied

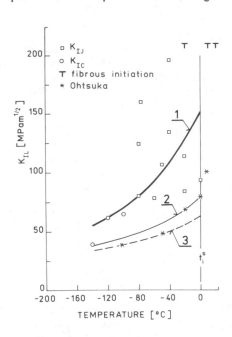

Figure 3. Fracture toughness vs temperature.

(2) K_{IC}; $K_{IJ} = K_{IL} = (K_{IC})_o + B \exp (-CT)$,

where T is the temperature in K, and $(K_{IC})_o$, B, and C are constants. On the basis of an analysis of experimental data with low alloy steels, $(K_{IC})_o$ was chosen 25 MPam$^{1/2}$. Then constants B and C can be determined by the regression analysis and a standard deviation (for a normal distribution) and the fracture toughness for the probability P = = 95 %, or P = 99 % can be determined.

In Fig. 3, the curve 1 represents the mean value of cleavage fracture toughness, curves 2 and 3 the values of fracture toughness for the probability P = 95 % and P = 99 %. Sumpter (1976), Landes and Shaffer (1980), and Ohtsuka (1983) have shown that the scatter decreased with increasing specimen thickness. In Fig. 3 the values K_{IC} (star points) found by Ohtsuka (1983) are plotted. Ohtsuka's data were determined using the 87,5 mm thick CT specimens machined from 90 mm thick plate of the steel 2 % Cr, 1 % Mo (practically the same steel as that in the present paper). The good agreement with the values K_{IL} determined with

25 mm thick SEN bend specimen for propabilities P = 95 %
can be seen.

The fracture toughness curves for P = 95 % and the temperatures t_i^s are given for different microstructures in
Fig. 4. These results are in coincidence with the Charpy
V-notch data. The microstructure with 25 % of ferrite has the lowest fracture toughness and the highest transition temperature
t_i^s.

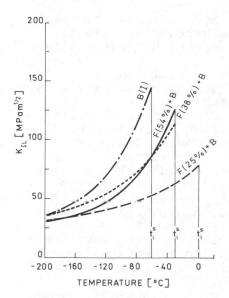

Figure 4. Comparison of
fracture toughness curves
for P = 95 %.

As shown by present authors
(Holzmann et al. 1983) the transition temperature behaviour of
mixed microstructure can be explained through the influence
of ferrite proportion on the
cleavage fracture stress and on
the value of yield stress and
its temperature dependence. Both
these parameters were, for 25 %
ferrite-bainite mixture, effected in such way that for fulfilling the conditions for initiation of cleavage fracture, only
small temperature decrease was
needed. The change of K_{IC} vs
temperature curve due to ferrite proportion results from the
model for cleavage fracture toughness of mixed microstructures
suggested by Strnadel et al.
(1985).

In Fig. 5 the dependence of
transition temperature t_i^s on
the yield stress is shown for
various microstructures tempered by the same manner. It is
clear that the ferritic-bainitic microstructures have the
highest transition temperatures.

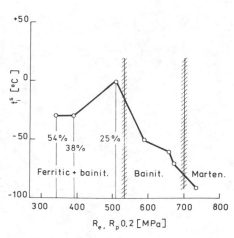

Figure 5. Transition tem-
perature t_i^s vs yield stress.

4 CONCLUSION

The proportion of ferrite in
the mixed ferritic-bainitic
microstructure influences mar-
kedly yield and tensile strength, the transition temperatures of Charpy V-notch specimens, the cleavage fibrous
initiation transition temperature of specimens with crack
and the dependence of fracture
toughness on temperature. The

largest shift of both the transition temperatures and the cleavage fracture toughness curve to higher temperature was found for 25 % proportion of ferrite in the microstructure.

REFERENCES

Helms, R., Kühn, H.J. & S. Ledwurski 1982. Assesment of Mechan. Behaviour of Mat. in the Notched Bar Impact Test. Research Report No. 82. Berlin, Bundesanstalt für Materialprüfung.
Holzmann, M., Vlach, B., Man, J. & E. Vyklická 1983. In Research Report No. VZ 537/646. Institut of Metallurgy-Czechosl. Acad. Sciences (in Czech.).
Landes, J.D. & D.H. Shaffer 1980. Statistical Characterization of Fracture in the Transition Region. In ASTM STP 700, p. 368-382. Philadelphia. Amer. Soc. for Test. and Mater.
Ohtsuka, N. 1983. An Interpretation of Scatter and Thickness Effect of Fracture Toughness in the Transition Region. In Mechanical Behaviour of Materials - IV, ICM 4 Vol. 2, p. 1133-1139, Carlson, J. & N.G. Ohlson (eds.) Oxford, Pergamon Press.
Strnadel, B., Mazancová, E. & K. Mazanec 1985. Some Problems of Cleavage Fracture Prediction in Engineering Steels. In Crack in Welded Joints, Vol. 1, p. 250-255, Bratislava, Institute of Welding.
Sumpter, J.D.G. 1976. The Prediction of K_{IC} Using J and COD from Small Specimen Tests. Metal Science 10:354-356.

Brittle fracture resistance of welded high pressure vessels

I.V.Gorynin, V.A.Ignatov, Yu.I.Zvezdin & B.T.Timofeev
Institute for Problems of Strength of the Ukr.SSR Academy of Sciences, Kiev, USSR

ABSTRACT

The paper summerizes experimental data on fracture toughness of heat resistant pressure vessel steels and their
welded joints for power equipment, and also, on the effect
of different process and experimental factors upon this
property. Suggestions to refine standard relationships and
to use them in the evaluation of high pressure vessel
strength are given.

1 INTRODUCTION

The main object at the stage of designing, production and
service is a reliable operation of welded high pressure
vessels. Its implementation necessitates a continuous refinement of the structure itself and the production process,
on one hand, and strength calculation methods, on the other
proceeding from the operational safety during the whole service life. It is the loss of tightness that is one of the
critical states in strength calculations of welded high
pressure vessels. Therefore, a verification estimate to ap-
praise brittle fracture resistance at the design stage is
needed (Calculated strength standards 1973) in view of fai-
ling to reveal defects at the production stage despite a
100 pct NDT flaw detection of 0.25 pressure vessel thickness
defects. A statistical analysis of test results in paper
(Gorynin 1985) for power equipment pressure vessels fabricated using the adapted production and assembly process has
indicated that the critical defect size is noticeably less
than the 0.1 pressure vessel wall thickness with more than
a 95 pct probability level. All the aforesaid necessitates
a close study of steel and welded joint brittle fracture
resistance using not only linear mechanics – based standard
techniques, but those involving small size defects and
test-to-fracture results for large specimens and prototypes.

2. TEST MATERIALS AND EXPERIMENTAL PROCEDURE

All the experiments were performed on heat resistant
15X2MФA and 15X2HMФA steels and welded joints to fit to
production conditions of power equipment. Different size
class base metal specimens 25...150 mm in thickness for
compact and from 50x100x450 mm to 150x300x1450 mm for three-
point bending were made from forgings. Fracture toughness
of weld metal was determined on automatic submerged arc,
electroslag and manual arc welds. The most information has
been obtained on circular welds made by an automatic welder
with Cb-10XMФT wire for Cr-Mo-V steel and Cb-08XГHMTA wire
Cr-Ni-Mo-V steel. Experimental data for 8...10 steel melts
(welding wire) are obtained with these welded joints and
base materials. The least information (2 to 3 wire melts)
has been obtained for electroslag and manual arc welds since
these welding techniques are less used in pressure vessel
fabrication (electroslag – for bottom blank welding and
manual – for production or assembly repair welding). In
order to evaluate welded joint fracture toughness mostly
the same specimen sizes were used as for base metal. Fracture toughness of weld metal deposited by manual arc welding
and of heat affected zone metal was investigated on three-
point bend tested specimens. In electroslag welding
Cb-13X2MФT and Cb-16X2HMФTA wire and in manual welding H-6
grade Cr-Mo-V and PT-45 grade Cr-Ni-Mo-V electrodes were
used. Fracture toughness tests of specimens were carried
out in accordance with the commonly adopted standards
(ASTM E399-81, Metod.Ukazanija 1982).

3. TEST RESULTS AND DISCUSSION

Until 1980 in our country when calculating brittle fracture
resistance, temperature criteria were used whereby the first
brittle strength diagram for power equipment (Calculated
strength standards 1973) was plotted as allowable stresses
vs. relative temperature using Robertson type specimen test
results as the base. A similar diagram (Pellini and Puzak
1963) is used in an ASME Code. Publications (Karzov et.al.
1982 and Shatskaya et.al. 1985) have shown that the crack
arrest criteria can not be physically grounded when designing power equipment vessels and pipeline with a high intr-
insic energy capacity. Therefore, in the second issue of
norms (Calculated strength standards 1973) it was found advantageous to shift to crack initiation criteria using linear fracture mechanics methods (Shatskaya 1984). The accumulation of experimental data on fracture toughness of heat
resistant steels and their welded joint metal was a requisi-
te for the application of this criterion in calculated high
pressure vessel strength. Figure 1 summerises the results
of brittle fracture resistance of 15X2MФA, 15X2HMФA steels
and their welded joints. It is evident from the figure that
there is a slight difference in fracture toughness of base
metal, welded joint and heat affected zone.
In the general form the temperature dependence of fracture
toughness of test materials is an exponent of the type

134

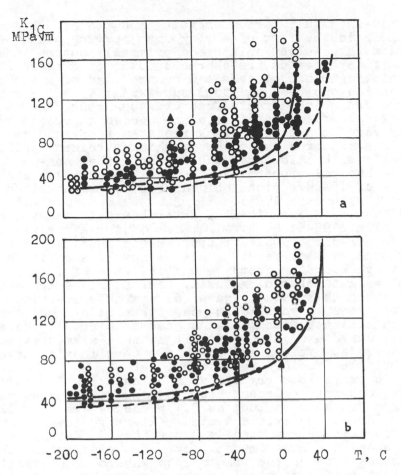

Figure 1. Fracture toughness of welded joints of
15X2MФA (a) and 15X2HMФA (b) steels: o - base metal,
● - weld, ▲ - HAZ.

$K_{1C} = A \cdot \exp(B \cdot T) + C$. For temperature dependence curves in
Fig.1 for 15X2MФA steel and welded joint metal. The equations take the form $K_{1C} = 50 \cdot \exp(0.032T) + 30$ and $K_{1C} =$
$= 32 \cdot \exp(0.024T) + 26$. For 15X2HMФA steel and welded joint
metal the equations are of the form $K_{1C} = 38 \cdot \exp(0.026T) +$
$+42$ and $K_{1C} = 47 \cdot \exp(0.023T) + 33$, respectively (**Timofeev**
et.al.1982). A mathematical representation of the above
equations corresponds to lower envelopes of experimental
data plotted to a 95 pct probability. It should be noted
that these equations incorporate also the effect on the
above materials fracture toughness of some process factors
which are the case at the stage of the production of high
pressure vessels in particular, a steel casting method
(acid open-hearth furnace, basic arc furnace), contamination of base metal and welded joints with deleterious (S and
P) and nonferrous (Cu,Sn,Sb,As,Bi) impurities, heat treatment conditions (basic heat treatment without and with additional tempers), and also the anisothropy of properties

of base metal (axial, circumferential and radial cutting
of specimens in the form of a ring from forgings) and welded joint metal. Welded joints and primarily automatic submerged multilayer welds are characteristic of a regional
inhomogeniety (weld root, top and centre) and of an inhomo-
geneity inside one bead due to a substantially different
microstructure which affects fracture toughness. Therefore
full thickness experimental data are needed to obtain reliable knowledge on a static crack resistance of such welded
joints. This requires up to 150 mm thick specimen tests.
Such results with allowance made for an actual non-uniformity of mechanical properties, chemical composition and
structures of 15X2MФA steel welded joints were obtained in
the paper (Kaplunenko 1985).
The application of experimental data obtained using standard
methods (ASTM E399-81 & Metod.Ukaz.1982)for a verification
calculation of the brittle fracture resistance can yield
a fairly conservative result since it involves an in-service
defect growth starting from the initial size of 0.25 pressure vessel wall thickness. Actually, defect sizes are noticeably less than this value, and with regard to low cyclic
loading of pressure vessels one can state to a high certainty that an ultimate size will not exceed 10 pct of the wall
thickness (Gorynin et.al. 1985). It is this size that can
be considered as typical and used in a verification calculation of brittle fracture resistance.
However, in order to carry out such calculation fracture
toughness temperature dependence plotting is required when
the fatigue crack is as long as 0.1 pressure vessel wall
thickness for all materials tested. Extensive investigations
on this problem are reported in paper (Gorynin et.al.1985).
Figure 2 shows the experimental data on the effect of the
fatigue crack size on the fracture toughness of one melt of
15X2MФA heat resistant steel in three-point bend testing of
50x100x450 mm specimens.

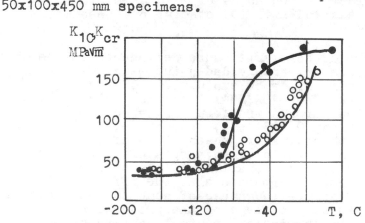

Figure 2. Fracture toughness test results of
15X2MФA steel specimens with 0.1 (●) and 0.5 (o)
crack depth.

It follows from the figure, that when the thickness is 10mm
the temperature dependence of this property shifts towards

136

a 80° lower temperature range as compared to the standard
crack depth (50 mm in this case). The discovered effect
was also noted for welded joints and base metal after a simulation of the operational action by a special heat treatment condition. This effect becomes stronger for a semi-elliptical crack which was revealed in both three-point bend
testing of 50x100x450 mm specimens and in tensile testing
of 150x600 mm cross-section specimens. The temperature dependence shift towards a lower temperature range with decre-
asing crack size in welded joints is less pronounced than
in base metal.
The data on the effect of anti-corrosive cladding on pressure vessel base metal (15X2MФA steel) and weld metal deposited by automatic welding with Cb-10XMФT wire are of interest.
In 50x100 mm cross section specimens with a 2 mm deep notch
(9 mm thick cladding) a fatigue crack was preliminary initiated in cyclic loading so that its tip was in perlitic
metal (base metal or weld metal) at a different distance
from the fusion line. Then the specimens were static three-
point bend loaded at different temperatures. The tests of
these specimens, if the cladding layer was damaged in service, more fully conformed to the practice, than the tests
of monometallic base metal specimen with short cracks. The
fracture toughness of base metal and welded joints, when
the crack is 0.1 specimen height, increases through damping
of plastic metal in austenitic cladding (figure 3).

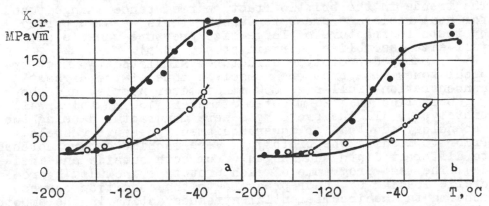

Figure 3. Fracture toughness of 15X2MФA steel of base (a)
and weld (b) metal having anticorrosive cladding with
0.1 (●) and 0.5 (o) depth of crack relative specimen
height.

The total effect of these factors results in the shift of
the fracture diagram plotted in the terms of the linear
fracture mechanics towards a lower temperature range by
80...150°C for base metal and by 30...50°C for welded
joints of heat resistant steels.
The analysis performed and the results obtained have shown
that in practice the fracture resistance of pressure vessels
with defects of 10 pct wall thickness features higher ultimate loads than it follows from data based on testing of
standard specimens with defects of 0.5 specimen thickness.
The result obtained with regard to a defect no greater than

0.1 wall thickness statistically revealed during NDT at the equipment production stage allows to raise the problem of a change in the standardizing approach to the evaluation of high pressure vessel brittle strength. Based on the experimental data, the relationships given in figure 4 are suggested to be used in static strength calculations from the brittle fracture point of view. The use of these relationships will permit to increase the service life of equipment and to appreciably simplify hydrotests of equipment at the production and operation stages.

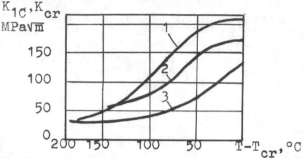

Figure 4. New temperature dependences of 15X2MФA (1), 15X2HMФA (2) steels and welds (3) fracture toughness for fatigue crack depth 0.1 relative pressure vessel wall thickness.

The trends in the brittle fracture resistance change found for heat resistant pressure vessel steels were supported by tests to fracture of large-size pressure vessels and full-size assemblies (Gorynin et.al. 1985).

. Damaged prototype structures simulated two most typical zones of a pressure vessel: a zone of the highest concentration (full-size 850 mm diameter nozzle) and the belt line (a 1.26 m outer diameter, 100 mm thick wall prototype). All the prototypes were manufactured using the technology adopted for power equipment pressure vessels. Prior to loading the prototypes were machine or arc notched to different depth and length (from both outside and inside). The test programme of each prototype included hydro-cyclic loading by inner pressure to the condition corresponding to the circumferential stresses acting in the smooth zone of a pressure vessel in hydropressing, and hydrostatic loading until fracture occurred. In the course of the experiment a cycle-to-cycle indication change was registered using notch edge displacement induction sensors and an acoustic-emission equipment. The results obtained in testing the prototypes were compared with those of testing specimens in the same steel melts and welding material lots. When a prototype was damaged in the base metal a sufficiently reliable correlation of results was obtained, whereas in the case of the prototype welded joint damage the difference in design values of the stress intensity factors of specimens and a prototype was greater, which is due to the neglect of the welding residual stresses when determining fracture toughness of the prototype. The results obtained

138

indicate that fracture mechanics criteria are applicable
for the analysis of high pressure vessels in heat resistant
steels with defects of 0.3...0.6 pressure vessel wall thickness in depth and loading below transition temperature T_{cr}.

4. CONCLUSIONS

1. Brittle fracture resistance of base metal and welded
joints used in the manufacture of high pressure vessels can
be fairly well described by the equation of the type
$K_{1C} = A \cdot \exp(B \cdot T) + C$. The parameters A,B,C of the exponential
equation have been determined to a 95 per cent probability
for 15Х2МФА, 15Х2НМФА heat resistant steels and welded joint
metal.
2. Fracture toughness of test steels increases with the defect size decrease, and also due to a damping action of the
anticorrosive cladding. The discovered effect shows up
both for through cracks and semi-elliptical ones in base
and weld metal, it being less pronounced for the latter.
3. The use of the new temperature dependence of fracture
toughness for a defect of 0.1 wall thickness permits to
raise the equipment service life and to significantly simplify hydrotests through the reduction of the environmental
temperature.
4. Brittle fracture resistance trends obtained have been
supported by tests to fracture of large size pressure vessels.

REFERENCES

ASTM E399-81. Standard test method for plane strain fracture toughness of metallic materials. Annual book of ASTM,
   Philadelphia, 1981, p.592-622.
Calculated strength standards of reactor elements, steam
   generators, vessels and pipelines of nuclear power plants,
   experimental and research nuclear reactors. Moscow, Metal-
   lurgiya, 1973, 408.
Gorynin I.V., Ignatov V.A., Zvezdin Y.I., Timofeev B.T.,
   Filatov V.M. Fracture resistance of thick welded pressure
   vessels of power equipment: Statistical analysis of defects and fracture resistance of pressure vessels. Problemy Prochnosti, 1985, N 11, p.3-11.
Kaplunenko V.G., Pokrovsky V.V., Timofeev B.T., Chernaenko T.A. Crack resistance of weld metal in 15Х2НМ A steel.
   Voprosy Sudostrojenija. Ser. Svarka, 1985, issue 39,
   p.26-33.
Karzov G.P., Leonov V.P., Timofeev B.T. Welded pressure
   vessels: strength and endurance. Leningrad, Mashinostroye-
   niye, 1982, 287p.
Metodicheskie ukazanija. Strength calculation and testing
   in mechanical engineering. Methods of metals mechanical
   testing. Evaluation of fracture toughness characteristics
   under static loading. RD50. 260-81. Izdatelstvo Standar-
   tov, Moscow, 1982, 96p.
Pellini W.N., Puzak P.P. Fracture analysis diagram procedures for the fracture safe engineering desing of steel

    structure. Report N 5920. Naval Research Laboratory, Wa-
    shington, 1963, 55p.
Shatskaja O.A. Main aspects of a new version of Calculated
    strength standards. Voprosy atomnoj nauki i techniky.
    Ser. Physika i technika jadernich reaktorov, 1984, issue
    1 (38), p.3-9.
Shatskaja O.A., Rivkin E.Y., Filatov V.M. Rated analysis of
    strength and service life of nuclear reactors. Mashino-
    stroyeniye, 1985, issue 4, p.3-9.
Timofeev B.T., Zherebenko A.C., Chernaenko T.A. Statistical
    approach to the evaluation of the quality and properties
    of welded joints. Leningradski Dom Nauchno-Tekhnicheskoi
    Propagandy, L., 1982, 22p.

Mechanical strength of large pressure-vessel materials in corrosive medium

F.F.Giginyak, Yu.I.Zvezdin, A.A.Lebedev & B.T.Timofeev
Institute for Problems of Strength of the Ukr.SSR Academy of Sciences, Kiev, USSR

1 INTRODUCTION

In accordance with statistical data (Klemin 1971) 31 to 41
percent of the total damage of the inner surface of high-
-pressure-vessel shells for power equipment is associated
with the formation of corrosion cracks, pittings, dimples
and other flaws. Hence, it is necessary to study thoroughly the resistance of base materials used in power equipment
as well as their welded joints to corrosion and fracture
under conditions of static and cyclic loading and durable
effect of an aqueous medium. It is of importance to conduct corrosive-mechanical testing of pearlite pressure-vessel materials irrespective of the fact that in service they
are protected from the contact with an aqueous medium by
the anticorrosive austenitic cladding since there exists
the possibility of the corrosion resistant layer damage
(Wimune 1966).

2 MATERIALS AND EXPERIMENTAL PROCEDURE

The authors studied the resistance of materials to low-cycle fatigue in an aqueous medium characterized by the parameters of high magnitude (high pressure and high temperature), as well as corrosion crack growth resistance under
static and cyclic loading. The materials studied were high-
-temperature large pressure-vessel steels 15Kh2MFA,
15Kh2NMFA, their welded joints made by automatic welding
under a layer of flux and the metal of the austenitic anticorrosive cladding made by automatic welding with the use
of Sv-07Kh25N13 strip electrode for the first layer and
Sv-04Kh20N10G2B electrode for subsequent layers. The last
layer was in contact with a heat-transfer medium.
Low-cycle fatigue tests were conducted on tubular specimens in biaxial stress state which was attained by constant
inner pressure (P = 6MPa, T = 270°C) and external tension-
-compression at different frequencies. During the tests
the loading rate varied by 20-fold and corresponded to the
frequencies of 0.00167...0.033 Hz. The water of high purity
with the content of oxygen up to 0.01 mg/kg and with the

141

boron-content control was used as a test medium.

To evaluate the corrosion-crack growth under static loading 16×18×180 mm-beam specimens were used tested in three point bending. Due to high plasticity of the materials tested the Cherepanov-Rice J -integral method was used for the estimation of the crack growth resistance. The experimental technique is described by Nikiforchin, Ignatov, Timofeev et al. (1984). The test basis was 10^3h, while for 15Kh2NMFA steel it was considerably higher and constituted one year. Different autoclaves were used for the evaluation of the corrosion-crack growth resistance under cyclic loading. The tests were conducted with a frequency of 0.0167 Hz, at which the maximum effect of the medium on the cyclic crack growth resistance was revealed. In the course of the crack growth studies the load cycle varied to obtain the stress ratio in a load cycle from R = 0 to R = 0.7.

In the present study 1 percent H_3BO_3 solution in distilled water with the addition of KOH was used to ensure pH8. The values of pH were monitored at a temperature of 25°C.

3 EXPERIMENTAL RESULTS AND DISCUSSION

As was shown by the experiments for the steels studied and their welded joints the effect of the medium is actually absent. This is supported by the test results presented in Figure 1 from which it follows that the resistance of a pearlite steel to fracture in air and in the water with boron-content control does not change at cyclic deformation up to 820 h over the life range from 10^2 to 10^4 cycles.

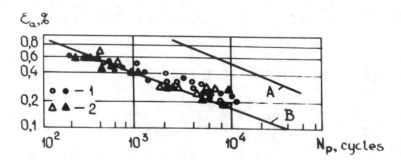

Figure 1. Resistance to low-cycle fatigue of steels 15Kh2MFA (1) and 15Kh2NMFA (2) in air (dark points), in an aqueous medium (light points) at 270°C. A,B curves obtained in air with 50 and 95 % of probability, respectively.

Similar effect was observed by Giginyak (1987). The absence of the aqueous medium effect upon fatigue strength of high-temperature pressure-vessel steels and their welds is accounted for by the formation of a protective magnetite film on the inner surface of the specimens tested. At low strains cyclic loading does practically no damage to the film. At essentially large strain amplitudes the magnetite film usually fractures but this practically does not effect

142

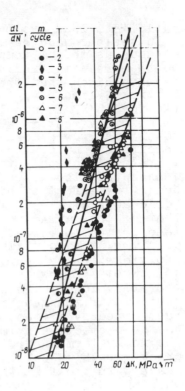

Figure 2. Fatigue crack growth
rate variation in reactor's waters at 250...300°C for high-temperature steels: 1 15Kh2MFA KP60,
R=0, T=250°C, δ =3 mm;
2 15Kh2MFA KP100, R=0, T=250°C,
 δ =3 mm; 3 15Kh2MFA KP60, R=0.7,
T=300°C; δ =25 mm; 4 15Kh2MFA
KP100, R=0.7, T=250°C, δ =3 mm;
5 15Kh2MFA KP60, R-0.7, T=300°C,
 δ =50 mm; 6 15Kh2MFA KP60,
R=0.2, T=270°C, δ =50 mm;
7 15Kh2NMFA KP60, R=0, T=250°C,
 δ =3 mm; 8 15Kh2NMFA KP100,
R=0, T=250°C, δ =3 mm.

the life of the material. The latter fact is associated
with rapid healing up of the film fractures in the medium
of neutral parameters as a result of which pits fail to form
on the specimen surface. The absence of a free access of
oxygen to the surface of the shell metal subjected to deformation is an additional guarantee that the corrosive
medium will not contribute to the reduction of life.
Investigation of the aqueous medium effect on the fatigue
crack growth rate in base materials and their welded joints
(automatic and electroslag welding) was carried out on disk
specimens which made it possible to observe visually the
crack extension. These experiments conducted on compact-
-tension specimens of the base-metal showed that the heat-
-transfer medium had little effect on the resistance to fatigue crack growth with a nearly repeating load cycle.
Generalized data for steels 15Kh2MFA and 15Kh2NMFA are given in Figure 2 where the experimental results are compared with those obtained in air (shaded stripe) and with
standard dependence from the ASME code (curve 1). It should
be noted that for 15Kh2MFA steel at the stress ratio in a
load cycle R=0.7 the fatigue crack growth rate in an aqueous
medium at ΔK = 30 MPa is an order of magnitude higher
than in air. These results were obtained in testing 25 mm-
thick compact tension specimens. With 3 mm-thick specimens such a reduction was not observed even at a high stress

ratio in a load cycle (Timofeev 1987).

In tests conducted under more severe conditions, i.e. at an aqueous medium temperature of 80°C, the medium was found to be more acidified at the crack tip in the 10 mm-thick beam specimen than the water in the autoclave. Thus, Dmitrakh, Timofeev et al. (1984) obtained pH3 at the crack tip for 15Kh2MFA steel and pH8 in the autoclave.

An improvement in the resistance of the materials studied to the fatigue crack growth with an increase in the medium temperature up to the maximum in-service one (270...290°C) is related not only to the changes in electrochemical processes but also to those in protective film properties. Under the effect of high-temperature water which contains boric acid a film is formed near the crack tip similar to that on the surface of the high-temperature steel specimens tested in low-cycle fatigue. With an increase in temperature, the solubility of magnetite reduces and the electrolyte stirring near the crack tip intensifies (Hartt, Henke, Fluet 1977) which hinders its local acidification and therefore lowers the intensity of local hydrogen embrittlement, while the penetration of hydrogen into the metal is restrained by protective oxide films. Therefore, fracture of the protective film and intensive local dissolution of the newly-formed surface are essential conditions for the crack growth under the influence of the medium. In accordance with the film theory these conditions are encountered when the deformation rate of the film does not exceed the rate of processes which cause its fracture. The data obtained by Scott, Truswell (1983) reveal the fact that under certain conditions a complete repassivation of the material at the crack tip may occur which causes a reduction in the crack growth rate.

For the tests conducted the stress ratio in a cycle was shown to have the largest effect on the crack growth rate in an aqueous medium as compared to all other service factors, such as frequency, temperature, stress ratio, the velocity of the medium. With an increase in the stress ratio from 0 to 0.7, the crack growth rate for the high-temperature pressure vessel steel in an aqueous medium can grow 15 to 20 times as compared to the same dependence in air. Basing on systematization of the experimental data on fatigue crack growth resistance in air and in an aqueous medium, the recommendations were given as to how to take into account this stage of fracture when calculating lives of structures with surface and internal manufacturing or in--service flaws. In this case it is practical to simulate the shape of the flaws which are revealed by nondestructive methods with plane defects outlined with the elliptical and circular contours as it was described in ASME Code (1978). To obtain the most conservative data on the crack extension under cyclic loading with the account taken of the location of flaws either an internal circular or surface semielliptical crack with a semi-axis ratio of 2:3 should be studied. Figure 3 presents the calculated fatigue crack growth versus conventional life curves which take into account the above-mentioned simulation procedure.

$$N_{con} = N(\Delta \acute{6})^m$$

where N is the actual life spent for the crack extension
from a_o = 1 mm up to a_f evaluated from the brittle
fracture condition; $\Delta\acute{6}$ is the range of the reduced stresses in MPa; m is the exponent in the Paris's equation.
According to Timofeev, Fedorova and Zvezdin (1987)
m = 2.47 for the materials tested.

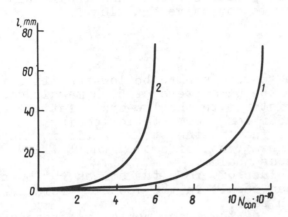

Figure 3. Calculated fatigue crack growth vs conventional
life curves for inner (1) and surface flaws (2) in pressure-vessel materials.

Curve 1 presented in Figure 3 is valid for the ΔK values varying over the range from 4 to 100 MPa$\sqrt{m}$ and can be
used without any correction for the evaluation of structures
lives in the stage of crack extention from the internal
flaw since there is no contact of a heat-transfer medium
with the metal in the locality of such a flaw. Curve 2
describing the surface flaw extension is plotted with the
account taken of the maximum correction for the corrosive
medium effect revealed under cyclic loading with the stress
ratio in a cycle R = 0.7.
No tendency to cracking was observed for high-temperature
steels, their welded joints and the metal of the austenitic
cladding at a temperature of 80°C when studying corrosion-
-crack growth resistance under static loading over a period
of 1000 h. However, for both the metal of the first layer
of the cladding made with the use of Sv-07Kh25N13 strip
and the metal of the heat-affected zone (HAZ) of the welded
joints for both steels tested under the conditions mentioned the crack extension was observed. Owing to a low crack
growth rate in the HAZ metal, its reliable monitoring in
the course of testing was impossible. Hence, the sensibili-
ty of a metal to the effect of a corrosive medium could be
estimated only after final fracture of specimens subjected
to the number of load cycles corresponding to the test basis. In this case the crack extension did not exceed 0.2mm.

At elevated temperatures (250-300°C) the tendency to crack extension for the materials tested in the aqueous medium under long-term static loading is far weaker than at a temperature of 80°C. All the data obtained mainly refer to 15Kh2MFA steel which was heat-treated to obtain different levels of strength. No subcritical crack growth was observed with 15 mm-thick compact-tension specimens tested for as long as 7000 h in autoclaves both with and without heat-transfer medium circulation. This testifies to the fact that high-temperature steels and their welded joints have no tendency to corrosive cracking.

CONCLUSIONS

The materials tested turn to be low-sensitive to the aqueous medium with the parameters of high magnitude in terms of both the resistance to the low-cycle fatigue and fatigue crack growth under long-term static loading.

Of all service factors the stress ratio in a cycle is found to have the largest effect on the fatigue crack growth rate in an aqueous medium.

At the same values of ΔK the stress ratio variation from 0 to 0.7 can result in the 10 to 15-fold increase in the crack growth rate.

Systematization of the experimentally obtained data on the fatigue crack propagation in high-temperature steels and their welded joints makes it possible to take into account the above stage of fracture in standard calculations for both internal and surface flaws.

REFERENCES

Klemin, A.I. & E.F.Polyakov. 1971. Quantitative analysis of fail-safe and reliable operation of atomic power stations (in Russian). Atomnaja tekhnika za rubezhom. 3: 3-11.

Wimune, E.A. 1966. How serious are vessel cladding failure. Power Reactor Techn.V.9, 12:101-109.

Nikiforchin, G.N., Ignatov, V.A., Timofeev, B.T. et al. 1984. Corrosion-crack growth resistance in large pressure-vessel high-temperature steels and their welded joints (in Russian). Fiz. Khim. mekh. mat. 4: 21-28.

Giginyak, F.F., Storchak, M.V., Timofeev, B.T. et al. 1987. Corrosive medium effect on cyclic creep in 15Kh2MFA steel tested under complex stress-state conditions (in Russian). Probl. Prochn. 1:31-35.

Timofeev, B.T., Fedorova, V.A., Zvezdin, Yu.I. et al. 1987. Resistance of high-temperature steels and their welded joints to corrosion-fatigue fracture (in Russian). Probl. Prochn. 1:25-31.

Ratych, L.V., Dmitrakh, I.N., Timofeev, B.T. et al. 1984. On electrochemical conditions at the crack tip in corrosion-crack growth resistance tests on 15Kh2MFA steel beam specimens in an aqueous medium (in Russian). Fiz. khim.

mekh. mat. 3: 31-42.

Hartt, W.N., Henke, T.E. & Y.E.Fluet. 1977. Acidification of electrolyte within corrosion fatigue cracks of 1018 steel in sea water. Corrosion V.33, 4: 151-152.

Skott, P.M. & A.E.Truswell. 1983. Corrosion fatigue crack growth in reactor pressure vessel steels in PWR primary water. Trans. ASME. J. Press. Vess. Techn. 105: 245-254.

ASME Boiler and pressure vessel code. Section XI. 1978. 232.

Analytical and numerical study of thermo-shock induced fracture by single and multiple crack growth

M.Kuna & H.G.Maschke

Institute of Solid State Physics and Electron Microscopy, Halle, GDR

H.A.Bahr & H.J.Weiss

Institute of Solid State Physics and Material Science, Dresden, GDR

1 INTRODUCTION

Following the approach of Stahn & Kerkhof (1977) and Emery & Kobayashi (1980), a generalized qualitative fracture-mechanical concept for the investigation of thermo-shock induced cracking was developed by Bahr & Weiss (1986a, 1986b). Based on the consideration of time-dependent energy release rates, this model takes into account stable and unstable crack growth processes, flaw size distribution, severity of the quenching temperature and mutual unloading of multiple cracks. The experimentally observed crack patterns (Bahr et al 1986b) and the residual strength of quenched ceramic samples could be explained. To substantiate the previous findings, in the present paper numerical calculations are performed for a heated long strip of width b with a single crack or periodically arranged edge cracks (length a, distance p) exposed to thermal shock on the cracked side. The essential features of this two-dimensional example problem should be transferable to more complex crack configurations under thermal shock. The calculation of the energy release rate G (or KI-factor) in dependence on time and crack length requires: i) evaluation of the transient temperature fields due to thermal shock, ii) solution of the resulting thermoelastic stress problem and determination of G.

2 TEMPERATURE AND STRESS FIELDS DUE TO THERMAL SHOCK

The cracks spread perpendicular to the cooled edge of a thin strip and lie parallel to the y-axis of the cartesian coordinate system (Fig. 4). In thermal shock the temperature on the quenched face suddenly drops from Ts by the amount Δ T.

$$T(\eta,\tau) = T_s + \Delta T \sum_{n=1}^{\infty} \frac{2 \sin \mu_n \cos(\mu_n \eta - \mu_n)}{\mu_n + \sin \mu_n \cos \mu_n} e^{-\mu_n^2 \tau^2} \tag{1}$$

Since the crack does not influence the heat flow, the
temperature field eq.(1) can be obtained by solving the
one-dimensional transient heat conduction equation (see
Bahr et al 1987), with η = y/b and τ = $\sqrt{\varkappa t}$ /b . The
constants μ_n are the positive eigen-values of the equation
μ_n= β cotμ_n . β =hb/k. Here, $\varkappa$, k and h denote the
thermal diffusivity, thermal conductivity and heat trans-
fer coefficient, resp. Newtonian (β =10) and ideal
(β = ∞) cooling were considered.

Normal to the crack the following stresses are induced
by the thermal field: (2)

$$\sigma_x(\eta,\tau) = E\alpha \left\{ T(\eta,\tau) - \int_0^1 T(\xi,\tau)d\xi - 12\left(\eta-\tfrac{1}{2}\right)\int_0^1 \left(\xi-\tfrac{1}{2}\right)T(\xi,\tau)d\xi\right\}$$

(E -Young's modulus, α -coefficient of thermal expansion)
For more details of the calculated temperature and stress
fields we refer to the paper of Bahr et al (1987).

3 CALCULATION OF STRESS INTENSITY FACTORS

The stress intensity factor KI for the mode-I crack problem at hand can be determined by means of the weight function method. For any thermal loading case an equivalent
crack face loading can be found which is equal but opposite to the tractions at the prospective crack line in the
body without crack. Then KI is found by integrating these
tractions given by eq.(2) over the crack :

$$KI(a,t) = \sqrt{\frac{2}{\pi}} \int_0^a M(a,y) \ \sigma_x (y,t)dy \qquad (3)$$

The weight function M(a,y) was derived by Bueckner
(1971) for the single edge cracked strip.

The application of FEM to this problem was accomplished
in three steps:
a)evaluation of the temperature field by solving the
transient heat conduction problem for the thermoshock;
b)solution of the associated thermoelastic stress problem;
c)determination of the stress intensity factor KI;

All steps were performed by means of the FEM-system
FRACTURE (Eisentraut & Kuna 1986) especially adapted to
the solution of plane, axi-symmetrical and spatial problems of thermoelasticity, fracture mechanics and heat
conduction. Fig. 1 illustrates the mesh generated for the
upper half of the sample with a single crack. The crack
tip region was modelled by quarter-point triangular elements . These elements can reflect the elastic crack tip
singularity and a constant thermal strain distribution.
The KI-factor was calculated from the FEM-results in two
ways: First, from the opening displacements of the nodes
at the crack face element . Second, the energy release
rate G = KI*KI/E was calculated using the technique of
virtual crack extension (Bass & Bryson 1983).

Next, the boundary element analysis code ATALANTE
(Maschke & Kuna 1985) for solving two-dimensional problems

in linear elastic fracture mechanics was applied. The
program is based on the direct boundary element method,
involving the free space Green's function and 3-noded
elements of quadratic isoparametric shape functions.
Special boundary elements were employed for modelling the
crack tip. The stress intensity factor is gained from the
BEM-solution utilizing either the crack face displacements
or the traction values of the crack tip node, see Bahr et
al (1987). By a special option of the BEM-system periodical structures can be treated, which is essential for
analyzing multiple crack configurations. Fig. 2 shows one
of the BEM-discretizations used for the analyses of periodically arranged cracks of two different lengths. Here,
originally thermal loads are transferred into the equivalent crack face tractions of eq.(2).

For a single crack, Fig.3 summarizes the results obtained by using weight functions, FEM and BEM at the time
τ =0.1 . The good agreement of all three solutions confirms the range of validity of the weight functions (up to
a=0.5 b) and demonstrates the applicability and quality of
both other numerical techniques. Compared at equal accuracy, the BEM requires much less expense of mesh creation
than the FEM does.

4 CRACK PROPAGATION SCENARIO

Fig.4 shows the normalized energy release rate plotted
versus crack length. Crack propagation starts as soon as G
exceeds a critical value Gc and continues as long as this
condition is met. There is a critical severity of shock,
Δ Tc, below which no cracking occurs. Just at Δ T = Δ Tc,
single crack propagation is observed, with the crack run-
ning unstably. Excess released energy may drive the crack
beyond the envelope of the family of G-curves (solid arrow
in Fig.4)

Higher Δ T means lower Gc-level in our normalized plot.
It is seen from Fig.4 that with increasing Δ T the delay
of crack initiation after shock is diminished. As a consequence, the unstable crack paths are shorter. Furthermore,
some of those initial flaws which are partly unloaded by
the growing cracks may still be able to start. This is due
to the fact, that their G increases as the cooling goes on
penetrating into the strip. This consecutive activation of
initial flaws has been analyzed for a periodic array by
means of BEM (Fig.2): By variation of the crack spacing p,
the minimum spacing is found where the remaining initial
flaws of length a2=a0 will just be able to start while
they would not do so with lower distance. This implies a
condition on the stress intensities of both growen cracks
(1) and initial flaws (2),

K1(a1, a2=a0, t, p) = Kc; Max {K2(a1, a2=a0, t, p)} = Kc.

It is found that larger Δ T results in smaller p (Fig.4,
inserts fit to scale). These conditions govern the establishment of the so-called parking order of cracks by

combined unstable and stable propagation (completed with the shaded arrow). `This is followed by stable propagation controlled by $KI(a1, a2=a1, t, p) = Kc$. As a consequence of progressive mutual unloading of the growing cracks, not all of them keep propagating: There is a bifurcation-type instability with every second crack popping off while the remaining ones are left behind (Bazant et al 1979). This occurs at $\partial K2(a1, a2=a1, t, p)/\partial a2 = 0$. The computations revealed two events of this kind for $\Delta T = 2.9 \ \Delta Tc$ (see inserts of Fig.4). Multiple crack growth finally stops at the decreasing branch of the envelope (contrary to single crack propagation, which may overshoot). Note that the envelopes related to single and multiple propagation differ (dashed lines in Fig.4).

The numerical results obtained for the quenched edge cracked strip support the recently proposed fracture-mechanical approach by Bahr & Weiss (1986a) to thermo-shock damage of brittle materials due to single and multiple crack growth. It allows to draw quantitative conclusions concerning the dependence of damage on severity of shock, initial strength and heat transfer.

REFERENCES

Bahr, H.A. & H.-J. Weiss 1986a. Heuristic approach to thermal shock damage due to single and multiple crack growth. Theor.Appl.Fracture Mech. 6: 57-62
Bahr,H.A., G. Fischer & H.-J. Weiss 1986b. Thermal shock crack patterns explained by single and multiple crack propagation. J. Mater. Sci. 21: 2716-2720
Bahr, H.A., M. Kuna, H. Liesk & H. Balke 1987. Fracture analysis of a single edge cracked strip under thermal shock. to appear in: Theor.Appl. Fracture Mech.
Bass, B.R. & J.W. Bryson 1983. Energy release rate techniques for combined thermo-mechanical loading. Int.J. Fracture 22: R3-R7
Bazant, Z.P., H. Ohtsubo & K. Aoh 1979. Stability and post- critical growth of a system of cooling or shrinkage cracks. Int.J.Fracture 15: 443-456
Bueckner, H.F. 1971. Weight functions for the notched bar. ZAMM 51: 97-109
Eisentraut, U.M. & M. Kuna 1986. A FEM-program for plane, axi-symmetric and spatial problems in fracture mechanics, elastostatics and heat conduction. (in german) Technische Mechanik 7: 51-58
Emery, E.F. & A.S. Kobayashi 1980. Transient stress intensity factors for edge and corner cracks in quench- test specimens. J.Amer.Ceram.Soc. 63: 410-415
Maschke, H.G. & M. Kuna 1985. A review of boundary and finite element methods in fracture mechanics. Theor. Appl. Fracture Mech. 4: 181-189
Stahn, D. & F. Kerkhof 1977. Bruchgefährdung von thermo-schockbelasteten Glasscheiben. Glastechn. Berichte 50: 121-128

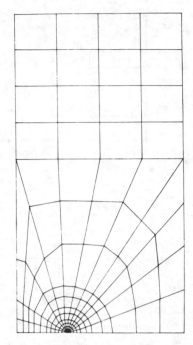

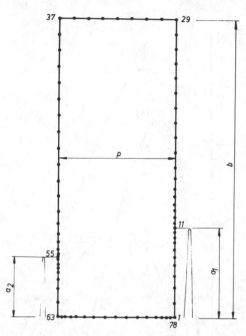

Fig.1 Finite-element mesh
    for the crack length
    of a=0.3b
(148 elements,475 nodes)

Fig.2 Boundary element mesh
    for periodical
    crack configuration

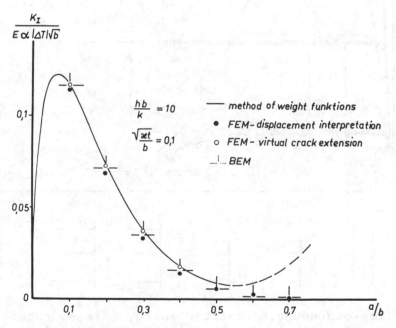

Fig.3 Comparison of K-factors derived from weight func-
    tions, finite elements and boundary elements

153

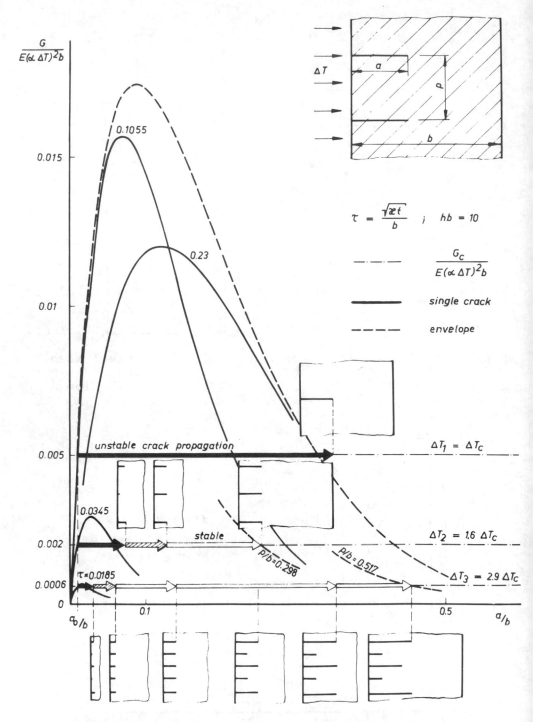

Fig.4 Time-dependent normalized energy release rates
versus crack length for finite heat transfer.
The crack propagation scenario is illustrated
by the arrows and the properly scaled inserts.

Fracture resistance of SA106B seamless piping and welds

B.Mukherjee, D.Carpenter & M.J.Kozluk
Ontario Hydro Research, Toronto, Canada

1 INTRODUCTION

Ontario Hydro has developed a leak–before–break (LBB) methodology for application to the large diameter heat transport piping for Darlington NGS A
because of the difficulties encountered in attempting to provide pipewhip
restraints and in recognition of the questionable benefits of providing such
devices. This approach has been developed for pipe sizes which are equal
to or greater than 21" nominal diameter (21" NPS).

Piping steels are expected to exhibit a stable crack extension behaviour
with rising load up to the point of instability in the ductile uppershelf
regime. Under these conditions, the application of structural integrity concepts that are based on crack initiation alone will be overly conservative.
Thus the margin against failure must be determined under elastic plastic
conditions by taking into account both crack initiation and stable crack
growth. The Ontario Hydro leak–before–break approach is based on the J/T
method.

The material value of J associated with crack extension which is also
known as material resistance curve or a J–R curve is required for stability
assessment. The nondimensionalised slope of a J–R curve at a given value
of J is defined as tearing modulus (T). The margin against instability is
determined from a J–T plot where J is plotted against T for the material
and the structure. In this Darlington leak–before–break material test
program J–R and J–T curves were determined from actual Darlington NGS–A
large diameter heat transport piping, forgings, associated welds and heat affected zones. A total of 91 J–resistance curve tests were conducted.
Sixty–three tensile tests were also performed to determine the stress–strain
behaviour accurately. This paper is a brief summary of the J–R curve test
results and major conclusions.

2 EXPERIMENTAL PROGRAM

The test program was designed to take into account the effect of various
factors which influence fracture properties. The objective was to identify
those factors which tend to lower resistance curves and apply the appropriate lower bound resistance curves that were obtained from this
program, for structural analyses. Factors that were included in this project
were:
   Pipe size;
   Heat of Piping Material;

Crack Plane Orientation;
Temperature;
Welding Effects.

Three different pipe sizes of seamless SA–106 Grade B were used (12, 22 and 24 inch NPS). Two SA–105 vesselet forgings were also tested. Vesselet is a trade name for a weld–in contour fitting. The test program utilized actual piping material from actual piping heats that were used in the construction of the Darlington heat transport system (Units 1 and 2). In the stability assessment of through circumferential (L–C) and longitudinal (C–L) flaws, J–Resistance curves in L–C and C–L crack plane orientations, respectively, are required. Specimens from SA106B and SA105 forgings were machined from both L–C and C–L orientation.

It is known that ductility of SA106B steel, as measured by the lateral contraction during a tensile test, shows a minimum value between 200°C and 300°C. Tensile tests were conducted in this temperature range and it was established that the lateral expansion of this material is minimum in the neighbourhood of 250°C. Therefore, 250°C was selected as the appropriate test temperature. These results will then be conservatively used for the appropriate operating temperature. In order to determine if there is any significant dependence of toughness on temperature over the temperature range, a temperature of 20°C was selected arbitrarily, as a lower bound of the lowest service temperature for the heat transport system.

Two different Ontario Hydro welding procedures (PN–108 and PN–232) were used for all welds within the scope of the LBB program. Each of these welding procedures for nuclear Class I piping, were included in the test program. These welding procedures are listed in Table I. Tests were conducted on the weld material and the heat affected zone (HAZ) material. The orientation of a crack in the weld and HAZ was in the L–C direction. Due to the V–shape of the weld preparation surface, the crack front in a HAZ specimen straddled the HAZ, parent and weld material. The measured HAZ toughness is therefore an average toughness.

3   TEST METHOD

Three different sizes of compact tension (CT) specimens, 1–1/4T, 1T and 3/4T, conforming to the geometry of the ASTM test procedure E813, Test for J_{lc}, a Measure of Fracture Toughness were used. The sizes of the specimens were determined by the largest CT specimens that could be machined without flattening from the 24", 22" and 12" NPS pipe, respectively. All specimens were precracked and sidegrooved after precracking to ensure a straight crack front. The R–curves produced with a straight crack front extension exhibit slopes (dJ/da) lower than those produced with a non sidegrooved specimen. Sidegrooving also appears to produce a slightly lower (more conservative) value of crack initiation toughness.

Regarding specimen geometry, existing data indicate that fracture toughness specimens having approximately the same thickness as the pipe wall and without sidegrooves tend to model actual pipe behaviour most accurately. However, sidegrooved specimens will provide an acceptable lower bound J–resistance curve/1/. Also note that C–T specimens are bend type specimens. J–R curves from bend type specimens defines the lower bound estimate of J–capacity as a function of crack extension, and has been observed to be conservative in comparison with those obtained with tensile loading specimen configuration/2/. The specimen thicknesses were approximately 80% of the respective pipe wall thicknesses from which they were machined. Therefore, this test program will provide an acceptable lower bound J–resistance curve.

A single specimen technique documented in /2/ was utilized to determine the J-R curves. In this test method, crack length and crack extension were determined from elastic compliance measurements. These measurements were taken on a series of unloading/reloading segments spaced along the load versus load-line displacement record.

A personal computer was used for test control and data acquisition. A computer program was developed to control the loading and unloading sequences. A separate computer program was used to analyze the acquired data and plot final graphs.

The calculation of J_{Ic} is according to ASTM E813–81. J_{Ic} is defined by the intersection of a linear regression fit to the data, that fall within the 0.15 and 1.5 mm exclusion lines, with the blunting line. A J-R curve is established by fitting a curve through the experimental data points using a power law equation in the form

$$J = C_1 \, (\Delta a)^{C_2}$$

This equation has been used extensively to establish J-R curves for ferritic steels. An examination of the data showed that this equation provides an excellent fit through the experimental J-Δa data points for each experiment in this test program.

4   DISCUSSION OF RESULTS

The objective of this test program was to develop a comprehensive material data base for all product forms, heats of steels and weld procedures that have been used in the construction of large diameter Darlington heat transport circuit. At the end of the test program, it can be stated with confidence that this objective has been achieved. All test results are stored in a computerized data base.

4.1   Parent

Specimens from four product forms were tested. These were 24", 22" and 12" NPS Sch 100 pipes and SA105 forgings. Toughness values and J-resistance curves of these product forms depend on temperature and orientation. However, at a fixed temperature and at a fixed orientation toughness value (Figure 1) and J-resistance curves (Figures 2) lie within a narrow scatterband.

A general observation from all parent material results is that all precracked parent material specimens showed stable and ductile crack extension with rising load. There was no systematic variation of toughness with product forms or pipe sizes. At both test temperatures and specimen orientations all parent material specimens showed high J_{Ic} and rising resistance curves with crack extension.

Toughness values of pipes depend on the direction of crack extension relative to the pipe geometry. Tests using conventional Charpy specimens have shown that toughness of SA106B steel, as indicated by energy absorption, depends on specimen orientation/3/.

At 250°C, J_{Ic} and J-resistance curves in the circumferential direction were higher than those in the axial direction for all product forms. Changes in J_{Ic} values and J-resistance curves were comparable for the 24" and 22" NPS Sch 100 pipes and SA105 forgings (Figure 3). The 12" NPS Sch 100 pipe specimens showed the least reduction of toughness with orientation.

Toughness behaviour of high toughness − low strength C−Mn steels such as SA106B and SA105 forgings show a transition with temperature. An examination of load displacement curves and Charpy energy values indicated that these steels show an upper shelf toughness behaviour at $20^{\circ}C$. Specifically, load versus displacement behaviour of all specimens were highly ductile. All specimens exhibited high crack initiation toughness, stable crack extension and steeply rising resistance curves beyond J_{Ic}.

J_{Ic} and J−resistance curves at $250^{\circ}C$ were lower than those at $20^{\circ}C$, which is characteristic of lower strength, high toughness structural steels (Figure 4). Reduction of material tensile stress with temperature will promote this behaviour.

4.2   Weld

Ontario Hydro has developed various welding procedures for the construction of Darlington PHT system. These procedures satisfy ASME Code requirements. The PN 108 and PN 232 weld procedures were used for all welds within the scope of the Darlington LBB project. Post weld heat treatment is required for these weld procedures. All tests on PN 108 and PN 232 weld and HAZ material at both test temperatures showed high J_{Ic}, stable crack extension and rising resistance curves, that is, general uppershelf toughness behaviour.

J_{Ic} values and J−resistance curves (Figure 5) for the weld and HAZ material of PN 108 and PN 232 procedures were higher than the corresponding parent material properties. Due to the geometry of weld preparation and the given thickness of pipes, the crack front in a HAZ specimen straddled the weld, HAZ and parent material. The measured HAZ J−resistance curves, therefore, lie between the weld and parent material resistance curves. Effect of test temperature on the toughness of PN 232 weld was similar to that observed in parent material tests in that J_{Ic} and J−resistance curves at $250^{\circ}C$ were lower than those at $20^{\circ}C$. An upper-shelf toughness behaviour was observed at both test temperatures.

5   VALIDATION OF TEST RESULTS

Since the material test program is pivotal to the Darlington LBB program, it was decided to validate the material test program by comparing six test results against tests conducted by an independent laboratory. Materials Engineering Associate of Lanham, Maryland was selected for conducting the validation tests because of their experience in conducting single specimen J−resistance curve tests for U.S. Nuclear Regulatory Commission [4].

Three 2.5 cm thick compact tension specimens were machined by MEA and were tested at $250^{\circ}C$. All specimens were machined in the L−C orientation. Yield and UTS values were provided to MEA by Ontario Hydro.

In summary, the J_{Ic} and J−resistance curves obtained by OH and MEA on parent and weld materials (Figure 6) were comparable. There were no systematic variation or discrepancy between the tests conducted by two independent laboratories. This comparison provides an independent but indirect validation of transducer calibration, test method, crack length and J calculation procedures.

6   CONCLUSIONS

1.   All product forms utilized in the construction of Darlington primary heat transport piping (24", 22" and 12" SA106B NPS Sch 100 pipes and SA105 forging) exhibited uppershelf toughness behaviour at 20°C and 250°C. All specimens from the above product forms showed high crack initiation toughness J_{Ic}, increasing resistance to crack extension and stable and ductile crack growth.

2.   Toughness (J_{Ic} and J–Resistance curve) of parent materials depends on specimen orientations. Toughness of a pipe in the circumferential orientation was higher than the corresponding toughness in the axial direction. The effect of orientation on toughness was quantified.

3.   For a given temperature and specimen orientation, toughness of various product forms was reasonably uniform.

4.   All welds and heat affected zone materials within the scope of the Darlington Leak Before Break project exhibited uppershelf toughness behaviour at 20°C and 250°C. All specimens from weld and HAZ material showed high crack initiation toughness, J_{Ic}, increasing resistance to crack extension and stable and ductile crack growth.

5.   The material test program was validated by comparing six test results against tests conduoted by Materials Engineering Associates of Lanham, Maryland. The J_{Ic} and J–resistance curves obtained by Ontario Hydro and MEA were comparable.

REFERENCES

1. Report of the U.S. Nuclear Regulatory Commission Piping Review Committee, Nureg–1061, Volume 3.
2. Standard Test Method for Determining J–R Curves, 12th Draft. A proposed test method developed by the ASTM E24.08 Subcommittee.
3. "PVRC Recommendations on Toughness Requirements for Ferritic Materials", Welding Research Council Bulletin No. 175.
4. "Structural Integrity of Water Reactor Pressure Boundary Components", Annual Report for 1984. Edited by F.J. Loss, Materials Engineering Associate, NUREG/CR–3228, MEA–2051, Vol. 3.

Table I.   Ontario Hydro Nuclear Weld Procedures

NPS (Inches)	Ontario Hydro Weld Procedure	Weld Process
24 22	PN 108	– GTAW root pass – SMAW fill up – for field and shop welding – PHWT performed
24 22	PN 232	– GTAW root pass – SMAW 2nd pass – SAW fill up – for shop welding only – PWHT performed

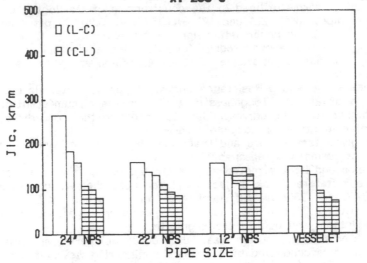

FIGURE 1

COMPARISON OF J_{IC} FOR VARIOUS PRODUCT FORMS

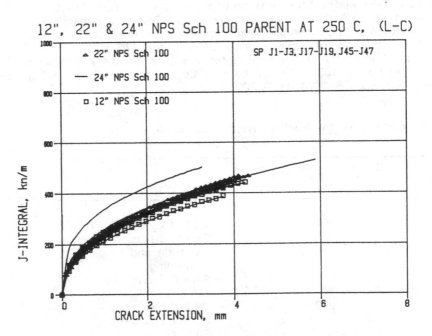

FIGURE 2

J-RESISTANCE CURVES FOR THREE PIPE SIZES

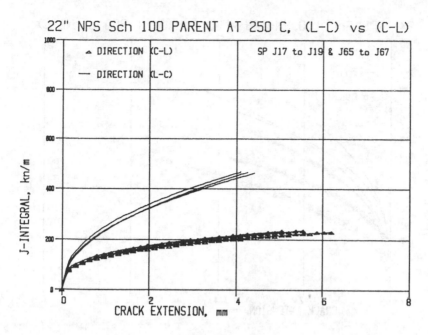

FIGURE 3

EFFECT OF ORIENTATION ON J-RESISTANCE CURVES

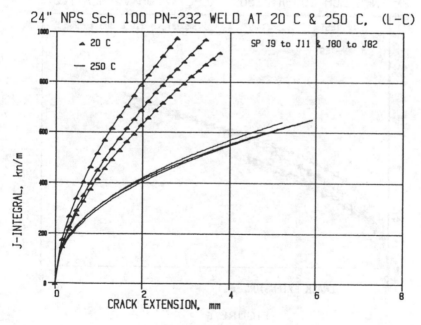

FIGURE 4

EFFECT OF TEST TEMPERATURE ON J-RESISTANCE CURVES

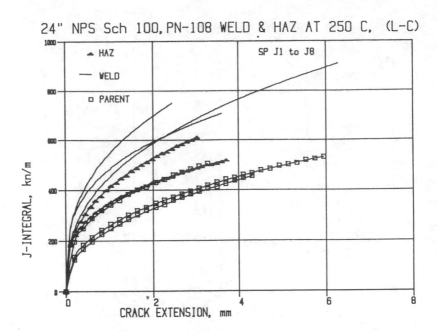

FIGURE 5

COMPARISON OF J-RESISTANCE CURVES FOR WELD,
PARENT AND HAZ MATERIALS

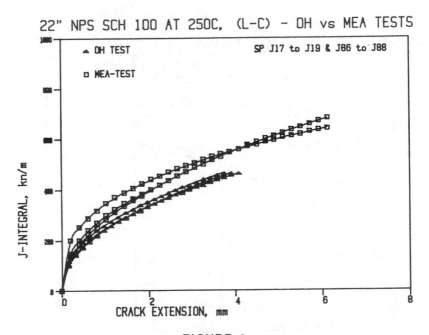

FIGURE 6

COMPARISON OF J-RESISTANCE CURVES BETWEEN
OH AND MEA TESTS

H Concrete and concrete structures

Verification of the design of a reinforced concrete structure by dynamic elastoplastic calculations

Y.L'Huby

Electricité de France, SEPTEN, Villeurbanne

J.Rivière

Commissariat à l'Energie Atomique, DAS, Fontenay-aux-Roses, France

1 INTRODUCTION

Some reinforced concrete structures used in nuclear power stations must be designed to withstand accidental dynamic loadings, such as aircraft impacts, explosions, earthquakes or pipe whippings. These loadings are phenomena with low occurence probability, and it is not necessary to design the structures to their resisting in the elastic behaviour range. French regulations allow the use, in a limited way, of plastic material capacities : a maximum strain rate of 0.8 % is permitted for the rein-forcing steel bars, a maximum strain rate of 0.35 % is permitted for concrete in compression, for nuclear structures under these accidental loadings.

Due to the evolution of calculation methods, it is now possible to reliably evaluate the elastoplastic behaviour of reinforced concrete structures. However these methods are extensive and time consuming and they are not fit for use in design offices. EDF was led to define sim-plified rules, by cutting down the maximum value of the load-function, in order to design with the classical elastic static methods. To vali-date this equivalent simplified method, EDF carried out dynamic elasto-plastic computations on several types of structures subjected to various accidental loadings. This method formed also the subject of a verifica-tion performed by the CEA which was representing the safety authorities.

The objective of this paper is to present the dynamic elastoplastic analysis of an internal structural wall of the PWR 1300 MWe reactor building, and to compare two approaches : one by EDF, the designer, one by CEA/DAS, the safety authorities. The purpose of this study is to prove that, under the combined loads due to LOCA (loss of cooling acci-dent) and SSE (safe shutdown earthquake), the safety criteria (above-mentioned strain rates) are respected ; hence it forms a part of the validation of the equivalent simplified design methods.

2 THE STRUCTURE AND THE LOADING

The structure under study is a rectangular plane wall of the internal structures of the PWR 1300 reactor building : it is a side-wall of the steam generator local. This 1 m thick wall is considered to be embedded in the two heavy floors at levels 6.56 m and 22.76 m ; one vertical edge of the wall is also embedded in the peripheral wall of the internal structures, the other side is bound to another symmetrical wall.

In case of pipe rupture or earthquake, this wall is impacted by the steam generator through a rectangular support (2 m x 0.80 m) at level 17.10 m. Under the effect of LOCA or SSE the steam generator induces a transient dynamic load on the support. The LOCA transient load depends on the localization of the pipe break ; the duration of the SSE transient load is long compared to the LOCA load. For the dynamic analysis of the wall, we have to take into account the most severe possible combination of LOCA and SSE. This has led us to phase the maximum of the force induced by the circumferential rupture of the cross-over leg and the higher peak of the force due to the SSE (figure 1). We have to keep in mind that the wall has been statically and elastically designed with a reduced equivalent force F_{eq}, combining the maxima F_{LOCA} of the LOCA transient effort and F_{SSE} of the SSE transient effort, according to the "quadratic combination rule" :

$$F_{eq} = \sqrt{F_{LOCA}^2 + F_{SSE}^2}$$

3 EDF ANALYSIS

3.1 Computer-code : DALTON

EDF carried out the dynamic elastoplastic computations for the steam-generator wall with the Dalton-program, developed by the Research Department of EDF. This program uses an explicit time-integration procedure. The finite elements are three-node triangular thin plate bending elements of Kirchoff theory. The integration over the element uses three Gauss points. The boundary conditions are represented by spring supports and restoring couples whose stiffness can be variable along the edges. We assume that the behaviour of the slab is orthotropic (according to the two orthogonal directions of reinforcement that are generally encountered in reinforced concrete slabs and walls) ; it is modelled by bilinear moment-curvature relations (one for each direction and each sense of bending). The first straight segment of the global behaviour law corresponds to elastic behaviour, the second part (half-line) corresponds to plastic behaviour, with kinematic hardening. We have adopted the Johansen's yield criterion (figure 2) with the associated flow rule according to the "normality law". Unloading is elastic, according to the first elastic slope.

In order to validate this program, many tests on reinforced concrete slabs have been performed. The tested slabs were simply supported or embedded, their reinforcement was spread over a large range. The loading force was generated by the dynamic crushing of steel pipes by a falling mass. The behaviour of these slabs have been calculated with Dalton and we have found a very good agreement between measurements and computations using Dalton.

3.2 Modelization

EDF modelled the structure as a rectangle, 7.50 m wide and 16.80 m high. The grid had 630 triangular elements, grouped in twos in regular 0.50 m x 0.80 m elementary rectangles (figure 3), and 352 nodes. The impact zone covered 4 rectangles. The wall was divided into six different zones, corresponding to different kinds of reinforcement, in order to better represent the real structure. This led to the definition of many different behaviour laws. As mentioned above, these laws are entered in the program as bilinear moment-curvature relations. These curves are calculated by a specific program (MOCO) and are simplified in the bilinear relations (figure 4) according to an "equivalent energy" method which had been validated by the computations for the tested slabs (§ 3.1) and other studies. In the case of the wall studied, 4 different laws have been defined for each zone, so the behaviour is orthotropic both in elastic and plastic domains.

4 CEA ANALYSIS

4.1 Computer-code : PLEXUS

CEA carried out the dynamic elastoplastic computations for the steam-generator wall with the Plexus-program, developed by the Mechanics and Thermics Department of CEA. This program also uses an explicit time-integration procedure. It is provided with a respectable library of elements and uses, in this case, four-node rectangular shell elements with four Gauss points for the integration. The boundary conditions are simple supports or embedments or can be represented by boundary elements. The behaviour is isotropic and is modelled by multilinear stress-strain relationships (of an homogeneous equivalent material). The first straight segment corresponds to elastic behaviour, the others correspond to plastic behaviour, with isotropic hardening. The Von Mises' yield criterion was adopted (figure 5), with the associated normality law. Unloading is elastic.
    This program Plexus had been validated by many tests on square slabs, with static and dynamic loadings. Some of the EDF tests had also been computed. A good agreement was found between tests and computations with Plexus.

4.2 Modelization

CEA modelled the structure as a 7.50 m x 16.80 m rectangle with 315 rectangular elements and 352 nodes, hence the grid is the same as EDF's. However, it has been divided into only three different zones with different behaviour laws (figure 6). These laws are defined as the stress-strain relations of an homogeneous reinforced concrete material. The original relation is calculated by a specific program (SAMSON) which takes into account the behaviour of concrete in compression and traction. This program is similar to the EDF's program MOCO and comparisons between the two computer codes have given the same behaviour curves. The stress-strain relation is also simplified in a multilinear curve (figure 7). In the case of the CEA's analysis, the behaviour is isotropic and symmetrical (this means that only one curve is employed for each zone) and the curves calculated with Samson have been averaged.

5 RESULTS AND COMPARISONS

In terms of displacements, the results for EDF and CEA seem a little different : the maximum value is reached at the same time and at about the same point (in the impacted zone), but has a value of 9.4 mm in the

EDF analysis and 12 mm in the CEA analysis (figure 8). However, the corresponding maximum tensile strain of steel (which is representative towards the safety criteria) is exactly the same in the two studies : ε_s = 0.50 %. Moreover, a computation, carried out by EDF, with a simplified reinforcement (only one zone with orthotropic behaviour) gave a maximum deflection of 11 mm in the impacted zone (the global mis-shape is the same as for the six-zones computation and the maximum curvature, corresponding to the maximum strain in steel, is a slightly lower). The plasticized zone (figure 9) has about the same pattern and the same extent in both the EDF and CEA analyses.

6 CONCLUSION

The computation results for EDF and CEA give deformations of the same value (steel strain $\varepsilon_s \leqslant$ 0.50 %), respecting the criteria defined by the regulations ($\varepsilon_s \leqslant$ 0.8 %) and justifying, for the studied example of internal structural wall of a nuclear power station, the simplified method used for designing. The other studies of various R.C. structures confirm the validity of the elastic static equivalent method.

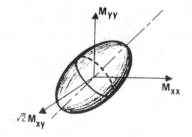

Figure 1. Loading (LOCA + SSE)

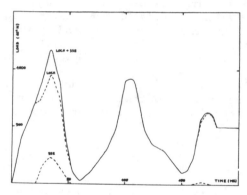

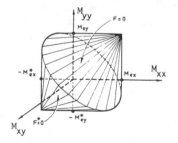

Figure 2. Johansen's yield criterion Figure 5. Von Mise's yield criterion

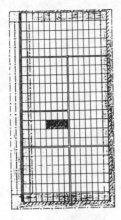

Figure 3. EDF modelization

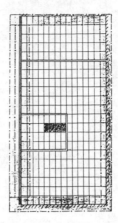

Figure 6. CEA modelization

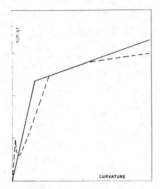

Figure 4. EDF schematization
of behaviour laws

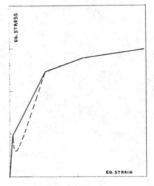

Figure 7. CEA schematization of
behaviour laws

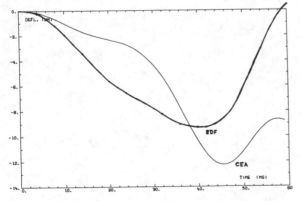

Figure 8. Deflection under the impact zone

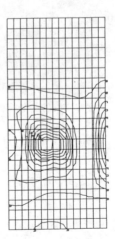

Figure 9. Plastic
deformations

169

Cross sectional design of reinforced concrete ducts for accommodation of emergency cooling water pipes

Y. Aoyagi & T. Endo
Central Research Institute of Electric Power Industry, Abiko-City, Chiba, Japan
K. Iida
Technology Institute of Taisei Construction Company, Yokohama-City, Kanagawa, Japan

1. INTRODUCTION

The Japanese reinforced concrete design code for civil engineering structures has been revised from the previous allowable stress approach (referred to as A.S.D) to that of limit state (referred to as L.S.D) since Oct. 1986.

It is expected that rational and economical results of design can be achieved by applying the new design method also in the case of nuclear power generation related civil engineering structures, such as reinforced concrete ducts for accommodation of cooling water pipes (referred to as D.C.W) ,cooling water intake pits, etc..

As a first step, loading tests were performed on two 1/2 scale models, which were designed by A.S.D and L.S.D,respectively ,and numerical simulations of these loading tests were carried out. This paper describes the main features of comparisons between the two design procedures,referring mainly to loading test results.

2. DESIGN PROCEDURES AND CROSS SECTIONS

Governing design loading was earthquake loading, which is shown in fig-1. The final sections designed by A.S.D and L.S.D are depicted also in fig-1.

The section for L.S.D had the same internal cavity as that for A.S.D. Seismic coefficient for horizontal and vertical directions were assumed as Kh=0.48 and Kv=0.24,respectively. The thickness of the walls for the former, however, was selected so as to achieve the maximum reinforcement ratio, allowed in the Japanese L.S.D code.

The amount of total reinforcement per unit length of D.C.W for L.S.D is 3.2 times as large as that for A.S.D. Volumes of concrete and excavation for the former are reduced to almost 50% and 60% of the latter, respectively. It may be conceivable that an optimum section exists between the above two extreme cases.

3. TESTING PROCEDURES

In the tests loads were applied to the specimens, simulating the loading conditions of underground water and earth pressures, inertial forces,self weight,etc. at the time of earthquake. The distributed

171

loads were approximated by a number of hydraulic jacks as shown in fig-2.

Two loading patterns were employed;
(1) Horizontal as well as vertical loads were increased in the fashion that similarities to the design loads were kept in terms of load distribution. Safety margins against total design loads could be judged in this loading stage.
(2) Only horizontal load was increased ,holding the vertical load constant at that of design during the subsequent loading. This loading pattern was intended to compare the ultimate load capacities of the D.C.W sections designed by A.S.D and L.S.D. This may be justified because the vertical loads are almost unchanged in the cases of ordinary and earthquake loading stages.

Applied loads, strains in reinforcement and concrete, displacement, deformations, local sectional radius, etc. were measured.

4. TEST RESULTS

4.1 Strength Capacities

In fig-3 , the relationships between total horizontal loads and the inward displacement at the mid-height of vertical walls are depicted. It was found that in the loading pattern (1) both A.S.D and L.S.D models showed no signs of reinforcement yielding as well as concrete crushing, even at the loads equal to 1.6 and 1.4 times the calculated ultimate loads for A.S.D and L.S.D, respectively.

In the subsequent loading pattern (2), since the ratio between axial force and bending moment at the mid-height section of vertical walls decreased as the horizontal load increased, the ultimate horizontal loads attained were 650 tons and 550 tons for A.S.D and L.S.D models, respectively. These strength capacities exceed the design earthquake loads by a safety margin of well over 3, which demonstrates ample safety against increment of earthquake load .

4.2 Deformational Behaviors

As may be clear from fig-3, the fact that the deformations at the mid-height of vertical walls are far more pronounced in the case of loading pattern (1) than those in the case of loading pattern (2) indicates the effects of difference in axial compressive force. The deformations of L.S.D model were two to three times as large as those of A.S.D model at the same loading level, which was attributed to the extremely low regidities of L.S.D model. Moreover, non-linear behaviors became distinct at a much lower loading level for L.S.D model than for A.S.D model.

4.3 Cracking Behaviors

Fig-4 shows the comparison of cracking patterns for A.S.D and L.S.D models at the loading stage equivalent to about three times the design load for the loading pattern (1). Only slight cracks could be observed in the central portion of vertical wall for A.S.D model, while cracks developed over the whole regions of L.S.D model .

172

Judging from strength, deformation and cracking points of view, it could be said that the A.S.D hitherto employed may result in unnecessary overdesign in terms of strength and serviceability requirements.

5. FINITE ELEMENT ANALYSIS

5.1 Basic Assumptions

Experimental data were compared with analytical ones, which were derived by applying a smeared crack non-linear finite element method (fig-5). Concrete elements were assumed to become orthogonally heterogeneous when cracked at the Gaussian integral points. Reinforcement elements were idealized by a linear member, and joint elements were inserted between concrete and reinforcing bars. Bond-slip characteristics were determined considering the experimental results.
  Kupfer's two-dimensional empirical formula was adopted as a constitutive equation for concrete, which was corrected based on the authors' experimental data. Reinforcement were formulated by a complete elasto-plastic bi-linear material.

5.2 Comparison of the Test Results with Analytical ones.

Comparisons of analytical and experimental horizontal inward deformations are shown in fig-6 and fig-7. Also, the analytical cracking patterns comparable to the experimental ones are illustrated in fig-4. Satisfactory analytical simulation could be found concerning deformations, cracks, strains in reinforcements, etc..
  Three cases of computed moment distributions along a vertical wall, which were obtained by non-linear solid, linear solid as well as elastic frame analyses, are compared in fig-8.
  Redistribution of moment at and around critical sections due to the non-linearity caused by cracking, material plasticity etc., could be quantitatively evaluated by the above mentioned FEM analysis. It was estimated that at the ultimate stage the maximum moment could be reduced by about 15% from the hitherto utilized elastic frame analysis.

6. CONCLUSIONS

(1) The reinforced concrete ducts for accommodation of cooling water pipes(D.C.W), which was designed by the conventional A.S.D procedures, had ample safety margin against failure, leading to an excessive over design.

(2) It was found feasible to design a rational D.C.W section by L.S.D, which satisfies both strength and serviceability requirements.

(3) Non-linear FEM may be useful as a basis for evaluating sectional forces for thick walled frame-like structures up to the ultimate loading stage.

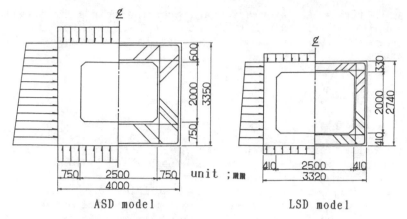

ASD model   LSD model

fig-1.   Dimensions and load distribution
          for models.

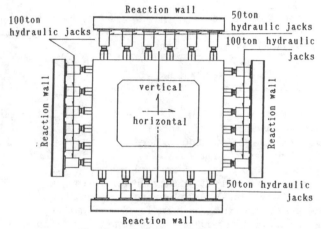

fig-2.   Loading scheme.

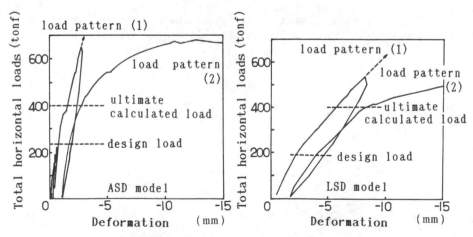

fig-3.   Relationships between total horizontal
          loads and inward horizontal deformations
          at the midheight of vertical walls.

174

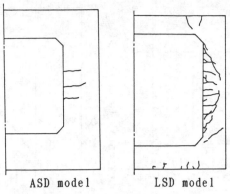

ASD model   LSD model

a) Experimental cracking pattern

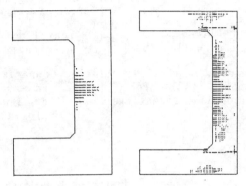

b) Analytical cracking pattern

fig-4. Cracking patterns of experiment
and analysis(at the load three
times the design level.)

Concrete elements   Reinforcement elements

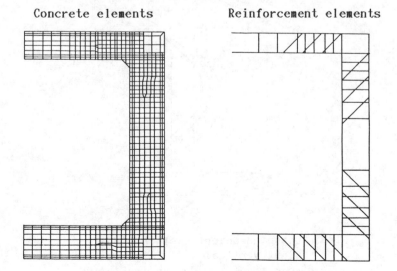

fig-5. Element mesh for FEM analysis

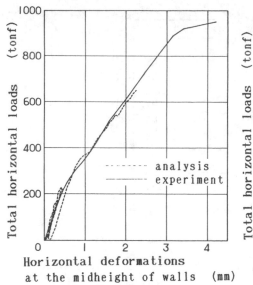

fig-6. Comparison with analysis
and experiment result for
inward horizontal deformations
of ASD model.

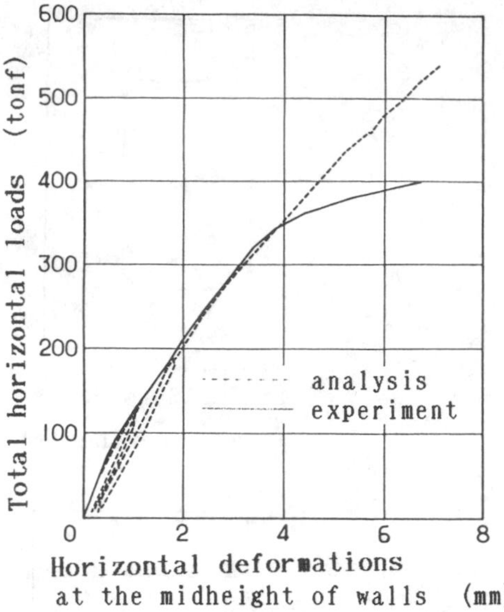

fig-7. Comparison with analysis
and experiment result
for inward horizontal
deformations of LSD model.

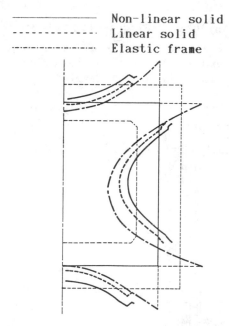

——————— Non-linear solid
------------ Linear solid
-·--·--·--·- Elastic frame

fig-8. Computed moment distributions
along vertical wall (at the loads
four times the design level for
ASD model)

Delayed phenomena analysis from French PWR containment instrumentation system

J.L.Costaz
Electricité de France, SEPTEN, Villeurbanne

J.Picaut & J.Chataigner
Coyne et Bellier, Paris, France

1. INTRODUCTION

French 900 MW reactors containments have been equipped with comprehensive instrumentation systems in order to provide strain and displacements measurements in concrete as well as stresses in presstressing tendons first during the preoperationnal air pressure test and then during the life of the structure.

More than 20 reactors containments have been so regularly monitored over several years of normal operating (since May 1977 for the Tricastin 1 unit, the first of these similar containments) and through one (or two for a few of them) air pressure test.

A large amount of measurements on concrete or tendons is now available; we intend to analyse them to study delayed phenomena such as concrete creep and shrinkage and steel relaxation of prestressing tendons.

2. DESCRIPTION OF MONITORING SYSTEM

The french PWR 900 MW reactor containment consists of a reinforced concrete raft, a prestressed cylindrical wall and a prestressed spherical dome; the inner face of the whole containment is covered with a steel liner which ensures air tightness.

The main dimensions of the structure are :

mean radius of cylindrical wall	:	18.95 m
height (from upper face of raft to dome)	:	60.35 m
raft thickness	:	3.5 m
cylindrical wall thickness	:	0.9 m
dome thickness	:	0.8 m

Basically, the instrumentation system consists of vibrating wire strain gauges (extensometers), pendulums for displacements measurements and dynamometers on prestressing tendons.

The more complete instrumentation systems have always been installed on the first containment unit of each project, the following units being more simply equipped.

Strains gauges are usually positioned in pairs, near the outer or inner surfaces of dome, containment wall or raft in typical or particular (hatches, prestressing ribs, ...) sections ; each of them is combined with a thermocouple which provides local temperature measurement.

Pendulums are positioned so that they allow the determination of the dimensional variations of two diameters, approximately perpendicular.

Four vertical tendons are fitted with dynamometers at their anchorages thus providing stress measurement in prestressing steel.

Measurements have been regularly (about four or five times a year) carried out on the twenty four units in operation at present time by a specialized service of Electricité de France and stored on computer files for further treatment. To eliminate temperature effects a statistical analysis (main components) is performed on the uncorrected values. The resulting values are then graphically displayed in regular reports; thus it is possible to check the behaviour of the structure as the plant is normally operating.

3. ANALYSIS OF MEASUREMENTS

As above mentionned the purpose of this study was to compare the strain variations actually measured by the instrumentation on reactor containments to the theoretical values for concrete shrinkage and creep proposed in french regulations for prestressed concrete structures design (BPEL 83 - Règles techniques de conception et de calcul des ouvrages et constructions en béton précontraint suivant la méthode des états limites).
To carry out these comparisons we shall first analyse the strains measured by vibrating wire strain gauges in selected typical sections of the standardized part of the cylindrical wall.

3.1. Strain gauges analysis

For this analyis, the selected sections are located approximately in the middle of the height of the containment (at a level of + 24 m or + 29 m depending on sites) in four different angular positions (42 gr., 176 gr., 242 gr. and 376 gr.); it must be noticed that selecting this level in the containment wall ensures us that the measurements are no longer disturbed by site conditions such as soil stiffness; we can thus compare the values obtained on units belonging to different sites.

On the other hand, we have usually choosen the results obtained from the instrumentation of the first unit of each site as they are generally more completely equipped (so we can more easily reject gauges suspected to be faulty).

To carry out a more valuable comparison between measured and predicted values for the cumulated actions of concrete creep and shrinkage we prefered to limit our study to the reactor containments for which measurements were available on a long space of time, in fact TRICASTIN 1, DAMPIERRE 1 and GRAVELINES 1, units which have been continuously followed for more than five years after their first air pressure test.

Theoretical strain variations have been calculated, using analytical formulas from BPEL 83 which take into account :
     ambient air moisture ratio
     wall thickness
     concrete age when prestress forces have been applied
     concrete elastic modulus
     normal stress in concrete.

Comparisons have been made for the two main stress directions in the containment wall, vertical and horizontal the design values being respectively 9.2. MPa (including dead weight) and 11.4 MPa.

Elastic modulus under instant loading have been determined from the results of containment air pressure tests and range from 28800 MPa (DAMPIERRE 1) to 36000 MPa (TRICASTIN 1). These elastic moduli have been used for theoretical calculation of creep.

The results of these comparisons are graphically displayed on fig. 1 to 10.
It appears that the predicted theoretical values fit well to the measured ones for practically all vertical strain gauges when calculations are carried out with a 55 % air moisture ratio ; on the other hand slight discrepancies appear for the horizontal gauges, the measured strains being generally higher by 10 % than the values given by the analytical formulas.
The above mentionned discrepancies relative to horizontal strains have probably their origin in the actual horizontal stress in concrete which can locally exceed the design values.

3.2. Diameter variations analysis

By mean of four pendulums, radial displacements are measured on two diameters; we analysed the results obtained from these devices at the same level on the containment wall as for the strains gauges; taking into account the relation between horizontal strain and radial displacement in a cylinder : $\varepsilon_T = U_r/R$

We were able to get an other evaluation of the horizontal strain variations versus time; it then appeared that the values estimated from the measurements were closer (for GRAVELINES 1) or quite similar (for DAMPIERRE 1) to the theoretical quantities (see fig. 11 to 12).

These results corroborate the hypothesis made about the slight differences we have noticed between theoretical predicted values and measurements for horizontal strain gauges; when measured in typical sections, diameter dimensional variations depend essentially, on mean horizontal stresses in containment wall; consequently, they are closer to the values we can get from analytical formulas based on the design mean stress.

3.3. Force losses in prestressing tendons

At the same time steps as for strain gauges, tensile forces in four vertical tendons are regurlarly measured using dynamometers set under their anchorage plates.
In order to compare the evolution of actual tensile force losses in these tendons to those we can calculate from the formulas given for

steel relaxation in french regulations BPEL 83 we analysed the results of measurements on dynamometers for several containment buldings on different sites.

To carry out this comparison we have first calculated from BPEL 83 steel relaxation losses which must have occured in the selected tendons since they are under tension and cumulated these stress.
Losses to those created by creep and shrinkage, the latter being evaluated directly from strain gauges located next to the tendons.
The total losses we have so obtained were then compared to the measured ones. As we can notice on the fig. 13 to 17, describing five such calculations made on GRAVELINES 1, DAMPIERRE 1 and TRICASTIN 1 units, predicted values are in good agreement which the measurements except for one cable (GRAVELINES 1, Tendon 20) were the estimated losses notably exceed the measured ones).

4. CONCLUSION

The analysis of the large amount of measurements which has been now gathered by Eectricité de France on its twenty two PWR 900 MW shows that the behaviour of concrete under creep and shrinkage effects is in good agreement with the values given as correct estimates by french regulations and taken into account for the design of nuclear prestressed structures.

None of the containment buildings studied here showed significant differences with the regulations theoretical values and consequently all the measurements remain in the field of the allowable strain variations used for design.

On the other hand, if the instant loading elastic modulus is clearly determined for each containment, and its effect on theoretical creep taken into account, it was not possible up till now to extract from measurements some particular effects such as type of concrete and agregates or climatic effects.

Strains : $e_t \times 10^{-6}$ Time : days

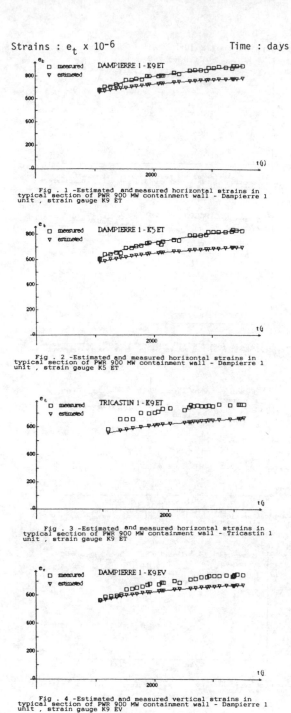

Fig . 1 -Estimated and measured horizontal strains in typical section of PWR 900 MW containment wall - Dampierre 1 unit , strain gauge K9 ET

Fig . 2 -Estimated and measured horizontal strains in typical section of PWR 900 MW containment wall - Dampierre 1 unit , strain gauge K5 ET

Fig . 3 -Estimated and measured horizontal strains in typical section of PWR 900 MW containment wall - Tricastin 1 unit , strain gauge K9 ET

Fig . 4 -Estimated and measured vertical strains in typical section of PWR 900 MW containment wall - Dampierre 1 unit , strain gauge K9 EV

181

Strains : e_t x 10-6   Time : days

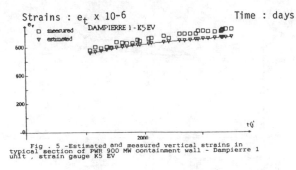

Fig . 5 -Estimated and measured vertical strains in
typical section of PWR 900 MW containment wall - Dampierre 1
unit , strain gauge K5 EV

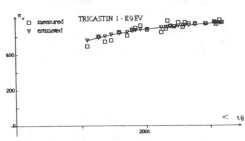

Fig . 6 -Estimated and measured vertical strains in
typical section of PWR 900 MW containment wall - Tricastin 1
unit , strain gauge K9 EV

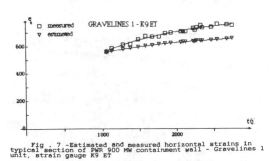

Fig . 7 -Estimated and measured horizontal strains in
typical section of PWR 900 MW containment wall - Gravelines 1
unit, strain gauge K9 ET

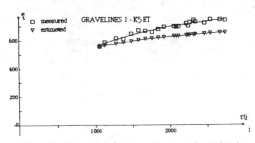

Fig . 8 -Estimated and measured horizontal strains in
typical section of PWR 900 MW containment wall - Gravelines 1
unit, strain gauge K5 ET

Strains : $e_t \times 10^{-6}$ Time : days

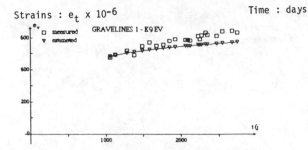

Fig . 9 -Estimated and measured vertical strains in
typical section of PWR 900 MW containment wall - Gravelines 1
unit, strain gauge K9 EV

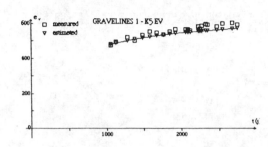

Fig .10 -Estimated and measured vertical strains in
typical section of PWR 900 MW containment wall - Gravelines 1
unit, strain gauge K5 EV

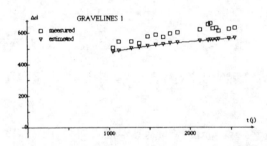

Fig .11 -Estimated horizontal strains in typical section
compared to measured values (computed from diameter variations)in
PWR 900 MW containment wall - Gravelines 1 unit ,level +26 m

DAMPIERRE 1

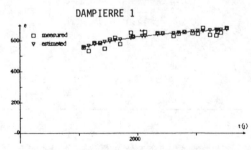

Fig .12 -Estimated horizontal strains in typical section
compared to measured values (computed from diameter variations)in
PWR 900 MW containment wall - Dampierre 1 unit ,level +26 m

183

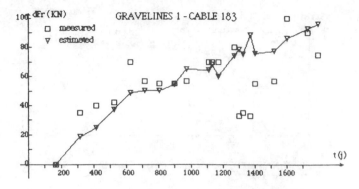

Fig .13 -Estimated and measured prestressing forces
losses (in KN) in vertical tendons   of PWR 900 MW
containment wall - Gravelines 1 unit , tendon 183

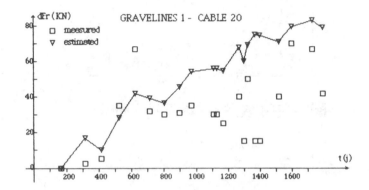

Fig .14 -Estimated and measured prestressing forces
losses (in KN) in vertical tendons   of PWR 900 MW
containment wall - Gravelines 1 unit , tendon 20

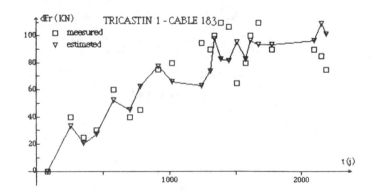

Fig .15 -Estimated and measured prestressing forces
losses (in KN) in vertical tendons   of PWR 900 MW
containment wall - Tricastin 1 unit , tendon 183

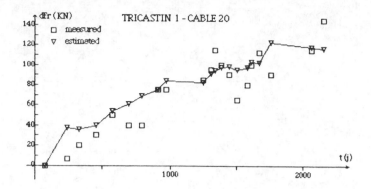

Fig .16 -Estimated and measured prestressing forces
losses (in KN) in vertical tendons   of PWR 900 MW
containment wall - Tricastin 1 unit , tendon 20

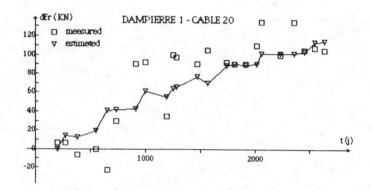

Fig .17 -Estimated and measured prestressing forces
losses (in KN) in vertical tendons   of PWR 900 MW
containment wall - Dampierre 1 unit , tendon 20

Fracture toughness determination of concrete and reactor surveillance procedure by use of breakoff tester and acoustic emission technique

T.Hashida & H.Takahashi
Research Institute for Strength and Fracture of Materials, Tohoku University, Japan

1 INTRODUCTION

With interest developing in the life extension of nuclear reactor systems, it is desired that procedures be developed which can be utilized to qualify nuclear safety related concrete structures for extended survice. It is known that the concrete in the nuclear reactor systems is subject to various potential degradation modes related to thermal and irradiation effects, which would cause a loss in strength or shielding efficiency (Naus 1985). In order to assure the integrity of the nuclear reactor concrete structures during future operation, the in-situ concrete strength in the structures must first of all be determined to assess the accumulated damage in the concrete. Breakoff test method is available to predict the in-place flexural strength of concrete (Johansen 1979). In the test method the breakoff tester is used as a loading apparatus.

It is well known that crack formation and growth play an important role in the performance of unreinforced and reinforced concrete. Hence, the fracture mechanics approach has been employed to analyze the cracking behaviors of concrete in many investigations. From this point of view, it will be advantageous if the fracture toughness of concrete in a structure can be determined by in-situ test as well as in the laboratory. Because the in-situ test on the concrete structure must be based on a small sample, the influences of specimen size have to be taken into account in the development of a suitable in-situ fracture toughness test of concrete.

The objective of the present investigation is to develop an in-situ fracture toughness test procedure by use of the breakoff tester. The J-integral approach and acoustic emission technique were employed in the fracture toughness determination. The validity of the breakoff fracture toughness testing procedure was discussed by comparing results of the standard fracture toughness test methods.

2 EXPERIMENTAL METHOD

The in-situ fracture toughness testing method proposed in this paper is illustrated in Fig. 1. The testing method uses a pair of concrete cores formed in the concrete structure, and determines the flexural strength and fracture toughness of the concrete. The two tubular slits surrounding the concrete cores are cut to come in contact with each other by use of a

187

special diamond drill. During the course of this investigation, the Norcem breakoff tester was employed as a loading device. The apparatus is illustrated in Fig. 2.

First, the conventional breakoff test is conducted on one of the test cores. A transverse force is applied at the end of the concrete core with the breakoff tester untill rupture occurs at the critical section on the bottom of the core. During the test the pressure reading of the manometer Pg is recorded. The force required to break off the core is taken as a measure of the flexural strength of the concrete. The load applied to the core F is calculated from the Pg using load calibration chart. The breakoff strength f_{bo} is determined as

$$f_{bo} = \frac{32 F_c L}{\pi D^3} \tag{1}$$

where Fc is the critical load, L is the distance between the loading point and the crack initiation point, and D is the core diameter.

Next, the other core is tested to determine the fracture toughness of the concrete. An artificial notch is introduced at the bottom of the core at a right angle to the core axis. F versus displacement at the end surface of the concrete core δ_M is monitored, and used to determine the fracture toughness of the concrete in terms of stress intensity factor K and J-integral. The K and J-integral values for the notched breakoff specimen are calculated through the following equations

$$K = Y' (\frac{L}{D}) \frac{F}{D^{1.5}} \tag{2}$$

$$J = \eta_{eq.} \frac{A}{B (D-a)} \tag{3}$$

where A is the area under a load-displacement curve and B is the notch front width. The value of Y' and $\eta_{eq.}$ in the above equations are given in Fig. 3. In this study the concrete core of the diameter 53 mm was used. The core length to diameter ratio (L/D) was 1.2 .

As supporting experiments, compact tension specimens (CT) and core based specimens were tested. The core based specimens have recently been proposed for the standard fracture toughness test of rock within the International Soceity for Rock Mechanics. The CT specimens of various sizes (1TCT-4TCT) were tested in accordance with ASTM Test Method for Plane-Strain Fracture Toughness of Metallic Materials (E399-78). The core based specimens were edge-notched round bend bar and shor rod (hereafter called BS and SRS specimen, respectively). The geometries and dimensions of the core specimens are shown in Fig. 4. The K and J-integral formula for the core based specimens can be found in the literatures (Ouchterlony 1982, Takahashi 1986).

The mix proportions of the concrete used were 1:3.6:4.4:0.75 by weight of cement:sand:coarse aggregate:water. Type I Portland cement was used. The maximum aggregate size was 20 mm. From the concrete slabs the breakoff, CT and core based specimens were mashined using a diamond wheel saw. The cylinders were used to conduct uniaxial compression tests. All the prepared specimens were tested within four days after approximately 70days of curing.

During the fracture toughness test, AE monitoring was conducted to detect the onset of crack growth at the notch tip. AE signals were detected using a broad band piezoelectric transducer (NF Corp. AE905S) with 80 gain. In the breakoff tests, the AE transducer was attached at the concrete surface close to the test core.

3 RESULTS AND DISCUSSION

The calculated values of breakoff strength f_{bo} are tabulated in Table 1. The primary objective of the conventional breakoff test is to predict the compressive strength of concrete based on the breakoff strength. In this regard the relationship between f_{bo} and compressive strength fc is plotted in Fig. 5. The fc values determined on the three cylinders of the concrete were 21.0, 23.2 and 24.5 MPa, respectively. The solid curve represents an empirical relationship measured by Johansen (1979). The apparent discrepancy in the figure indicates that the relationship between f_{bo} and fc is not unique but parameter dependent. Recently Yener and Chen (1985) have reported similar results ; their data are also plotted in the figure. It is desirable, therefore, to exploit a suitable evaluation method for determining the fracture strength of concrete.

Typical load-displacement record of the notched breakoff specimen is shown in Fig. 6. The concrete used shows the nonlinear deformation well below the maximum load. In order to correlate AE characteristics with the fracture behaviors, AE energy of events E_{AE} was determined with respect to time and the summation of the AE energy ΣE_{AE} is shown as a function of displacement in Fig. 6. It is seen that the AE signals are detected at an early stage of loading and their activity increases rapidly with the increasing nonlinear deformation. The ΣE_{AE} is plotted as a function of J-integral in Fig. 7. In view of the nonlinear deformation behaviors, J-integral is chosen as a loading parameter. The ΣE_{AE}-J curve can be divided into two regions . The first region, of lower slope, denotes microfracture processes with little acoustic activity. The second region of the curves has a high value of slope, and consists of AE events of much higher amplitude than those detected earlier in the test. Thus, it can be concluded that the resulting bilinear nature of the ΣE_{AE} vs J curves indicates the advance of crack at the point shown by arrows. The critical J-integral corresponding to the knee point is denoted by J_{iAE}.

The J_{iAE} values are given for all the fracture toughness specimens tested in Fig. 8. Although some degree of scatter is observed in the plots, it is noted that a fairly constant value of J_{iAE} is obtained even for ligament lengths as small as 20 mm. Thus, the J_{iAE} is considered to be a inherent material property of the concrete. Based on those experimental results , the fracture toughness K_{IC} of the concrete is determined from the J_{iAE} using the relationship,

$$K = \sqrt{E' \ J} \qquad (4)$$

where E'(=30 GPa) is the effective Young's modulus. The results are summarized in Fig. 9. The K_{iAE} values, which are calculated based on fracture load at the crack initiation point and initial notch length, are given for comparison with the J-based toughness results. The K_{iAE} values are dependent on specimen size and increase with specimen size. It is noted that the K_{iAE} approaches the converted fracture toughness K_{IC} as the specimen size becomes large. This experimental evidence indicates that the J-integral approach makes it possible to determine the valid fracture toughness value by use of the small breakoff specimen.

4 CONCLUDING REMARKS

In this study an in-situ test method was developed for determining the fracture toughness of existing concrete structural components, which uses the notched core specimen of 53 mm diameter established into the concrete. The J-integral based test combined with the acoustic emission technique was shown to determine the valid fracture toughness of the concrete even for such a small specimen. The experimental results give a firm basis for the development of a surveillance procedure of nuclear reactor concrete structure after long term service.

REFERENCES

Johansen, R. 1979. In Situ Strength Evaluation of Concrete- The Break-Off Method, Concrete International: Design & Construction, 1:44-51.
Naus, D. J. 1985. Concrete Material Systems in Nuclear Safety Related Structures-A Review of Factors Relating to Their Durability, Degradation and Evaluation, and Remedial Mearsures for Areas of Distress, EPRI Report NP-4208, August :B1-60.
Ouchterlony, F. 1982. Fracture Toughness Testing of Rock, SveDefo Report DS 1982:5.
Takahashi, H., Hashida, T. & Fukazawa, T. 1986. Fracture Toughness Tests by Use of Core Based Specimens, GEEE Research Report, No. T-002-86,Tohoku Univ.
Yener, M. & Chen, W.F. 1985. Evaluation of In-Place Flexural Strength of Concrete, ACI Journal, 82:788-796.

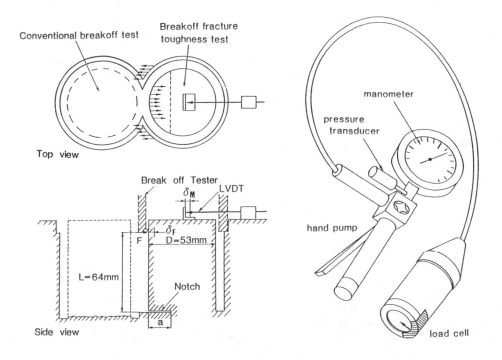

Fig. 1. Principle of breakoff fracture toughness testing method

Fig. 2. Norcem breakoff tester

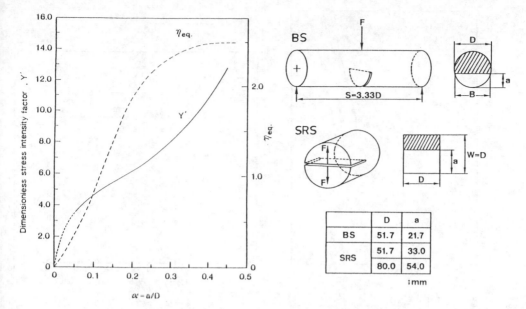

Fig. 3. Y' and $\eta_{eq.}$ as a function of relative crack length

Fig. 4. Core based specimens

	D	a
BS	51.7	21.7
SRS	51.7	33.0
	80.0	54.0

;mm

Table 1. Result of conventional breakoff tests

Specimen No.	Pg , bar	Fc , kN	$f_{b.o.}$, MPa
BO–A1	101.5	1.67	7.84
BO–B1	88.8	1.46	6.86
BO–C1	91.8	1.52	7.42

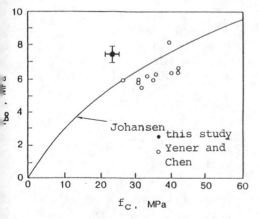

Fig. 5. Breakoff strength versus compressive strength

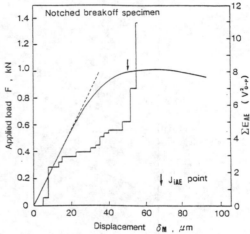

Fig. 6. Load versus displacement curve for notched breakoff specimen

191

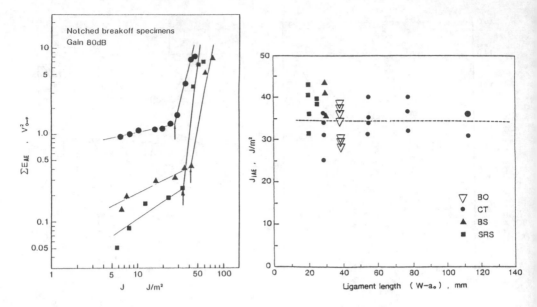

Fig. 7. ΣE_{AE} versus J-integral   Fig. 8. J_{iAE} versus specimen size
for notched breakoff specimens

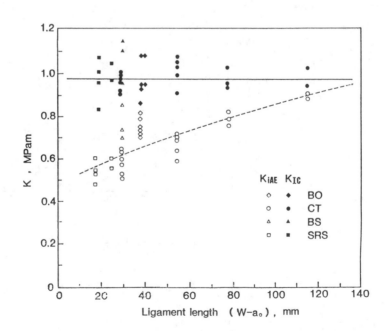

Fig. 9. K_{IC} and K_{iAE} versus specimen size

An experimental investigation of biaxial and triaxial compressive concrete strength and deformation

C.-Z.Wang, Z.-H.Guo & X.-Q.Zheng
Department of Civil Engineering, Tsinghua University, Beijing, People's Republic of China
J.-Y.Qu & R.-M.Zheng
Tsinghua University, Beijing, People's Republic of China

1. INTRODUCTION

Since A. Foppl of Germany tested the biaxial strength of mortar in 1900, many scientists in different countries have investigated the biaxial and triaxial strength and deformation of various concretes. Many criteria and constitutive equations have been proposed.

Tsinghua university started to test the biaxial and triaxial compressive concrete strength and deformation since 1983. Test results of biaxial and triaxial compressive concrete strength are analyzed. A concrete failure criterion is suggested and is compared with others. Biaxial compressive stress-strain relationships under different load ratios are investigated.

2. TESTS

The tests were taken in the pressure chamber of the triaxial loading device which consists of a biaxial loading pressure chamber to exert lateral loadings and an ordinary testing machine to exert vertical loading can reach about 1.2 MN as shown in fig.1. The lateral loads can reach about 800 KN. The three directions of loading are independent and can be in radom ratios. The load ratios were synchorized by using electro-hydraulic valves.

Two sheets of 2 mm teflon with grease of molybdonum sulfide (MoS_2) were used as the measure for reducing friction on surface of specimens.

225 specimens of 70.7 mm cube with concrete strength 10 N/mm^2, 20 N/mm^2 and 40 N/mm^2 were tested for biaxial and triaxial compressive strength.

52 specimens of 100x100x25 mm plates, 18 specimens of 70.7 mm cube and 15 specimens of 100 mm cube with concrete strength 20 N/mm^2 were tested for biaxial strength and deformation.

Electrical strain gages were used to measure biaxial compressive stain. Part of the specimens were measured the third direction of deformation by using transducers, placed on the diagonal of the free surface. For cubes, strain gages were embedded into the channels 12 mm width for 100 mm cube and 10 mm width for 70.7 mm cube and 3 mm depth. Expansive cement paste was used to fill into the channels.

3. FAILURE MODES OF THE SPECIMENS

Failure modes of the specimens of plates and cubes under different load

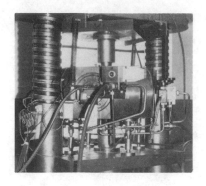

Fig.1 A triaxial testing
device

Fig.2 Failure mode of specimens
under biaxial compression(plate)

Fig.2 (cube)

Fig.3 Failure modes of specimens
under triaxial compression

ratios of biaxial compression are shown in Fig.2 and cubes under
different load ratios of triaxial compression are shown in Fig.3.
Three different kinds of failure modes can be classified.

1. Failure cracks were parallel to the free surface of biaxial plate
or cube specimens and parallel to the min. principal stress for
triaxial cube specimens.

2. Failure cracks were formed in angles with the free surface in
biaxial specimens and in angles with the min. principal stress in
triaxial specimens.

3. When the lateral compression loading were high in triaxial
conpression, the high confinement caused the cube to fail in flow and in
very high deformation.

4. CHARACTERISTICS OF BIAXIAL AND TRIAXIAL CONCRETE STRENGTH

From the test results some characteristics are shown to effect the
biaxial and triaxial concrete strength.

4.1 The effect of intermediate stress

Fig.4 is shown the effect of intermediate stress σ_2 to the strength
when concrete strength is 20 N/mm^2. When σ_3/σ_1 are 0, 0.1, 0.2, all the
intermediate stresses σ_2 effect the biaxial and triaxial strength. When σ_2/σ_1
is less than 0.5, the effect is to increase the strength and then
starts to decrease.

4.2 The effect of concrete strength

Fig.5 is shwon the test results of the biaxial compressive strength under concrete strength 20 N/mm^2 and 40 N/mm^2. Lower the concrete sttrength gives the higher increasing biaxial compressive value. Fig.6 is shown the triaxial strength in octahedral coordiantes under concrete strength 10 N/mm^2 and 20 N/mm^2. Under different meridian curve and some effect for tensile meridian curve. Lower the concrete strength gives the higher τ_{oct}/R_a.

4.3 Effect of reducing friction measure

Fig. 7 and 8 are shown the biaxial and triaxial compressive strength with dry steel platen and with 2 sheets of 2 mm teflon and grease of as reducing friction measure. It is shown that the friction has obvious influence of the biaxial and triaxial strength.

4.4 Effect of shape of the specimen

Fig.9 is shown that there is no clear difference to the biaxial compressive strength for plate and cube specimens.

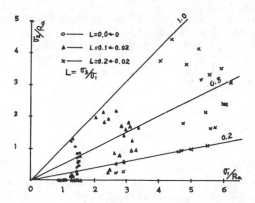

Fig.4 Test results of triaxial concrete strength ($L = \sigma_3/\sigma_1$, R_a uniaxial concrete strength)

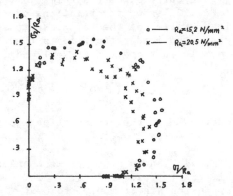

Fig.5 Test results of biaxial compression under different concrete strength

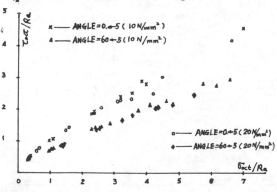

Fig.6 Test results of triaxial strength expressed in meridian planes

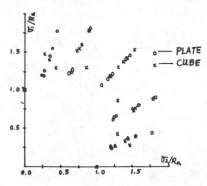

Fig.9 The effect of shape to biaxial strength

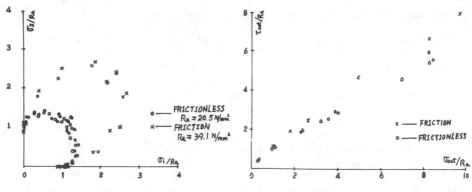

Fig.7 The effect of friction
to biaxial strength

Fig.8 The effect of friction to
triaxial strength

5 BIAXIAL COMPRESSIVE STRESS-STRAIN RELATIONSHIP

Fig.10 and 11 are shown two typical biaxial compressive stress-strain curves under different load ratios. Fig.10 is the stress-strain curve under different load ratios of plate specimen and fig.11 is the stress-strain curve of cibe specimen.

Fig.12 shows the relations of strain at peak stress with different load ratio of the test results and the other author's. It is shown that the load ratio effects the strain at peak stress. A regression formula is suggested as following:

$$\varepsilon_{1P} = (2.16 + 7.25\alpha^{0.8} e^{-3.71\alpha}) \times 10^{-3}$$

where ε_{1p} is the strain at peak stress
       α is σ_2/σ_1.

Fig.13 shows the volume changes of the biaxial compression under different load ratios. The turning point from compression to expansion are from 0.90 to 0.95 R_a where R_a is the uniaxial strength of concrete strength.

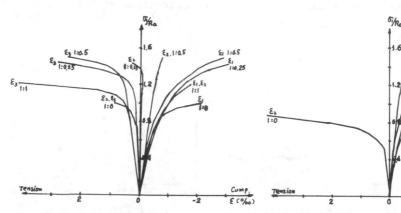

Fig.10 Biaxial compressive
stress-strain curve(plate)

Fig.11 Biaxial compressive
stress-strain curve (cube)

196

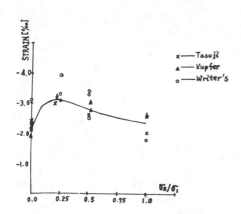

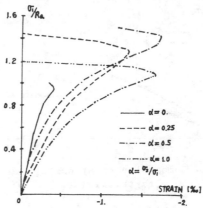

Fig.12 Relation of strain at peak stress to load ratio

Fig.13 Volume change of the biaxial compression

6 FAILURE CRITERION

Experimental results indicate that the failure criterion should satisfy the following general characteristics.

The failure curve in the deviatoric plane is smooth and convex. It is nearly triangluar for small hydrostatic pressures and becomes increasingly bulged for high hydrostatic pressures. The failure meridian curves are curved smooth and convex.

A four-parameters failure criterion is suggested as following:

$$\lambda = (sin \tfrac{3}{2}\theta_o)^{(0.8+\tfrac{3}{2+\theta_o})}$$

$$\overline{\tau}_{oct} = a\left[(\frac{\overline{\sigma}_{oct} + b}{\overline{\sigma}_{oct}'+2a})^\alpha(1-\lambda) + \lambda(\frac{\overline{\sigma}_{oct} + b}{\overline{\sigma}_{oct}'+3a})^\beta\right]$$

where $\overline{\tau}_{oct} = \tau_{oct}/R_a$, $\overline{\sigma}_{oct} = \sigma_{oct}/R_a$

λ is an affected factor for angle of simility

a, b, α, β, are parameters determined by tests.

Based on the test results of concrete strength 20 N/mm² four characteristics conditions to determin the parameters:

1. uniaxial compressive strength R_a
2. uniaxial tensile strength $R = 0.1R_a$
3. biaxial compressive strength $R_{bc} = 1.28R_a$
4. triaxial compressive strength $\theta_o = 0°$, ($\overline{\sigma}_{oct}, \overline{\tau}_{oct} = 2,1.649$)

From the above conditions, a=13.35, b=0.736, α=0.796, $\beta = 0.818$

Fig.14 and 15 are shown the comparison of biaxial test results and the test results of compressive and tensile meridian-curves with the theoretical curves calculted from the failure criterion.

Fig.16 is the comparison of the test results with the suggested criterion, Bresler - Pister, William - Warnke, Ottosen and Kotsovos criteria. It is shown that the suggested criterion satisfies the required characteristics for the failure criterion and is more simple.

7 CONCLUSIONS

1. Biaxial and triaxial compression increase the strength value under various conditions.
2. The intermediate stress effects the triaxial concrete strength.

197

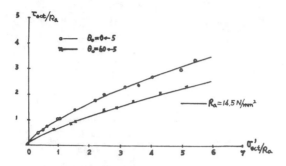

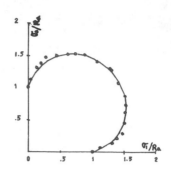

Fig.14 Comparsion of Strength
Criterion with triaxial test
data

Fig.15 Comparsion of strength
criterion with biaxial test
data

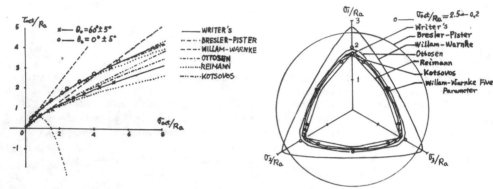

Fig.16 Comparsion of strength criteria

3. The concrete strength effects the biaxial and triaxial strength, lower the concrete strength gives the higher value.

4. The friction on the surface of the specimen has obvious influence of the biaxial and triaxial strength.

5. The load ratio effects the strain at peak stress.

6. A four-Parameters failure criterion is suggested and satisfies the requires characteristics for the failure criterion.

REFERENCES

Wang Chuan-zhi et al, 1985. Experimental investigation of biaxial and triaxial compressive concrete strength. Research Report, Tsinghua University, Beijing.

Qu Jun-yi, 1985. Experimental investigation of concrete strength under biaxial and triaxial compressive stresses. Master thesis, Tsinghua University, Beijing

Zheng Ru-mei, 1987. an experimental investigation of biaxial compressive concrete strength and deformation. Master thesis, Tsinghua University, Beijing

Gerstle, K. H. et al, 1980. Behaviour of concrete under multiaxial states. Journ. of ASCE EM6

Chen, W.F., 1982. Plascity in reinforced concrete. McGraw-Hill Book co.

J Extreme loading and response of reactor containment

Design and analysis of the Mark I Full Scale Test Facility

D.J.Drag, M.F.Samara, G.F.Fanous & A.Walsenko
Santa Fe Braun Inc., Alhambra, Calif., USA

INTRODUCTION.

The Mark I Full Scale Test Facility (FSTF) was designed and
constructed by Santa Fe Braun (formerly C F Braun) under contract
to the General Electric Company (GE). GE served as the Project
Manager on the behalf of the 16 utilities who owned and operated
Mark I BWR Nuclear Power plants. The FSTF was designed to simulate
both saturated water and saturated steam breaks. The test
facility consisted of a full-scale section of a GE Mark I wetwell
and the peripheral equipment required to provide complete
simulation of the hydrodynamic and structural response of the
wetwell to a Loss of Coolant Accident (LOCA).

The FSTF wetwell is a 22.5 degree segment with the exact
dimensions of a Mark I Nuclear plant constructed in the United
States and has a volume of 12,000 cubic feet. The two major
supporting pieces of equipment were the steam vessel, a flash
boiler capable of operating at reactor conditions (550 degrees F
and 1,050 psia) and the drywell, a vertical pressure vessel with a
volume equal to 1/16 of the drywell volume, 6,900 cubic feet,
of the operating Mark I.

During the simulation of either a saturated water or saturated
steam break, fluids flow from the steam vessel through a blowdown
line to the drywell and then from the drywell through vent lines
to the wetwell. A LOCA simulation is initiated by rupturing
discs located in the blowdown line. The flow rate is controlled
by a critical flow nozzle located in the blowdown line. During
the LOCA test, some 256 data points were recorded at a rate of up
to 1,000 times a second.

SYSTEM DESCRIPTIONS.

The configuration of the FSTF wetwell consisted of the 22.5 degree
segment with a 4'-3" extension on either side. This provided
the attenuating distance necessary to eliminate any local
discontinuity stresses in the wetwell that are caused by the flat
head end closures. Inside each extension was a sealed and
reinforced cylinder that extended back to the mitered junction of

201

the wetwell and extensions. This arrangement allowed the 22.5 degree wetwell section to completely simulate the effects of a blowdown. Figure 1 shows the construction details of the wetwell arrangement.

Fourteen hinged pipe struts connected the 4'-3" extension on either side of the 22.5 degree wetwell to the concrete abutment structure on each side. The hinged struts allowed both radial and axial displacements of the wetwell to be properly simulated.

The concrete abutments were multi-cell box-like structures with a flat vertical face to which the pipe struts were joined. The entire system consisting of the wetwell, extensions, pipe struts, and concrete abutments was supported on a 5 ft thick rigid mat foundation poured directly on lean concrete covered bedrock.

The FSTF configuration described above had to be designed and have both a static and dynamic response equal to that of one of GE's Mark I Nuclear plants in operation. To obtain this correlation, numerous configurations of the wetwell, abutments, foundation, and underlying rock were completely analyzed to determine the applicable stiffnesses and masses of the respective structural components.

A schematic view of the wetwell system is represented in Figure 2. A complete wetwell (360 degrees) as well as the FSTF wetwell of 22.5 degrees were modeled and their frequencies compared. The analyses accounted for the water contained in the wetwell using Westergard's method of hydrodynamic mass (no sloshing).

DYNAMIC ANALYSIS OF WETWELL.

A Spring-Mass model was first developed to define, size, and develop paramatric evaluations for various wetwell configurations. After the various configurations were evaluated, a complete 3-dimensional static and dynamic finite element analyses were performed on the FSTF configuration as well as the complete wetwell of the operating GE Mark I nuclear power plant. The mode shapes and the frequencies (Figures 3, 4, 5) as well as the time history behavior of the FSTF wetwell were compared to those of the complete wetwell. FSTF wetwell parameters were further refined to obtain a very close level of correlation with the behavior of the complete wetwell system.

DYNAMIC ANALYSES AND EVALUATIONS OF ABUTMENT CONFIGURATIONS.

In developing the required abutment configuration, numerous structural configurations were designed with static and dynamic analyses being performed. In each configuration, the stiffness and associated mass properties were varied in an attempt to develop an abutment system that when coupled to the FSTF wetwell, the entire configuration would yield a wetwell static and dynamic responses identical to those of the complete wetwell in operation. Figure 6 depicts some of the configurations considered in this process.

During this phase of the FSTF design it became apparent that since the foundations were being founded on bedrock covered with lean concrete, the influence of soil-structure interaction was negligible, and it would not need to be considered in the final FSTF system analyses.

FSTF SYSTEM ANALYSES.

In this series of static and dynamic analyses, the 3-dimensional FSTF wetwell model was joined to the 3-dimensional model of the concrete abutments and mat foundation.

Table I shows the comparison of natural frequencies between the complete wetwell and those of the FSTF wetwell system. The frequencies listed are grouped into local ovalling modes, overall wetwell modes, and wetwell bouncing modes. A review of Table I shows that the dynamic response of the two systems is in close correlation as desired. Although the results of some of the local ovalling modes were not in as close a correlation as the other modes, these modes were not of interest in the study and nor do they affect the results of the FSTF tests that were performed.

CONSTRUCTION.

The project was constructed on a site owned by Wyle Laboratories at Norco, California, U.S.A. The site is underlain with bedrock at a shallow depth. After excavating for the foundations and removing the loose material present, a lean concrete cover was poured to fill the surface fractures and to level the site. During the construction of the large mat foundation and abutments, 4" diameter pipes were embedded in the structures to permit the placement of explosive charges. This was done to permit easier demolition of the structures after the testing program was completed. All vessels were shop-fabricated, shipped to the site, for placement and final erection.

The entire project, engineering, procurement, and construction was completed and ready for testing in eight months.

FSTF TESTING PROGRAM AND RESULTS.

Numerous dynamic analyses of the various test conditions were performed on the FSTF wetwell system to predict results prior to actual testing. Results of these analyses were within 3% of tested results at the FSTF facility. This testing program demonstrated that the structural behavior and response of the Mark I wetwell under LOCA conditions met all US NRC requirements.

As a result of constructing this facility and the performance of these LOCA testing conditions, millions of dollars of potential Mark I modifications on existing plants were found not to be necessary.

TABLE I   COMPARISON OF NATURAL FREQUENCIES
BETWEEN THE COMPLETE WETWELL AND THE FSTF WELL SYSTEM

COMPLETE WETWELL (Hz)	FSTF WETWELL (Hz)	REMARKS
2.54	3.35	Local ovalling modes
3.03	4.79	
3.46	5.99	
4.18	6.45	
15.11	15.04	Overall wetwell modes
16.95	17.10	
17.43	18.46	
17.95	18.90	
27.07	27.31	Bouncing modes
27.20	28.05	
27.85	28.72	

ACKNOWLEDGMENT

The authors wish to acknowledge the assistance of Steve Dorsey
and Iman Phillips in the preparation of the numerous figures
required for this paper. The figures were prepared using Braun's
3-D CADD PDS system.

204

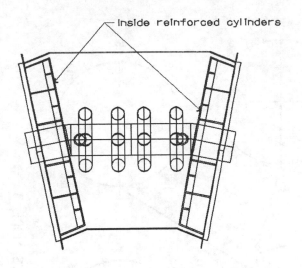

Figure 1. Wetwell arrangement

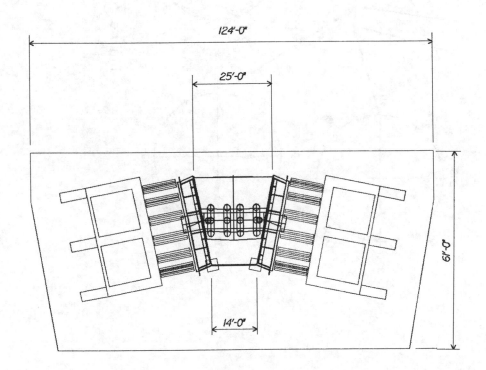

Figure 2. Wetwell - Abutment system arrangement

205

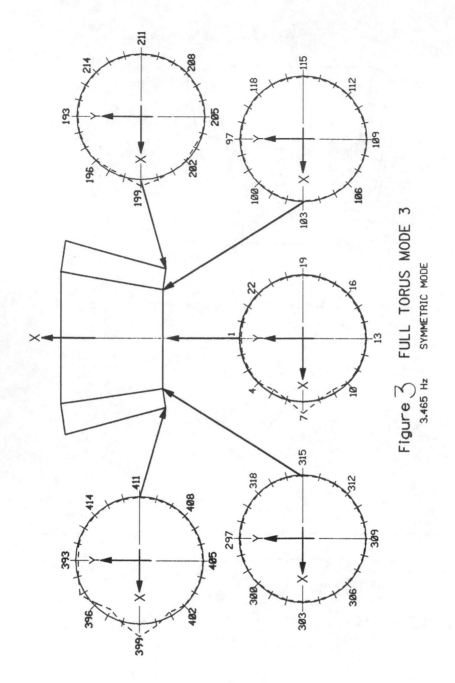

Figure 3 FULL TORUS MODE 3
3.465 Hz SYMMETRIC MODE

206

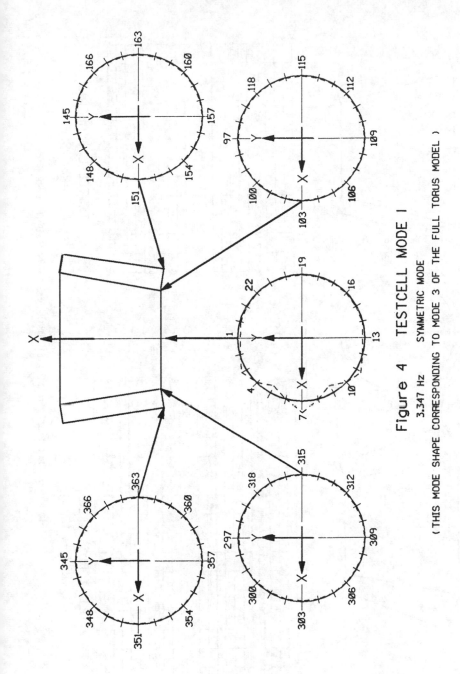

Figure 4   TESTCELL MODE I

3.347 Hz   SYMMETRIC MODE

(THIS MODE SHAPE CORRESPONDING TO MODE 3 OF THE FULL TORUS MODEL)

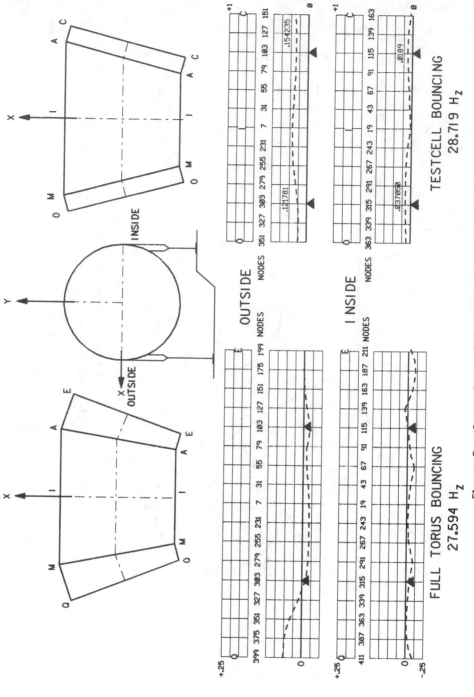

Figure 5. Comparison of Full Torus vs. Testcell Bouncing Mode

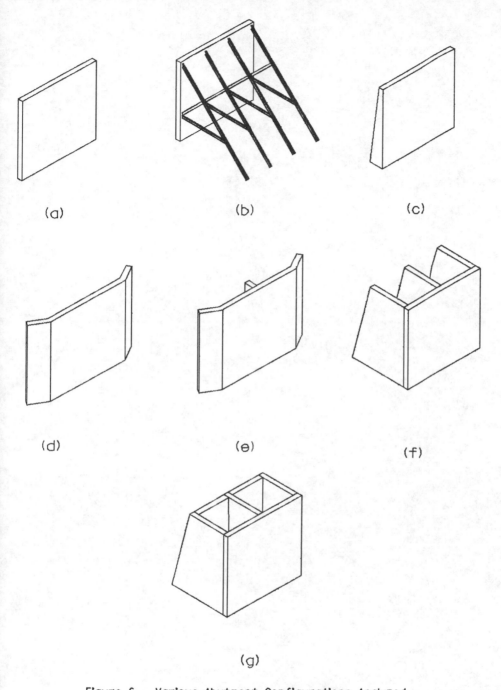

(a) (b) (c)

(d) (e) (f)

(g)

Figure 6. Various Abutment Configurations Analyzed

Dynamic response of beams on elastic foundations to impact loading

B.B.Prasad
M.I.T.Muzaffarpur, Bihar, India
B.P.Sinha
Science & Technology, Bihar, India

1. INTRODUCTION

The dynamic response of beams on elastic foundation to impact loadinghas been presented in this paper. The beam considered is a Timoshenko beam in which the effects of rotatory inertia and shear deformations are included and the foundation model consists of Winkler-Zimmermann type having Hookean linear elastic springs. Prasad et el (1975-1984) have presented the response of Euler-Bernoulli beam on elastic foundations to a class of dynamic loads. But for beams having large cross-sectional area dimensions in comparison to their length as well as for beams vibrating with higher modes, the inclusion of rotatory inertia and shear deformation provides a better approximation to the true behaviour of the beam on elastic foundation. In general the effect of shear deformation and rotatory inertia is to reduce the frequencies of vibrations. In the first mode the reduction is small but for higher modes the reduction is very much pronounced. Fig.1 shows the variation of natural frequencies with foundation modulus in non-dimensional form where in the foundation constant has been represented as $KL^4/EI \pi^4$. Modal analysis is being applied for predicting the response. The impact loading under consideration is that of a striking mass with uniform velocity. The effect of external damping is not considered.

2. DYNAMIC RESPONSE

The governing differential equation for the dynamic response of an uniform simply supported Timoshenko beam $(0 \leqslant x \leqslant L)$ on an elastic foundation (Winkler) without damping may be expressed as

$$EI \frac{\partial^4 W(x,t)}{\partial x^4} + PA \frac{\partial^2 W(x,t)}{\partial t^2} + \frac{P^2 I}{SG} \frac{\partial^2 W(x,t)}{\partial t^4}$$

$$-PI(1+E/SG)\frac{\partial^4 W(x,t)}{\partial x^2 \partial t^2} +KW(x,t) = 0 \ldots \ldots \quad (1)$$

Where in $W(x,t)$ = Total Dynamic Deflection, S = Constant depending upon shape of cross-section, I = moment of inertia of the cross-section, P = density of the beam, A = Cross-sectional area of the beam, K = Foundation Modulus.

The solution of eq.(1) may be expressed as

$$W(x,t) = \sum_{n=1}^{\infty} X_n(x) \cdot (A_n \cos w_n t + B_n \sin w_n t) \quad \ldots\ldots(2)$$

But for pinned-pinned boundary conditions

$$X_n(x) = \sin \frac{n\pi x}{L} \quad \ldots\ldots\ldots\ldots\ldots \quad (3)$$

Introducing $\quad \dfrac{EI}{PA} = a^2 \quad$ and $\quad \dfrac{I}{A} = r_g^2$

The eq.(1) takes the form

$$a^2 \frac{\partial^4 w}{\partial x^4} + \frac{\partial^2 w}{\partial t^2} + r_g^2 \frac{P}{SG} \frac{\partial^4 w}{\partial t^4} - r_g^2 (1 + \frac{E}{SG}) \frac{\partial^4 w}{\partial x^2 \partial t^2}$$

$$+ \frac{KW}{PA} = 0 \quad \ldots\ldots\ldots \quad (4)$$

Substituting eq.(2) in eq.(4), the frequency w_n may be obtained as

$$w_n = (\frac{EI}{PA})^{1/2} \frac{n^2 \pi^2}{L^2} \left[\left\{ 1 - \frac{\pi^2 \cdot r_g^2}{2(L/n)^2} \right\}^2 + \frac{\lambda'}{n^4} \right]^{1/2} \quad \ldots \quad (5)$$

Where in $\quad \lambda' = \dfrac{KL^4}{EI \pi^4}$

Having obtained the value of w_n, the initial conditions to incorporate impact may be considered as

$$W(x,0) = 0 \quad \ldots\ldots\ldots\ldots\ldots \quad (6)$$

$$\left. \begin{array}{l} \dfrac{\partial W}{\partial t} = 0 \text{ for } t = 0 , \quad x \neq L/2 \\[3mm] \dfrac{\partial W}{\partial t} = v_o \text{ for } t = 0, \quad x = L/2 \end{array} \right| \quad \ldots\ldots\ldots \quad (7)$$

Thus evaluating the constants A_n, B_n of eq.(2), the dynamic response is expressed as

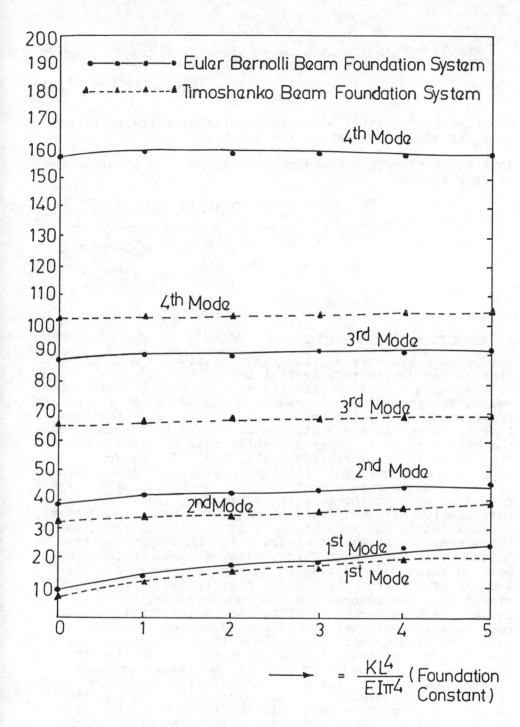

Fig. 1.   Variation of Natural Frequency with Foundation Modulus.

$$W(x,t) = m_2 v_0 \sum_{n=1}^{\infty} \frac{1}{w_n} \; \text{Sin} \frac{n\pi x}{L} \; \frac{\text{Sin } n\pi/2}{\frac{PAL}{2}+m_2 \text{Sin}^2 \frac{n\pi}{2}} \quad ..(8)$$

Simillarily for other boundary conditions suitable value of X_n in eq.(2) will yield the result.

Thus for a free-free boundary conditions, the solution may be expressed as

$$W(x,t) = m_2 v_0 \sum_{n=0}^{\infty} \frac{1}{w_n} \frac{X_n(x). \, X_n(L/2). \, \text{Sin} w_n t}{PAL \frac{\text{Cos}^2 \alpha_n L}{\text{Sin}^4 \alpha_n L}(\cosh \alpha_n L - \cos \alpha_n L) + m_2 X_n^2(L/2)} \quad ...(9)$$

3. RESULTS AND DISCUSSIONS

The dynamic response of Timoshenko beam $(0 \leq x \leq L)$ On Winkler foundation to a central impact $(x=L/2)$ by a moving mass m_2 with a uniform velocity v_0 have been expressed in eq.(8) and eq.(9) for pinned-pinned boundary conditions and free-free boundary conditions. In Fig.2 the deflection-time trace and in Fig.3 strain-time trace have been compared with that of an Euler-Bernoulli beam on elastic foundation as reported by Prasad et el(1975). The natural frequency of vibrations of Timoshenko beam on elastic foundation is smaller than that of Euler-Bernoulli beam. The agreement between theoretical predictions even after taking into considerations effects of rotatory inertia and shear deformation is not that good. This is primarily due to three dimensional stress distribution in the immediate neighbourhood of impact point. The response of Timoslenko beam on elastic foundation is in better agreement. With that of Euler-Bernoulli beam on elastic foundation. The effect of shear deformation and rotatory inertia is to reduce the frequencies of vibration in the first mode the reduction is small but for higher modes the reduction is very much pronounced. Fig.1 shows thev variation of natural frequencies with foundation modulus for the Euler-Bernoulli beam on elastic foundation and for the Timoshenko beam on elastic foundation. For n = 1 the difference between values of natural frequencies for Euler-Bernoulli beam and the Timoshenko beam is rather small but for n = 4, the difference is large. However for the both types of beams the effect of foundation modulus is to increase the value of natural frequencies of vibration.

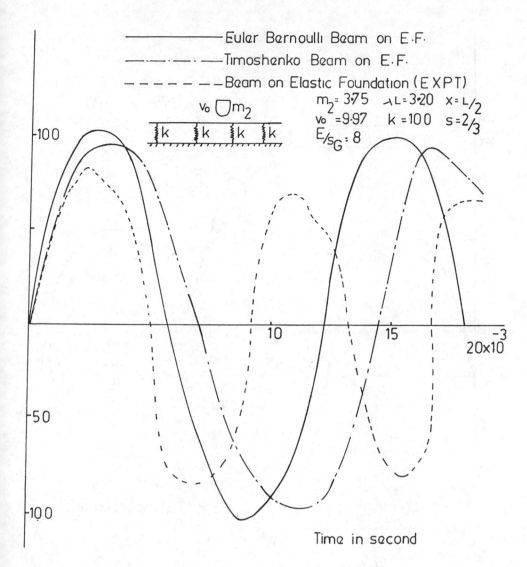

Fig. 2.   Variation of Deflection with time (x=L/2).

4. CONCLUSIONS

The present analysis is very useful for predicting the
dynamic response of structural components of Aircraft or
Nuclear Reactors or even runways if that component may be
mathematically idealized as a beam on elastic foundation.
The effect of rotatory inertia and shear deformation is
very much pronounced and hence should not be neglected in
solving such impact problems. In general the effect of
foundation modulus is to further increase the values of
frequencies of vibrations.

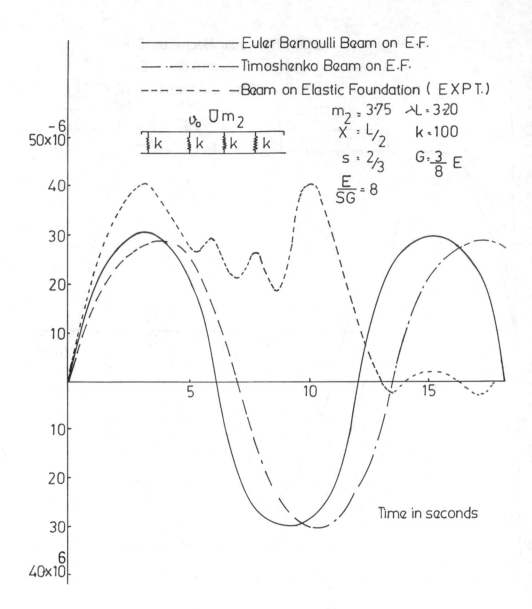

Fig. 3.   Variation of Strain with time (x=L/2).

REFERENCES

B.B.Prasad & N.S.V.Kameswar Rao, 1975. Impact loads on
Elastic Foundations, 3rd SMIRT conference, London, J5/8.

B.B.Prasad, 1978. Propagation of Flexural wave in beams
on elastic foundations. Proceedings 6th Symporium on
earthquake, UOR Roorkee, p. 395-400.

B.B.Prasad & N.S.V.Kameswar Rao, 1979. Beam Foundation
Interaction to impact loads, J.Indian Geotechnical,
p.257-278.

B.B.Prasad & N.S.V.Kameswar Rao, 1980. Response of beams on elastic foundations to dynamic loading , J.ASCE EMD, p.179-185.

B.B.Prasad, 1983. Soil-structure in Nuclear structures, ʾroc. International workshop on Soil-structure Interactions, UDR Roorkee, p.194-197.

B.B.Prasad, 1984. Dynamic stability of beam on elastic foundation. Proc. 5th ASCE speciality conference wyoming, U.S.A, p.1476-1479.

Goldsmith, W 1962. Impact Arnold Edward Publications, London.

K Seismic response analysis of nuclear power plant systems

Influence of uncertainties about depth and epicentral location of earthquakes on seismic hazard analysis

M.Chavez

Istituto de Ingenieria, UNAM, Mexico D.F.

1 INTRODUCTION

It is a common practice in seismic hazard studies for a site or region to use raw data about the location parameters of seismic events from world catalogs of events, this data is used as input to seismic hazard models from which the seismic hazard at a site is estimated. Recent studies have shown that there is a systematic mislocation of shallow events in subduction zones from teleseismic data, Singh and Lermo (1986). As the location parameters play an important role in the determination of the ground motion intensities at a site, through the attenuation relations used in the study, its mislocation has an effect on the estimated hazard for a site.

A procedure is proposed in this paper to incorporate the uncertainties about epicentral location E, and depth, H, of earthquakes on seismic hazard analysis. The procedure makes use of bayesian statistics, Monte Carlo simulation and point estimate techniques, to include the mentioned uncertainties into the seismic hazard model. Parametric studies and a real case problem are included. From these studies the influence of the uncertainties about the location parameters of earthquakes on the seismic hazard estimated for a site is highlighted.

2 DATA

In a recent paper, Singh and Lermo (1986) compared the locations of several mexican large shallow events and their aftershocks, determined from field seismographs or from particular studies (which in what follows we will call them local events) with those reported in PDE and ISC bulletins. They concluded that in general both bulletins reported epicenters shifted tens of kilometers from their actual locations; also the depths given in those bulletins showed a disagreemet of that order with respect to the ones of the locally determined events. Those authors suggested that the mislocations are probably due to the higher velocity of the Cocos plate below Mexico. These type of mislocations has also been reported for other subduction zones, for example in Japan (Utsu, 1971), Tonga (Mitronovas et al (1969) and the Aleutians (Engdahl et al., 1982).

In table 1 it is shown the epicentral locations and depths of the 33 out of 35 events used by Singh and Lermo (1986). Also their surface magnitudes are included in the table; Some of those magnitudes were converted from the body wave magnitudes reported by those authors. The information about the locally determined events (Local in Table 1) was taken as the actual one for the events and it was assumed (after Singh and Lermo, 1986) that the corresponding to the PDE and ISC bulletins had error on E and H. It was also assumed that there were not errors in the magnitudes included in Table 1. The errors in the epicentral locations (Δe) and in the depths (Δh) were computed as the absolute values of the differences between the values reported by the Local and the PDE, and the former and the ISC catalogs. From the values of Δe and Δh for the PDE and ISC catalogs, their mean ($\overline{\Delta e}$, $\overline{\Delta h}$) and standard deviations (σe, σh) values were computed and provided the following results: for PDE, $\overline{\Delta e}$=37.12 km, σe=18.64 km, $\overline{\Delta h}$=17.50 km, σh=10.82 km and for ISC, $\overline{\Delta e}$=45.90 km, σe=31.07 km, $\overline{\Delta h}$=24.64 km, σh=24.33 km. Also, histograms were computed for the Δe and Δh of each catalog. In order to find which probability distribution best fitted those histograms, several test were performed and the lognormal distribution provided that fitting. Further results are given elsewhere (Chavez et al, 1987).

3. PROCEDURE

After a site and the region were the former is located are known, and a catalog has been selected, the following procedure can be applied to incorporate the effect of the uncertainties about E and H in the seismic hazard analysis: 1) Determine the probability distribution of the error in E and H implicit in the catalog (Chavez et al, 1987); 2) Obtain a random sample of Δe and Δh values for each of the events of the catalog from their respective distributions, sample size $N \geq 100$; 3) Generate a sample of epicentral distance, E', and depth, H', for each of the events of the catalog by randomly adding or subtracting to their respective E and H, the Δe and Δh generated in 2). 4) Obtain the ground motion intensities associated to the Δ', H' and Ms of the sample of each event, by using an appropriate attenuation relation; 5) Compute the exceedance rates for each of the events of the catalog; 6) Calculate for each of the exceedance rates the probability of exceedance of their associated ground intensities by following Gumbel's criteria. This implies that the accumulated probability distributions of the ground intensities is of the extreme type, and also that each of those intensities correspond to independent events; 7) Select for each of the exceedance rates the ground intensities with the same value of the probability of exceedance of interest; 8) Compute, for the data obtained in 7) the rate of occurrence $\nu(y)$ with the expression

(1) $$\nu(y) = a \; y^{-b}(1-(y/y_1)^\varepsilon)$$

where y is the ground motion intensity of interest, a, b, c, depend on the seismicity of the region where the site is located and y_1 is the maximum y which may occur at the site. The estimation of those parameters can be achieved by using bayesian statistics and point estimate techniques, and therefore one is able to actualize the $\nu(y)$ when new data is available (Esteva and Chavez, 1982); 9) compute the expected $\nu(y)$ when the uncertainties about the attenuation relation used are taken into account, Esteva and Chavez (1982).

4. PARAMETRIC STUDY

A parametric study was performed with the procedure proposed in 3. the objectives of the study were the following: a) to compare the effect on the seismic hazard for a site of the uncertainties on E and H implicit in the PDE and ISC catalogs, with respect to the seismic hazard calculated for the local catalog (Table 1); b) To compare the same effect with respect to the effect of the uncertainties related to the attenuation relation used in the study.

The chosen site has as coordinates $17.5°\,N$, $98.75°\,W$ and is located in the region enclosed by the latitudes $15°$ to $19°\,N$ and longitudes $96°$ to $104°\,W$. All the events of table 1 occurred in the region. The y of interest is the peak ground acceleration, computed by the relation propose by Esteva and Villaverde (1973)

$$(2) \qquad y = 5600\ e^{0.8M}(R+40)^{-2}\quad (cm/s^2)$$

where $R = \sqrt{E^2 + H^2}$, $M = M_s$. A y_1 value of 300 gals was assumed and N = 1000 was used in the simulation step 2. The correction of $\nu(y)$ associated to the uncertainties on the expression (2) was performed as proposed by Esteva and Chavez (1982).

The results of the study are shown in Figures 1 to 3. In Figure 1 the data for the local catalog and the corresponding to the PDE_{10} and ISC_{10} data. The subindex 10 means that there is a 10 percent probability of exceedance of the peak ground accelerations associated to a particular ν(step. 5 of the procedure). In Figure 2 the $\nu(y)$ resulting from applying the step 8 to the data sets of Figure 1 are shown. From this Figure it can be concluded that for a given ν the peak ground acceleration y are larger for the PDE_{10} and ISC_{10} data sets. This effect is stressed for certain ranges of the y's and is larger for the ISC_{10} data. Finally, in Figure 3 it is shown that the effect of the uncertainties in E and H, of the ISC catalog, on the seismic hazard at the site are of the same order, than the ones related to the uncertainties on the attenuation relation used in the study.

5. APPLICATION TO A REAL CASE

The proposed procedure was applied to a site with coordinates $18.2°\,N$, $102°\,W$, located in a region defined by the latitudes $14°$ to $21°\,N$ and longitudes $98°$ to $106°W$. Notice that about 50 percent of the area of this region coincides with the one used in 4. The catalog used was from the ISC. Equation (2) was utilized to calculate the y's. N=1000 for the simulation of E' and H'. By using the seismotectonic information available for the region a y_1 value of 700 (cm/s^2) was used. The results of the application are shown in Figure 4. From this Figure it can be concluded of that the effect of the uncertainties about E and H on the seismic hazard of the site are as important as the related to the attenuation relation.

6. CONCLUSION

1) The effects of the uncertainties about E and H on the seismic hazard estimated for a site are (at least) as important as the associated to the uncertainties related to the attenuation relation. 2) This effect can be incorporated in the hazard analysis by using a methodology as the one proposed in this paper. 3) Efforts should be dedi-

223

cated to the determination of probability distributions of the errors on E and H for the seismic region of interest.

REFERENCES

Chavez, M., R. Vega, G. Jimenez, R. Monroy. 1987. Uncertainties about epicentral distance and depths of mexican subduction earthquakes and their influence on the evaluation of seismic risk at a site. Report No   Instituto de Ingenieria, UNAM (inpress)
Engdahl, E.R., J.W. Dewey & K. Fujita. 1982. Earthquake locations in island arcs. Phys. Earth Planet. Interiors. 30 : 145 - 156.
Esteva, L., M. Chavez. 1982. Analysis of uncertainty on seismic risk estimates. Proc. 3rd Int. Earthquake Microzonation Conf. III: 1273-1283.
Esteva, L., R. Villaverde. 1973. Seismic risk, desigh spectra and structural reliability. Procc. 5th WCEE Rome II: 2586 - 2597
Mitronovas, W., B. Isacks & L. Seeber 1969. Earthquake locations and seismic wave propagation in the upper 250 km of the Tonga island arc. Bull. Seism. Soc. Am. 59 : 115 - 1135
Singh, S.K. & J. Lermo. 1986. Mislocation of mexican earthquakes as reported in international bulletin. Geofisica Internacional. Vol 24 Num 2 : 333 - 352.
Utsu, T. 1971. Seismological evidence for anomalous structure of Island arcs with special reference to the japonese region. Rev. Geophys. Space Phys. 9 : 839 - 890.

Table 1. Epicentral locations, depths and surface magnitudes (M_s) of the 33 events used in this study (after, Singh and Lermo, 1986)

Date	Local Lat N (°)	Local Long W (°)	Depth (km)	PDE Lat N (°)	PDE Long W (°)	Depth (km)	ISC Lat N (°)	ISC Long W (°)	Depth (km)	M_s
730130	18.39	-103.21	32.0	18.48	-103.00	43.0	18.53	-102.93	48.0	7.5
730210	18.41	-103.63	11.0	18.89	-103.55	33.0	18.78	-103.79	42.0	5.6
781129	16.00	-96.69	18.0	16.01	-96.59	18.0	16.07	-96.55	23.0	7.7
781202	15.53	-96.68	13.0	15.79	-96.48	50.0	15.81	-96.47	36.0	4.6
781202	15.57	-96.73	12.0	15.85	-96.49	33.0	16.08	-96.39	21.0	4.0
781202	15.48	-96.73	10.0	15.75	-96.52	33.0	15.83	-96.48	23.0	4.8
781202	15.73	-96.82	13.0	16.02	-96.44	33.0	16.07	-96.39	50.0	4.4
781205	15.72	-97.30	11.0	16.06	-96.98	33.0	16.10	-96.93	31.0	4.3
781205	15.60	-96.75	24.0	15.95	-96.58	33.0	15.91	-96.54	34.0	4.7
781208	15.80	-96.78	19.0	15.67	-96.52	33.0	15.70	-96.49	53.0	3.2
781211	15.50	-96.85	15.0	15.75	-96.62	33.0	15.70	-96.69	19.0	3.2
790314	17.46	-101.46	20.0	17.81	-101.28	49.0	17.76	-101.30	3.0	7.6
790314	17.40	-101.40	16.0	17.71	-101.08	61.0	17.80	-100.90	104.0	4.2
790316	17.34	-101.38	25.0	17.99	-101.15	33.0	18.00	-100.70	106.0	4.2
790318	17.42	-101.10	25.0	17.55	-100.99	33.0	17.72	-100.89	61.0	5.4
790320	17.34	-101.44	30.0	17.53	-101.29	51.0	17.57	-101.26	56.0	4.8
790322	17.74	-101.65	30.0	17.96	-101.54	76.0	18.02	-101.52	77.0	5.0
790328	17.41	-101.16	30.0	17.14	-101.04	42.0	17.20	-100.60	99.0	4.3
790406	17.45	-101.63	14.0	16.76	-102.12	51.0	17.40	-101.50	100.0	4.6
811025	17.75	-102.25	16.0	18.05	-102.08	33.0	18.18	-102.01	28.0	7.3
811028	17.89	-102.35	15.0	18.46	-102.48	33.0	16.30	-102.90	33.0	3.6
820607	16.38	-98.38	20.0	16.61	-98.15	41.0	16.51	-98.25	19.0	6.9
820607	16.48	-98.55	15.0	16.56	-98.36	34.0	16.58	-98.34	20.0	7.0
820608	16.40	-98.39	38.0	16.37	-98.36	33.0	15.90	-98.41	61.0	3.9
820609	16.59	-98.44	23.0	16.66	-98.33	33.0	16.86	-98.38	52.0	4.8
820609	16.36	-98.51	15.0	16.57	-98.28	33.0	16.54	-98.22	53.0	4.4
820613	16.16	-98.44	20.0	16.18	-98.40	33.0	16.00	-98.50	33.0	3.9
820613	16.51	-98.40	25.0	16.26	-98.44	6.0	16.28	-98.42	7.0	3.9
820613	16.50	-98.40	25.0	16.13	-98.39	33.0	16.10	-98.42	33.0	3.4
820613	16.56	-98.44	27.0	16.18	-98.37	34.0	16.30	-98.29	41.0	3.6
820614	16.36	-98.30	26.0	16.60	-98.05	40.0	16.55	-98.05	46.0	4.7
820615	16.55	-98.27	24.0	16.30	-98.10	33.0	15.90	-98.00	33.0	3.8
820615	16.63	-98.47	30.0	16.46	-98.38	38.0	16.65	-98.36	38.0	3.6

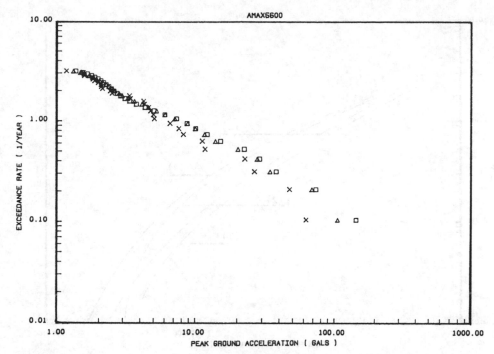

Figure 1. Rate of exceedance of peak ground accelerations for the local (X), PDE_{10} ($\triangle$) and ISC_{10} ($\square$) data, the later are for the ten percent probability of exceedance of the ground intensities.

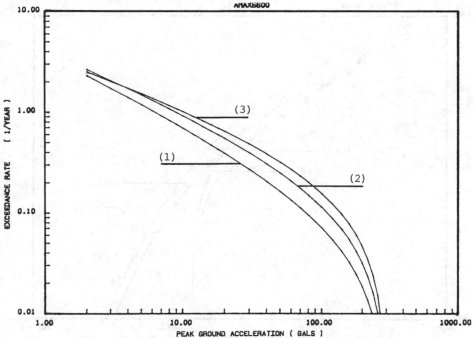

Figure 2. $\nu(y)$ computed for each of the data sets shown in Figure 1. Local (1), PDE_{10} (2), ISC_{10} (3); see Figure 1 for symbology

225

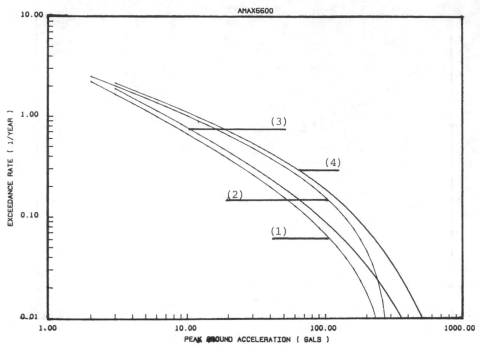

Figure 3. $\nu(y)$ computed for the original ISC catalog (1), for the ISC$_{10}$ (2) and their corresponding $\nu_c(y)$ corrected by uncertainties on the attenuation relation used, original ISC(3), ISC$_{10}$(4), see Figure 1 for symbology.

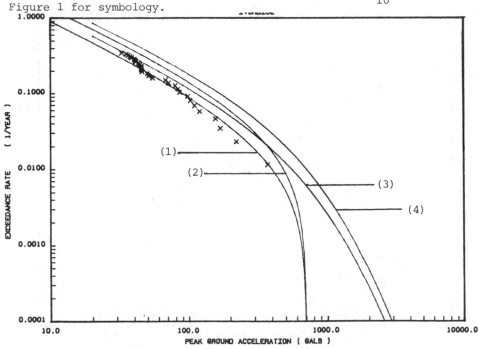

Figure 4. $\nu(y)$ computed for a real case: data (X), original ISC catalog (1), ISC$_{10}$(2), and their corresponding $\nu_c(y)$ correct by uncertainties on the attenuation relation used, original ISC(3), ISC$_{10}$(4).

The power spectral density method for seismic response analysis and its application to the improvement of the response spectrum method

M.Livolant

CEA-CEN Saclay, DEMT, Gif-sur-Yvette, France

INTRODUCTION

Two methods are widely used to calculate the structural response to seismic movements in the linear range: the response spectrum method (RS) and the time history method (TH).

The RS method starts from a ground response spectrum which is generally a standardized design response spectrum, sometimes adapted to the site. Its main defect is that it does not give correctly the response of a complex structure vibrating significantly on several modes.

The time history method starts from time representations of the earthquake. As the earthquake movement is random in nature, the use of a sufficient number of ground time histories is necessary: typically, the reduced standard deviation is 20% for the response of a given parameter to various time histories targeting the same ground spectrum. To obtain a reduced standard deviation of 6%, ten time histories are in principle necessary. The use of one time history with a spectrum enveloping the design ground response spectrum is not at all a guarantee that the parameters which involve a combination of vibrational modes are conservatively calculated.

Using the well-known results of the random vibrations theory and some more recent detailed studies on the peak values probabilities, it is possible to build a third method, which works well for complex structures, and gives directly results useful for risk analysis. That method is more complex in its presentation than the two others, but with the help of some light programming on a personal computer, it may be used easily.

METHOD PRESENTATION

For a given ground motion $\gamma(t)$, the response as function of time of a parameter y which is a linear combination of the structure displacements may be calculated by:

$$y(t) = \sum_i b_i \, \alpha_i(t) \qquad y_{max} = max \, [y(t)]$$

with $\alpha_i(t)$ solution of:

$$\alpha''_i(t) + 2\beta_i \, \omega_i \, \alpha'_i(t) + \omega_i^2 \, \alpha_i(t) = \gamma(t)$$

$\omega_i = 2\pi t_i$. f_i: resonant frequency of the ith mode

β_i : equivalent viscous damping value

b_i : participation factor of the ith mode in y.

The basic idea of the method is to consider $\gamma(t)$ as a transient random function, characterized by its mean frequency content, mathematically represented by a power spectral density $F(f)$, and its envelope $E(t)$. From that input, one has to calculate the statistics of the maximum value y_{max} of the response parameter y: mean, mean plus one standard deviation, probability of exceedance of some value...

FIRST STEP: Response standard deviation as a function of time

For simplicity, the presentation is made for a constant envelope: $E(t) = 1$ for $0 < t < T$. The time varying envelope is shortly presented after.

1) General expression

From the usual formulas for the response of linear systems to random input extended (12) to transient response:

$$\sigma_y^2(t) = 2 \int_0^\infty H(f,t) \; H^*(f,t) \; F(f)df$$

with $H(f,t) = \int_0^t h(\tau) \, e^{-2i\pi f\tau} \, d\tau$

$h(\tau)$: impulse response of the system.

2) Application to modal analysis:

For an harmonic oscillator, the impulse response is:

$$h_i(\tau) = \frac{1}{\omega'_i} \, e^{-\beta_i \, \omega_i \, \tau} \, \sin \omega'_i \, \tau \qquad \text{with} \quad \omega'_i = \omega_i \sqrt{1-\beta_i^2}$$

Consequently, for the y parameter:

$$h(\tau) = \sum_i b_i \, h_i(\tau) \qquad H(f,t) = \sum_i b_i \, H_i(f,t)$$

$$\sigma_y^2(t) = 2 \sum_i b_i^2 \int_0^\infty H_i \, H_i^* \, F(f)df + 2 \sum_{j>i} b_i \, b_j \int_0^\infty (H_iH_j^* + H_jH_i^*) \, F(f) \, df$$

228

So the standard deviation of the response in the sum of two types of terms: diagonal terms, involving only one resonance, and cross terms, for the interaction between resonances. These cross terms are important only when the resonances are sufficiently close to each other. Due to the time truncation, the exact expression of $H_i(f,t)$ is not single, but usually it is without interest to take all its complexity. For practical uses, the following approximative expression gives good results:

$$\tilde{H}_i(f,t) \approx \frac{\sqrt{4\,\beta'_i\,\omega_i^{\,3}}}{\omega_i^{\,2}-\omega^2 + 2i\beta'_i\,\omega\,\omega_i} \; \times \; A_i(t)$$

with

$$\beta'_i = \sqrt{\beta_i^{\,2} + \left(\frac{1}{\omega_i T}\right)^2}$$

$$A_i^{\,2}(t) = \frac{1 - e^{-2\beta_i\,\omega_i t}}{4\,\beta_i\,\omega_i^{\,3}}$$

PRACTICAL FORMULAS

$$\sigma_y^{\,2}(t) = \sum_i b_i^{\,2}\,A_i^{\,2}(t)\,I_i + 2\sum_i \sum_{i>i} b_i\,b_j\,A_i(t)\,A_j(t)\,I_{ij}$$

with

$$I_i = 2\int_0^{\infty} G_i(f)\,F(f)df$$

$$G_i(f) = \frac{4\,\beta'_i\,\omega_i^{\,3}}{(\omega_i^{\,2} - \omega^2)^2 + 4\,\beta'_i\,\omega^2\,\omega_i^{\,2}} \qquad \beta'_i = \sqrt{\beta_i^{\,2} + \left(\frac{1}{\omega_i T}\right)^2}$$

Remark: For $F(f) = F$ constant, $I_i = F$. That property can be use to derive an approximation when $F(f)$ is a smooth spectrum: $I_i \simeq F(f_i)$. In the general case, I_i has to be numerically calculated:

$$I_{ij} = \sqrt{I_i\,I_j}\;R_{ij} \qquad \text{with } R_{ij} = \frac{C_{ij}}{\sqrt{C_{ii}\,C_{ij}}}$$

$$C_{ij} = \frac{4(\beta'_i\,\omega_i + \beta'_j\,\omega_j)}{(\omega_i'^{\,2}-\omega_j'^{\,2})^2 + 2(\omega_i'^{\,2}+\omega_j'^{\,2})\,(\beta'_i\,\omega_i+\beta'_j\omega_j)^2 + (\beta'_i\,\omega_i + \beta'_j\,\omega_j)^4}$$

That interaction term is similar to that obtained by other authors, but it takes care of a reduced time effect on apparent damping by the use of β' insted of β.

Standard deviation maximum: $\sigma_{y\,max} = \sigma_y(T)$.

229

TIME VARYING ENVELOPE

In the preceding formulas, the duration T can be considered as the strong motion duration. If an envelope E(t) is given, a valuable estimation of T is:

$$T = \frac{1}{E_{max}^2} \int_0^\infty E^2(t) \ dt$$

For precise evaluations of the time variation of $\sigma_y^2(t)$, the $A_i^2(t)$ function may be obtained from the differential equation:

$$\frac{d}{dt} A_i^2(t) + 2 \ \beta_i \ \omega_i \ A_i^2(t) = \frac{E^2(t)}{2 \ \omega_i^2}$$

In that case, the maximum $\sigma_{y \ max}$ has to be determined numerically.

PEAK FACTORS FORMULAS FOR A SET OF OSCILLATORS

For a stationary random signal, the peak factor r in a time interval is the ratio between the maximum value in the interval and the standard deviation. Considering several time intervals with the same duration Θ gives the statistical behaviour of the peak factor $r(\Theta)$, which may be presented under the following form:

$r(p,\Theta)$: peak factor value with p as probability of non-exceedance during the time Θ.

The peak factors probability distribution depends of the signal type: narrow band or broad band for example, and there is no general method to obtain it. Practically, empirical formulas have been deduced in some cases from numerical simulations. The formulas presented below correspond to the stationary random response of a set of oscillators to a stationary random input.

$$r(p,\Theta) = \sqrt{1 + Log \ (1 + 0.7 \ N_e + N_e^2)}$$

$$N_e = 2 \ F_m \ \Omega \ \Theta \ \frac{Log \ 2}{Log \ 1/p}$$

$$F_m = \sqrt{\frac{\sum_i \sigma_i^2 \ f_i^2}{\sum_i \sigma_i^2}} \qquad \qquad \Omega = \sqrt{\frac{(0.05 + 10 \ \beta)^2}{(1 + (10 \ \beta)^2}}$$

$$\beta = \frac{1}{\sum_i \sigma_i^2} \left[\sum_i \sigma_i^2 \ \beta_i'' + 2 \ \sum_{j>i} |\sigma_i \ \sigma_j| \sqrt{\beta_i'' \ \beta_j''} \ (1-R_{ij}) \right]$$

$$\beta_i'' = \sqrt{\beta_i^2 + (\frac{0.05}{f_i \Theta})^2}$$

230

σ_i : standard deviation of the ith oscillator response.

R_{ij}: interaction term as defined precedently.

That set of formulas gives much better results for the response peak factor of a set of oscillators than the general formulation given in ref. (12). They are an improvement of the ref. (15) formulas.

MAXIMUM RESPONSE ESTIMATION TO A SEISMIC INPUT

The preceding formulas are not directly applicable, because the response standard duration of the set of oscillators is not constant. An approximate way to make the estimation is to use the maximum standard deviation value $\sigma_{y\ max}$, and to define an equivalent duration time Θ_e, as follows:

$$\Theta_e(r(p)) = \int_0^{t_{max}} e^{-\frac{r^2(p)}{2} \left(\frac{\sigma_{y\ max}^2}{\sigma_y^2(t)} - 1\right)} dt$$

That equivalent duration time is dependent of $r(p)$, which makes necessary to iterate to solve the overall system. A first approximation, generally conservative, is given by:

$$\Theta_1 = \Theta_e(2) = \int_0^{t_{max}} e^{-2 \left(\frac{\sigma_{y\ max}^2}{\sigma_y^2(t)} - 1\right)} dt$$

the y value with the probability p of non-exceedance is then:

$$y(p) = r(p, \Theta_e)\ \sigma_{y\ max}$$

Due to the shape of the amplitude distribution, which is not far from a gaussian distribution, the mean maximum value is approximately equal to $y(\frac{1}{2})$, the mean plus one standard deviation approximately equal to $y(0.84)$.

NUMERICAL APPLICATION

To verify the validity of the method, numerical simulations have been made and the results compared to those obtained by the DSP method and the response spectrum method with the usual SRSS combination rule, and the improved complete quadratic combination (CQC).

In the comparisons presented in this paper, earthquake movement was simulated by time histories with a constant envelope, a 10 s duration fitting a white noise spectrum in the frequency range of interest. To obtain a good statistical estimation, 200 time histories are used ; which gives a standard deviation of the result of the order of 1%.

231

The results are presented in the tables I, II, III:

Table I   : two oscillators with closed frequencies and the same
             coefficient.

Table II : two oscillators with closed frequencies and coefficients
             of opposite sign.

Table III: various combinations of oscillators with largely spaced
            frequencies, with coefficients selected to give the same
            displacement response standard deviation.

| Resonant Frequencies: | 0.52 | 1.17 | 1.89 | 3.54 | 7.32 |
Coefficients	1	3	6	16	47
Case 1	x		x		
Case 2		x	x		
Case 3			x	x	
Case 4			x		x
Case 5	x	x	x	x	x
Case 6	x				x
Case 7	x	x			
Case 8	x			x	

These tables show the improvement of the PSD method compared to the
CQC results for largely spaced resonances.

These tables show clearly the defects of the SRSS rule:

- For closed frequencies, the interaction effect on the standard
  deviation is ignored. This point is well corrected by the CQC rule.

- For largely spaced frequencies, the interaction effect on the peak
  factor is ignored, which gives an underestimation of the result. An
  extra-term has to be added to the CQC to give correct results
  (ref. 16).

The PSD method gives good results. One can notice a tendency to
overestimate the interaction effect for very largely spaced resonances
(Example: 0.52 Hz and 7.32 Hz).

CONCLUSION

In parallel with the usual response spectrum method and the time histories methods, the power spectrum method (PSD) can be used to calculate the response of mechanical systems to seismic movements. Its
main interests are to give directly the statistics of the response, to
treat correctly the interaction between resonances, and to allow floor
response spectra calculations. For its practical use, some works is
still needed, for example to define a standard PSD, corresponding to
the standard RS. Theoretical work is also necessary to improve the
peak factor estimation methods.

REFERENCES

(1) USAEC - Regulatory guide 1.60. Design Response Spectra for Seismic Design of Nuclear Power Plants.

(2) Ohsaki, Y., Iwasaki, R., Ohkawa, I., Masao, T. - Phase Characteristics of Earthquake Accelerogram and its Application. Ve SMIRT, Berlin, Août 1979, K1/4.

(3) Vanmarcke, E., Gasparini, D. Simulated Earthquake Ground Motions. IVe SMIRT, San Francisco, Août 1977, K1/9.

(4) Jeanpierre, F., Roullier, F. Caractéristiques des signaux sismiques pour l'analyse des structures. Cycle de Conférences CEA-EDF Jouy 1979.

(5) Scanlan R., Sachs, F. Earthquake Time Histories and Response Spectra. Journal of the engineering mechanics division, Août 1974, p. 635 à 655.

(6) Preumont, A. Vibrations aléatoires. Rapport BELGONUCLEAIRE 8404-03 (1984).

(7) Vanmarcke, E. Random Fields: Analysis and Synthesis. The MIT Press Cambridge Mass (1983).

(8) Livolant, M., Gantenbein, F., Gibert, R.J. Méthode statistique pour l'estimation de la réponse des structures aux séismes. Mécanique - Matériaux - Electricité n° 394-395, Oct. 1982.

(9) Combining Modal Responses and Spatial Components in Seismic Response Analysis. USNRC. Regulatory guide 1-92 (Fev. 1976).

(10) Rosenblueth, E., and Elorduy, J., 1969. Responses of Linear Systems to Certain Transient Disturbances, Proc. 4th WCEE, Santiago, A1-185.

(11) Maison, B.F., Neuss, C.F., Kasai, K. The Comparative Performance of Seismic Response Spectrum Combination Rules in Building Analysis. Earthquake Engineering and Structural Dynamics, Vol. 11, 623-647 (1983).

(12) Vanmarcke, E. Structural Response to Earthquakes. Ch. 8 in seismic risk and engineering decision (1976).

(13) Vanmarcke, E.H., 1975. On the Distribution of the First-passage Time for Normal Stationary Random Processes. J. Appl. Mech., 42 (Ser. E): 215-220.

(14) Gibert, R.J., Livolant, M., Fakhfakh, T. Analyse probabiliste du maximum de la réponse à une sollicitation sismique d'un système linéaire à plusieurs degrés de liberté. 1ere Conférence Association Française de Génie Parasismique, St-Rémy les Chevreuse (1986).

(15) Livolant, M. Structural Analysis in Time and Frequency Range Eurochina Joint Seminar on Earthquake Engineering, June 9-14, 1986, BEIJING, China.

(16) Fakhfakh, T., Gibert, R.J. Statistical Analysis of the Maximum Response of Multi-degree Linear System to a Seismic Excitation, CEA-CEN Saclay, IRDI-DEMT 91191 Gif sur Yvette Cedex, France.

233

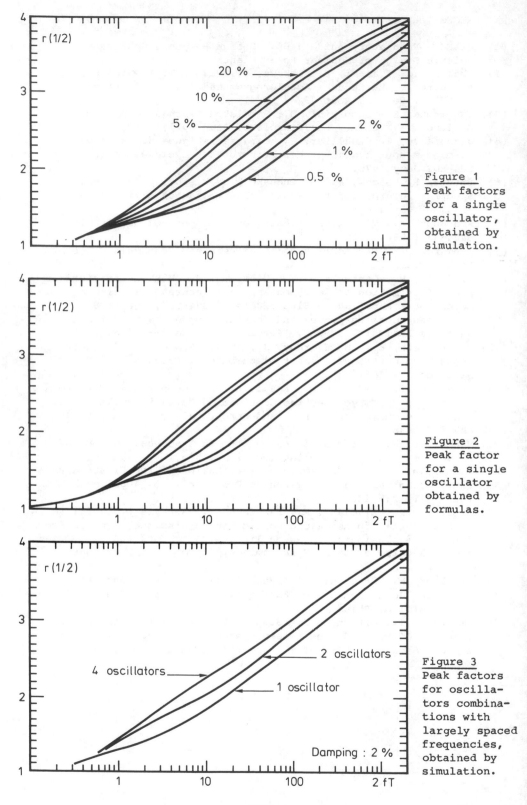

Figure 1
Peak factors for a single oscillator, obtained by simulation.

Figure 2
Peak factor for a single oscillator obtained by formulas.

Figure 3
Peak factors for oscillators combinations with largely spaced frequencies, obtained by simulation.

Table I. 2 oscillators with closed frequencies - Damping 2%
Duration 10 s - Same coefficient

Δf/2f	0	0.5	1	1.75	2.5	3.75	5	7.5	12.5
Simulation	0.350	0.346	0.339	0.324	0.310	0.291	0.277	0.261	0.245
PSD	0.356	0.353	0.347	0.330	0.310	0.291	0.279	0.263	0.248
CQC	0.348	0.344	0.331	0.308	0.288	0.266	0.252	0.237	0.222
SRSS	0.248	0.246	0.244	0.242	0.240	0.237	0.233	0.228	0.218

Table II. 2 oscillators with closed frequencies - Opposite sign coefficients - Damping 2% - Duration 10 s.

Δf/2f	0	0.5	1	1.75	2.5	3.75	5	7.5	12.5
Simulation	0	0.039	0.075	0.117	0.148	0.180	0.198	0.216	0.221
PSD	0	0.052	0.100	0.157	0.192	0.221	0.231	0.240	0.240
CQC	0.	0.055	0.102	0.152	0.180	0.204	0.213	0.218	0.218
SRSS	0.248	0.246	0.244	0.242	0.240	0.237	0.233	0.228	0.218

Table III. Oscillators combinations with largely spaced frequencies.

Case	1	2	3	4	5	6	7	8
Simulation	0.102	0.111	0.143	0.164	0.243	0.144	0.088	0.125
PSD	0.104	0.107	0.145	0.170	0.256	0.164	0.088	0.138
SRSS	0.089	0.100	0.127	0.143	0.194	0.129	0.079	0.111

Some conservative aspects in nuclear power seismic design

G.E.Howard & G.I.Yessaie
ANCO Engineers, Culver City, Calif., USA
R.F.Oleck
Niagara Mohawk Power Corporation, Syracuse, N.Y., USA

1 INTRODUCTION

Realistic prediction of the seismic response of nuclear
power facilities is particularly important for existing
nuclear plants subject to design review with respect to
seismic inputs more severe than used in plant design.
Design review using the conservative methods typical of the
design process can be expected to indicate that some form of
retrofitting is required to meet the increased seismic
inputs.

Nonlinearities in the dynamic behavior of structures and
equipment are potentially significant sources of
conservatism with respect to nuclear power plant seismic
response. This paper presents preliminary results of a study
of the potential influence of nonlinear reinforced concrete
shear wall effects in reducing the seismic loads imposed on
plant internal equipment for ground motions exceeding the
original design condition.

The focus of the study is the reactor building of a
Boiling Water Reactor (Mark-1, USA), as represented by a USA
nuclear power station that began commercial operation in the
late 1960's. A simplified nonlinear model provides the
basis for a parametric study to examine the potential
significance of reactor building nonlinear behavior with
respect to reduction of dynamic loads imposed on internal
equipment.

2 PHYSICAL DESCRIPTION

The modeled reactor building was rectangular in plan,
160 ft x 160 ft (49 m x 49 m), and 180 ft (55 m) in height.
The bottom 130 ft (40 m) of the building is a shear-wall
type structure, enclosing the reactor and it's associated
'lightbulb-shaped' drywell shell and the toroidal
suppression chamber. A steel superstructure occupies the
upper 50 ft (15 m) of the building.

The main Lateral Load Resisting System (LLRS) of the building consists of: a centrally located shield wall of thickness ranging from 3 ft - 9 ft (0.9 m - 2.7 m); and box wall of four peripheral walls forming a box around the shield wall. The wall thickness ranges up to 3 ft (0.9 m) with reinforcing ratios of less than 0.5%. Additional modest resistance is provided by internal shear walls.

The building is founded on a 6 ft (1.8 m) thick foundation resting in turn at approximately 60 ft (18 m) below grade in a rock socket on sandstone with a shear wave velocity exceeding 8000 ft/sec (2440 m/sec). The gap between the rock socket vertical walls and the reactor walls forms an annular region 5 ft (1.5 m) in width by 45 ft (14 m) high that is filled with very low shear wave velocity (order of 600 ft/sec [180 m/sec]) material; the remaining 15 ft to grade is soil backfill.

Input motion criteria used in plant design was a modified version of the Helena, Montana earthquake trace obtained at a rock site, with a peak acceleration for the subject plant site of 0.11 g.

3 MODEL DESCRIPTION

A simple multi-mass stick model derived from the original plant design model provided the basic model used in the parametric studies reported below, and included both translational and rotational inertial properties associated with the structure and internals.

A critical parameter in performing the elasto-plastic simulations was the assumed point of the onset of nonlinearity of the reinforced concrete structure. Because shear deformations for the building are significantly greater than bending deformations, shear stress was used to define departure from linearity in the model response. Figure 1 depicts the model used in this study, with the onset of nonlinear response denoted by Vy, corresponding to the onset of web shear cracking in squat shear walls. Following cracking onset, the shear stiffness decreases to 10% of the original stiffness.

Ductility is a common parameter used to quantify inelastic behavior, where for this study interstory ductility is defined as [Dmax / Dyield], also referred to as 'local interstory ductility' since it only applies to certain portions of the structure. Dmax is the maximum interstory drift and Dyield is that drift resulting in first yielding. System ductility is a measure of inelastic behavior throughout the structure and can be related in a fashion to local ductility; Kennedy (1) has reported system ductility of about one-half of local ductility for a multi-degree-of-freedom soil/structure model.

238

Ground motion characteristics - frequency content, duration, energy - are of particular importance in evaluating the severity of response of an elasto-plastic model. Decreases in apparent frequency of response can, for example, result in the structure shifting into higher or lower spectral acceleration regimes of the input motion; duration of strong motion is important with respect to structural degradation through the number of inelastic stress reversals.

For this study, two different earthquake motions were used: a time-history prepared by URS/John Blume (2) that bounds a US Nuclear Regulatory Commission-approved response spectrum; and a site specific time-history generated by Dames and Moore (3) for a western New York rock site on Lake Ontario. The URS/John Blume (JB) input is more severe than Dames and Moore (C-M for Cornell-McGuire) in the frequency region of significance for the reactor building. However, the C-M motion is believed to be representative of the expected site motions. The 5% damped response spectra for both inputs are shown in Figure 2.

Figure 3 shows the 'Energy' of the C-M and JB inputs, where 'Energy' is defined as the integral over time of the acceleration squared. Housner and Jennings (4) have demonstrated that the above defined 'Energy' multiplied by pi/2 may be interpreted as the frequency ensemble work and is a measure of the capacity of ground motion to do work on an idealized population of structures of all natural frequencies. Although the peak accelerations are about the same, the JB input imposes about an order of magnitude more 'Energy' than the C-M input and the duration (time to achieve approximately 90% of the total Energy') of strong motion is 7 sec for C-M vs 16 sec for JB.

4 COMPARISON OF FLOOR RESPONSE SPECTRA: ELASTIC VS INELASTIC
   MODELS

First estimates of influence on floor response spectra - which provide the loads imposed on internal equipment -can be obtained from single degree of freedom models for the JB and C-M inputs. The results indicated a suppression in floor response spectra by a factor of 2 near the natural frequency of the oscillator for a ductility of 4.5, as anticipated.

The multi-mass models (elastic and elasto-plastic) were likewise subjected to amplitude-scaled versions of CMNS to achieve various levels of interstory ductility, with the onset of yielding taken to occur at shear stresses at 4 x (f_c'raised to the 0.5 power), where f_c' is the compressive strength of concrete (3500 lb/ft-ft for the structure modeled here). The building models were fixed-base since

239

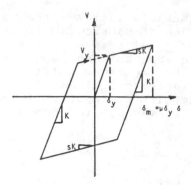

FIGURE 1 THE NON-LINEAR RESISTANCE FUNCTION USED IN THE ELASTO-PLASTIC ANALYSIS

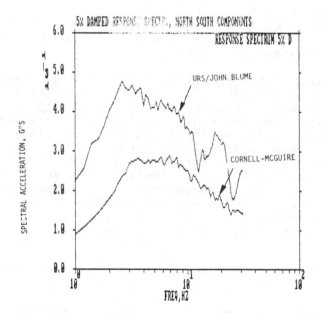

FIGUIRE 2 5% DAMPED RESPONSE SPECTRA FOR THE CORNELL-MCGUIRE AND
URS/JOHN BLUME GROUND MOTIONS.

the fundamental frequency was reduced by only 3% when 'soil
springs' were included to model the foundation medium. The
first mode at 2.7 Hz was the response of the steel
superstructure; the second and third modes at 6.3 Hz and
18.8 Hz were the primary modes of interest, with the 6.3 Hz
dominating due to the relatively low spectral input at the
higher frequency.

    The floor response spectra at the operating floor (142
ft or 43 m above the foundation) resulting from C-M
excitation of the elastic and elasto-plastic
multi-degree-of-freedom models are shown in Figure 4 with a

240

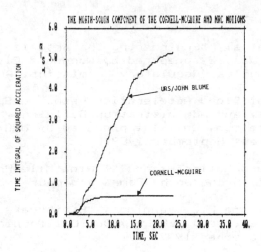

FIGUIRE 3 INTEGRAL OVER TIME OF ACCELERATION SQUARED FOR THE
CORNELL-MCGUIRE AND URS/JOHN BLUME GROUND MOTIONS.

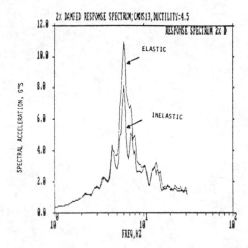

FIGUIRE 4 2% DAMPED FLOOR RESPONSE SPECTRA AT THE OPERATING FLOOR
LEVEL USING THE ELASTIC AND IN-ELASTIC SHEAR WALL RESISTANCE
FUNCTIONS.

maximum imposed local ductility of 4.5 occurring at the
lower elevations. The 2% floor response spectrum for the
elasto-plastic model is reduced relative to the elastic
result by a maximum of about 30% at the dominant response
frequency (f_N), with varying amounts of spectral suppression
over the range of approximately $f_N/1.5$ to $1.5 \times f_N$. A
ductility of 4.5 is expected to correspond to the onset of
'significant' structural damage in the concrete shear wall
structure.

For a maximum local ductility of up to 2.75, no
significant floor spectral suppression was obtained.

REFERENCES

1. R. Kennedy, et al, "Engineering Characterization of Ground Motion", Vol. 2, prepared by Woodward Clyde, NUREG/CR-3805, Nuclear Regulatory Commission, August 1986.

2. "Generation of Flood Acceleration Response Spectra for Reactor, Turbine, and Administration Buildings, Nine Mile Point Nuclear Station, Unit 1", prepared by URS/John A. Blume & Associates, September 1983.

3. "Seismic Ground Motion Hazard at Nine Mile Point Nuclear Station, Unit 1", prepared by Dames and Moore, July 1984.

4. Housner, G.W. and P.C. Jennings, "Earthquake Design Criteria for Structures", EERL 77-06, California Institute of Technology, November 1977.

Seismic design of the HDR/VKL-piping system without snubbers and preliminary results of building shaker tests

L.M.Habip & W.Schrader
Kraftwerk Union AG, Offenbach, FR Germany
L.Malcher
Kernforschungszentrum Karlsruhe GmbH, Projekt HDR-Sicherheitsprogramm, FR Germany

1 EXPERIMENTAL DYNAMIC LOADS

The reactor building of the decommissioned HDR-plant, originally housing an experimental superheated-steam reactor system, has recently been subjected to intense dynamic loads in a series of tests which induced cyclic base excitation of existing mechanical components and piping systems.

A large, eccentric mass, coast-down shaker was installed on the operating floor of the building, as shown in Figure 1, for this purpose. The design and operation of the shaker and the excitation levels that were reached are summarized elsewhere (Malcher et al. 1987).

Horizontal floor accelerations were measured in the vicinity of the VKL-piping system to be considered here in the course of various tests with different initial shaker coast-down frequencies. The corresponding response spectra are shown in Figure 2, where they can be compared with an enveloped safety-earthquake spectrum at a relevant location of the reactor building of a modern PWR-plant in the FRG. Measured vertical floor accelerations were an order of magnitude smaller.

2 SELECTION OF PIPING SUPPORT CONFIGURATION

The dynamic response of the VKL-piping system shown in Figure 3, for two different seismic support configurations which did not make use of snubbers, is of particular interest: a) the so-called HDR-configuration with only two lateral sway struts (H4, H5) and b) the KWU-configuration with 5 struts (H4, H5, H9, H10, H11). The 4 nozzle connections at the ends of the piping system served as anchor points. Intermediate supports in the vertical direction (not shown) consisted of 4 constant force hangers, two spring hangers and a threaded rod.

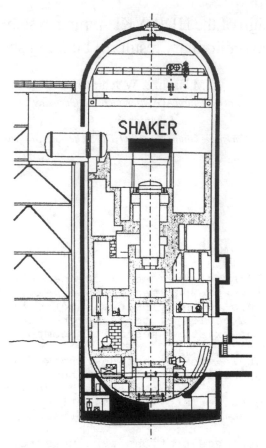

Figure 1. Location of shaker.

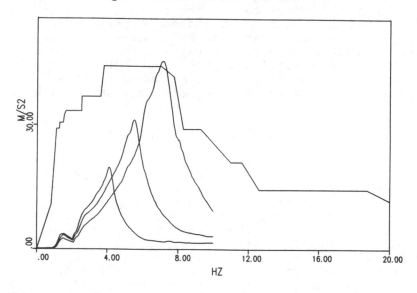

Figure 2. Response spectra of measured horizontal
floor accelerations.

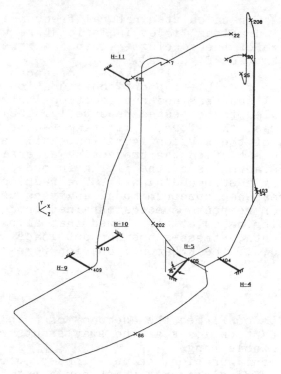

Figure 3. Routing of VKL-piping
system with sway struts.

Table 1. Natural frequencies, Hz.

Support configuration	
HDR	KWU
1.8	2.2
2.6	3.2
3.2	4.1
3.6	4.6
4.1	5.8
4.6	6.4
5.5	7.0
6.5	7.3
6.6	7.8
7.1	8.6

Most of the lower calculated natural frequencies of these two configurations listed in Table 1 are well within the range of the forcing frequencies generated during the tests (see Figure 2).

The lateral struts were already part of another, dynamically less flexible support configuration which included several snubbers and one additional strut and were scheduled to be tested repeatedly. Their rated load capacity was, therefore, a limiting factor in the seismic design of the KWU-configuration. Still another condition to be satisfied was preventing excessive horizontal displacements of the piping in order for it to avoid hitting a surrounding wall. The results of accompanying response spectrum modal analyses using horizontal floor response spectra obtained from preliminary building calculations showed that allowable pipe displacements, stresses and support loads would not be exceeded, if the KWU-configuration were chosen.

3   TEST RESULTS

The test results confirmed the adequacy of both designs and that the reaction loads at the sway struts remained within the allowable range.

For the HDR-configuration, calculations predicted partly excessive displacements which could not be verified experimentally, because displacements were not measured.

Thus, the present series of full-scale dynamic tests during which resultant acceleration levels of up to 4.5 g were recorded on parts of the VKL/KWU-piping - a system with several natural frequencies in the range of the forcing frequencies - has shown that a support configuration with lateral sway struts and no snubbers is indeed a good solution in designing piping systems against base excitation.

Another series of dynamic tests in which the piping would be excited directly at selected locations is planned for 1988. This should allow a thorough in situ investigation of the seismic load-carrying ability of a piping system with multiple supports.

ACKNOWLEDGMENT

This work was part of the project HDR/SHAG sponsored by the BMFT and the USNRC, Office of Research.

REFERENCES

Malcher, L., H. Steinhilber, D. Flade & D. Schrammel. 1987. Earthquake investigations at the HDR facility. Paper K9/1. 9th International Conference on Structural Mechanics in Reactor Technology, Lausanne.

Stochastic response of WWER Core Barrel to seismic ground motion

L.Pečínka & M.Čechura

Škoda Works Plzeň, Power Engineering Division, Czechoslovakia

1 INTRODUCTION

Earthquake resisting design of nuclear power plants ranks among the important problems of nuclear safety. There are two possible methods of analysis: the deterministic approach and the stochastic approach. In the former method the time histories of the motion of structure floors are calculated by the numerous available analytical or numerical techniques, some adequate strong motion records being considered as the input earthquake ground motion.

In this study the stochastic response of thin cylindrical shell (core barrel - CB) to random earthquake support motion is considered. The formal general solution in time domain is at first derived. The statistically stationary model of earthground acceleration in the form of filtered white noise is briefly reviewed and the general expression for the mean-square response is derived.

2 EQUATION OF MOTION

The equation of motion of thin cylindrical shell is given in the form

$$Au = Bu + f_g + f_i \tag{1}$$

A ... matrix of Flügge differential operators

$$B = -\frac{\varrho_s R_s (1-\nu^2)}{Eh} \frac{\partial^2}{\partial t^2} E \qquad E \text{ ... unit matrix}$$

$$f_g = \left[0; 0; -\frac{\varrho_s R_s (1-\nu^2)}{Eh} \ddot{u}_g(t) \right]^T$$

$\ddot{u}_g(t)$ ... horizontal earthquake acceleration

$$f_i = \left[0; 0; p_w \right]^T \frac{1-\nu^2}{Eh} R_s \qquad \text{... vector of fluid - CB interaction}$$

p_w ... radial pressure generated by fluid - CB interaction

Using the method derived in /1/, we obtain the differential equation as follows

$$\ddot{q}_\alpha + 2D_\alpha \omega_\alpha \dot{q}_\alpha + \omega_\alpha^2 q_\alpha = $$
$$= \frac{1}{1+\delta_\alpha} \frac{1}{\int_o^L F_m^2 dz} \int_0^L \int_0^{2\pi} \ddot{u}_g(t) F_m(z) \cos m\varphi \, d\varphi dz \tag{2}$$

$F_m(z)$ - beam function for given boundary condition

δ_α - added mass coefficient

The radial displacement is supposed as

$$w(z,\varphi,t) = q(t) F_m(z) \cos m\varphi \tag{3}$$

Because of only radial component of $\ddot{u}_g(t)$, e.g. $\ddot{u}_g(t) \cos\varphi$ takes the place in (2), it will be received as a result

$$\ddot{q}_\alpha + 2D_\alpha^* \omega_\alpha \dot{q}_\alpha + \omega_\alpha^2 q_\alpha = -\frac{\pi}{1+\delta_\alpha} \frac{\int_o^L F_m(z)\,dz}{\int_o^L F_m^2(z)\,dz} \ddot{u}_g(t) \tag{4}$$

$$\alpha = (1, m)$$

The general solution in the time domain has the following form

$$q_\alpha(t) = -\frac{\pi}{1+\delta_\alpha} \frac{1}{\omega_\alpha} \frac{\int_o^L F_m(z)dz}{\int_o^L F_m^2(z)dz} \int_0^t e^{-D_\alpha^* \omega_\alpha(t-\tau)} \times$$

$$\times \sin \omega_\alpha(t-\tau) \ddot{u}_g(\tau)\, d\tau \tag{5}$$

3 RESPONSE TO STATIONARY GROUND MOTION

If the ground motion $\ddot{u}_g(t)$ is given in the form of power spectral density $S_{\ddot{u}_g}$, then the mean-square of $q_{1,m}(t)$ is defined by

$$\langle q_{1,m}^2 \rangle = \int_0^{+\infty} S_q(\omega)\, d\omega \tag{6}$$

with

$$S_q(\omega) = \lim_{T \to \infty} \frac{1}{2T} q(\omega) \hat{q}(\omega) \tag{7}$$

248

$$\hat{q}(\omega) = \mathcal{F}\{q(t)\} = C \frac{\hat{\ddot{u}}_g(\omega)}{\omega_{1,m}^2 - \omega^2 + 2i D_{1,m}^* \omega_{1,m} \omega} \qquad (8)$$

Substituting (8) and (7) into (6) we obtain after some calculations

$$\langle q_{1,m}^2 \rangle = \sum_{m=1}^{M} C_m \int_0^{+\infty} \frac{S_{\ddot{u}_g}(\omega)}{(\omega_{1,m}^2 - \omega^2)^2 + 4 D_{1,m}^{*\,2} \omega_{1,m}^2 \omega^2} \, d\omega \qquad (9)$$

$$C_m = -\frac{\bar{\eta}}{1 + \delta_{1,m}} \frac{1}{\omega_{1,m}} \frac{\int_0^L F_m(z)\, dz}{\int_0^L F_m^2(z)\, dz}$$

If $S_{\ddot{u}_g}(\omega)$ is suggested by Zeman /2/ as

$$S_{\ddot{u}_g}(\omega) = \sum_{i=1}^{\bar{I}} S_{0g,i} \frac{\omega_{g,i}^2 + 4 D_{g,i}^{*\,2} \omega^2}{(\omega_{g,i}^2 - \omega^2) + 4 D_{g,i}^{*\,2} \omega_{g,i}^2 \omega^2}$$

the general form of $\langle q_{1,m}^2 \rangle$ is given by

$$\langle q_{1,m}^2 \rangle = \sum_{i=1}^{I} C_{1,m} \int_{-\infty}^{+\infty} \frac{S_{0g,i}}{(\omega_{1,m}^2 - \omega^2)^2 + 4 D_{1,m}^{*\,2} \omega_{1,m}^2 \omega^2} *$$

$$* \frac{\omega_{g,i}^2 + 4 D_{g,i}^{*\,2} \omega^2}{(\omega_{g,i}^2 - \omega^2)^2 + 4 D_{g,i}^{*\,2} \omega_{g,i}^2 \omega^2} \, d\omega \qquad (10)$$

The computation of $\int_{-\infty}^{+\infty} \ldots \, d\omega$ will be carried out using the rezidue theorem. The zero points of denominator of (10) are

$$Z_{1,2,3,4} = \pm \omega_{1,m} \sqrt{1 - D_{1,m}^{*\,2}} \pm i D_{1,m}^* \omega_{1,m} \qquad (11a)$$

$$Z_{5,6,7,8} = \pm \omega_{g,i} \sqrt{1 - D_{g,i}^2} \pm i D_{g,i}^* \omega_{g,i} \qquad (11b)$$

Because of

$$\operatorname{rez} f(z_1) = \frac{\omega_{g,i}^2 + 4 D_{g,i}^{*\,2} z_1^2}{\prod_{j=2}^{8} (z_1 - z_j)} \qquad (12a)$$

249

$$rez\, f(z_2) = \frac{\omega_{g,i}^2 + 4D_{g,i}^{*2}\, z_2^2}{8 \prod\limits_{\substack{j=1 \\ j \neq 5}} (z_2 - z_j)} \tag{12b}$$

$$rez\, f(z_6) = \frac{\omega_{g,i}^2 + 4D_{g,i}^{*2}\, z_6^2}{8 \prod\limits_{\substack{j=1, j \neq 6}} (z_6 - z_j)} \tag{12c}$$

$$rez\, f(z_5) = \frac{\omega_{g,i}^2 + 4D_{g,i}^{*2}\, z_5^2}{8 \prod\limits_{\substack{j=1, j \neq 5}} (z_5 - z_j)} \tag{12d}$$

the resulting formula of $\int_{-\infty}^{+\infty} \ldots d\omega$ takes the form of

$$\int_{-\infty}^{+\infty} f(z)\, dz = 2\pi i \left[rez\, f(z_1) + rez\, f(z_2) + \right.$$
$$\left. + rez\, f(z_5) + rez\, f(z_6) \right] \tag{13}$$

Substituting (11) and (12) into (13) and rearranging, we obtain the mean-square of $\langle q_{1,m}^2 \rangle$ as

$$\langle q_{1,m}^2 \rangle = -4\pi C^2 S_{0g,i}\, \omega_{g,i}^2 \left[\frac{H_1 E_1 - F_1 G_1}{E_1^2 + F_1^2} + \frac{H_3 E_3 - F_3 G_3}{E_3^2 + F_3^2} \right] \tag{14}$$

with

$E_1 = A_1 C_1 - B_1 D_1$ $\qquad\qquad$ $F_1 = B_1 C_1 + A_1 D_1$

$A_1 = - 8\omega_1^3\, D_1^{*2}$ $\qquad\qquad$ $B_1 = 8\omega_1^3\, D_1^*$

$C_1 = (\omega_{g,i}^2 - \omega_1^2)^2 - 4\omega_1^2 (D_1^{*2}\omega_1^2 - D_{g,i}^{*2}\, \omega_1^2 -$

$\qquad - D_{g,i}^{*\,2}\, \omega_{g,i}^2)$

$$D_1 = 4D_1^* \omega_1^2 (\omega_1^2 - \omega_{g,i}^2) + D_{g,i}^{*2} \omega_{g,i}^2$$

$$G_1 = \omega_{g,i}^2 + 4D_{gi}^* \omega_1^2$$

$$H_1 = 8 D_1^* D_{g,i}^* \omega_1^2$$

$$E_3 = A_3 C_3 - B_3 D_3 \qquad\qquad F_3 = B_3 C_3 + A_3 D_3$$

$$A_3 = -8 D_{g,i}^{*2} \omega_{gi}^3 \qquad\qquad B_3 = 8 D_{gi}^* \omega_{gi}^3$$

$$C_3 = (\omega_1^2 - \omega_{gi}^2)^2 - 4\omega_{gi}^2 (D_{gi}^{*2} \omega_{gi}^2 - D_1^{*2} \omega_1^2)$$

$$D_3 = 4 D_{gi}^* \omega_{gi}^2 (\omega_{gi}^2 - \omega_1^2) + D_{gi}^{*2} \omega_{gi}^2$$

$$G_3 = \omega_{gi}^2 (1 + 4 D_{gi}^*)$$

$$H_3 = 8 D_{gi}^{*2} \omega_{gi}^2$$

CONCLUSIONS

In our calculations, the 3σ level of radial deflection, e.g.

$$3\left[\langle W_{1,m}^2 (z,\varphi)\rangle\right]^{0.5} = 3\left[\langle q_{1,m}^2 \rangle \, \bar{F}_m^2 (z) \cos^2\varphi\right]^{0.5}$$

has been, for the given accelerogram, compared with deterministic approach $W_{1,m}(z,\varphi,t) = q_{1,m}(t) \, F_m(z) \cos\varphi$ obtained by direct integration of equation (4). It has been determined that the 3σ level is a conservative estimate of the maximum response. The $1,5\sigma$ level is more realistic. This fact is in a good agreement with /3/.

REFERENCES

Pečínka,L. & Čechura, M. 1986. The Improved Method of Calculation of Core-Barrel Response During Blow-down. ŠKODA Works Internal Rept.(only in Czech).
Zeman, Vl. 1985. The Mathematical Model of WWER - type Reactor. Internal Rept. of Czechoslovac Academy of Sciences, Institute of Theoretical and Applied Mechanics, Division Plzeň.
Newmark, N.M. & Rosenblueth,E. 1971. Fundamentals of Earthquake Engng. Prentice-Hall.

Confirmation of soil radiation damping from test versus analysis

J.M.Eidinger, G.S.Mukhim & T.P.Desmond
Impell Corporation, Walnut Creek, Calif., USA

1 INTRODUCTION

The validity of using high analytical soil damping values in soil-structure interaction analyses of nuclear plants has been the subject of much debate. In many instances, engineers and regulatory bodies have mandated the use of low soil damping simply to ensure conservatism. This paper shows that high radiation damping in soils do occur and that there is good correlation between analytical and test results.

The work was performed to demonstrate that soil-structure interaction effects for nuclear plant structures can be accurately (and conservatively) predicted using the finite element or soil spring methods of soil-structure interaction analysis. Further, the work was done to investigate the relative importance of soil radiation versus soil material damping in the total soil damping analytical treatment.

The analytical work was benchmarked with forced vibration tests of a concrete circular slab resting on the soil surface. The applied loading was in the form of a suddenly applied pulse load, or snapback. The measured responses of the slab represent the free vibration of the slab after the pulse load has been applied. This simplifies the interpretation of soil damping, by the use of the logarithmic decay formulation. To make comparisons with the test results, the damping data calculated from the analytical models is also based on the logarithmic decay formulation.

An attempt is made to differentiate the observed damped behavior of the concrete slab as being caused by soil radiation versus soil material damping. It is concluded that both the traditional soil radiation and material damping analytical simplifications are validated by the observed responses.

It is finally concluded that arbitrary "conservative" assumptions traditionally made in nuclear plant soil-structure interaction analyses are indeed arbitrary, and not born out by physical evidence. The amount of conservatism introduced by limiting total soil damping to values like 5% to 10% can be large. For the test slab sizes investigated in this paper, total soil damping is about 25%. For full size nuclear plant foundations, total soil damping is commonly in the 35% to 70% range.

Therefore, the authors suggest that full soil damping values (the combined radiation and material damping) should be used in the design, backfit and margin assessment of nuclear plants.

2 FORCED VIBRATION TEST

The forced vibration test described in this report was performed by the Southern California Edison Company, with technical direction by the Woodward-McNeill Company

(Woodward-McNeill, 1974). The tests were performed in 1972 for various slab configurations and embedments. The discussion in this paper is limited to a surface-founded slab (i.e. no embedment).

A concrete slab of 10 feet in diameter and 5 feet thick was placed on a horizontal soil surface. Steel lugs were embedded in the concrete block to allow the application of external loads. Instrumentation consisted of velocity geophones placed on the slab such that the motions of the slab in the horizontal direction and in the vertical direction would be recorded. Figure 1 shows the slab geometry and instrument locations. Locations No.1 and No.2 record the slab motions in the horizontal direction and in the vertical direction, respectively. Transient motions to the slab were induced through an external load applied by tensioning a cable and weak link until the weak link failed. The tensile strengths of the weak links vary from 3 to 16 kips. The actual breaking force was not recorded.

Soil conditions beneath the concrete slab are characterized as San Mateo sand. Very low strain shear wave velocities near the surface of this sand were estimated to be in the range of 900 to 1200 feet per second based upon cross hole and down hole seismic velocity tests. The soil profile has this sand layer extending for over 500 feet beneath the surface. This soil's profile is therefore characterized as a uniform halfspace. Dynamic triaxial shear tests were performed for this soil to determine its shear modulus and soil material damping properties (Woodward-McNeill, 1974). Shear modulus versus soil shear strain, and the soil material damping versus soil shear strain curves are presented in Figure 2. From the tests, it was determined that the frequency of the slab-soil system in the horizontal direction is about 20 Hz.

3 SOIL DAMPING

Soil damping can be separated into two categories, namely, radiation damping and material damping. In some manner, soil radiation and soil material damping combine to result in total soil damping. In this paper, we assume that this combination is a simple addition, i.e.,

Dtotal = Dradiation + Dmaterial, where

Dtotal = equivalent viscous damping

Dradiation = radiation (or geometric) damping

Dmaterial= material (or hysteretic damping)

Soil material damping is most commonly determined by calculating the energy lost per cycle in a dynamic triaxial soil column test. Results of this type of calculation are available in the literature (Seed and Idriss, 1970). The adequacy of calculating soil material damping in this manner has often come to question. For seismic events which can produce soil shear strains on the order of 1%, soil material damping can be calculated as high as 15% to 25%. Because dynamic soil triaxial tests may not totally duplicate field soil conditions, and the normal scatter often found in triaxial tests, the soils engineer will often recommend a range of soil material damping on the order of ±50% of the average soil material damping value. This apparent scatter in damping value raises concern for both the analyst, and the regulatory reviewer, with often the most conservative approach taken, i.e., to use the lower bound soil material damping curves.

The choice of value for soil radiation damping has at least as much variance in range as soil material damping. Frequency-independent soil-structure interaction techniques were developed in the 1960's and 1970's, using the lumped parameter soil-spring approach. In this manner, many investigators calculated soil radiation damping using textbook formulas (Richart, et al, 1970). By this method, soil radiation damping values, for the vertical direction of response, are typically in the 40% to 70% range, for nuclear plant reactor buildings. In the horizontal direction, the damping values are in the 25-40% range. For the rocking direction, lower damping values, about 10%, are typical.

The structural analyst is often presented with a dilemma with these soil radiation damping values. First, they are extremely high, as compared to traditional structural materials (steel, concrete) damping values. Second, they are known to be frequency-dependent, yet most soil-structure interaction computer codes of the 1960's and 1970's could not handle frequency-dependent parameters. Third, the damping values vary considerably upon direction. Finally, and perhaps most importantly, these analytically derived soil radiation damping values were based upon many analytical assumptions, and not all were supported by field test data. Therefore, in the prudency of timeliness for design, and the lack of widespread acceptance of these high damping values, compromises have often been taken, to limit the total soil damping value between 5% to 20%. This limitation is in contrast to computed soil damping values often in the 40% to 70%+ range.

4 ANALYTICAL MODEL

To determine the accuracy of these high analytically-computed soil damping values, an analytical model was developed using the SASSI computer code (Lysmer, et al, 1971). In very concise terms, SASSI is a finite element analysis program based in the frequency domain. Soil impedances are calculated over the frequency range of interest. The soil may be represented as either a layered halfspace, or as discretized soil elements. By careful subtraction of the impedances of soil elements from that of a layered halfspace, SASSI allows both soil layers and soil elements to be combined.

The model is three-dimensional and is shown in Figure 3. A horizontal analysis was performed since the focus of the paper is on soil damping during horizontal translational motion. One quarter of the slab is modeled with one plane of symmetry and one plane of anti-symmetry. The concrete slab is modeled by four elements. A lateral ramp load with a peak value of 10 kips was applied to induce motion in the slab. The loading starts with a zero value, slowly increasing to 10 kips in 0.15 seconds, constant at 10 kips for 0.06 seconds, then a sudden drop to zero in 0.003 seconds followed by zero load for the rest of the analysis. This loading function is considered to be a good fit to the test conditions, even though the actual forcing function was not recorded. Up to 12 feet of the soil half-space are explicitly modeled by finite elements to obtain soil strains. Two layers of soil, each 3 feet thick, below the concrete slab are included, having the same discretization as the concrete slab. Below these two layers are two more layers, also 3 feet thick each, comprising of one element at each layer at the center of the slab. The reduction of soil elements with increase in depth is done to improve computational efficiency. To the sides of these soil elements and under the bottom soil element are frequency-dependent springs representing the soil half-space.

The soil shear modulus is taken as 2100 ksf, with the soil unit weight of 0.13 kcf and Poisson's ratio of 0.35. Soil material damping was taken as 2 percent. The soil shear modulus of 2100 ksf was applied to all soil layers and was obtained by trial and error until the analysis frequency matches the frequency of 20 Hz obtained from the field tests. The value of shear modulus of 2100 ksf and the material damping of 2 percent were later confirmed for compatibility with field-determined soil properties.

5 CALCULATION OF DAMPING

Total damping from the analysis, can be computed as follows:

$$D_{total} = \frac{\ln (v_n / v_{n+m})}{(2\pi m)} \quad ...[1]$$

where,
D_{total} = total damping (percent of critical damping)
v_n = peak velocity, cycle n, during free vibration from SASSI analysis
v_{n+m} = peak velocity, cycle n + m, during free vibration, from SASSI analysis

255

To obtain an average damping value, applicable over several cycles, a logarithmic decay curve for an ideally damped single degree of freedom oscillator can be fitted to the time history of response. This curve can be expressed with the following formula (Clough and Penzien, 1975):

$$v = p e^{-D w t}, \qquad \qquad ...[2]$$

where,
v = velocity at time 't'
p = constant, found by best fit
D = percentage of critical damping, found by best fit
w = frequency (in radians per second)
t = time in seconds

Equations [1] and [2] are generally expressed in terms of displacement rather than velocity, however since the slab can be treated as a single degree of freedom system in the horizontal direction with frequency 'w' (radians/second) the displacement term is substituted by the velocity term 'v'. This implies that the velocity time history has the same rate of decay as the displacement time history.

Using the horizontal direction velocity time history obtained from the SASSI output, at the center of the slab, the total damping value was calculated by using the log decay approach. As discussed above, we assume that the total amount of damping is the algebraic sum of the material and radiation amounts of damping. The average total damping is calculated by enveloping the velocity time history by the exponential decay curve, over the first few cycles of motion.

6 STRAIN-COMPATIBILE PROPERTIES

In order to validate the correctness of soil parameters used, the analytically computed soil strains were used to calculate strain compatible properties. The time histories of soil strains for all soil elements were calculated in the SASSI analysis. The effective values were calculated from the peak values in the analysis by multiplying by 0.65, to approximate the effective maximum shear strain. The values are averaged for each soil layer. The material damping values are shown in Figure 2. Table 1 summarizes the effective soil shear strain at the two top layers.

Table 1. Strain-compatible properties

Layer	Depth (ft)	Effective shear strain (10^{-3}%)	Strain-compatible material damping (%)	Strain-compatible shear modulus (ksf)
1	0.0 to -3.0	1.4	4.0	1900
2	-3.0 to -6.0	0.5	3.0	3200

From the strain-compatible properties shown in Table 1, assumed values of shear modulus of 2100 ksf and assumed soil material damping of 2 percent appear to be reasonably accurate. Owing to the low values of strain, iteration on soil properties was felt to be not necessary, as properties are not largely changed. Damping results are not expected to change significantly with subsequent iterations.

7 RESULTS AND CONCLUSIONS

Assumed soil parameters match reasonably well with strain-compatible properties, and the SASSI analytical frequency of about 20.5 Hz closely matches the field-observed frequency of 20 Hz.

Test damping values for the translational mode are observed to vary with different tests. The average observed damping value is about 29%. Using the frequency-independent soil spring damping formula (Richart, et al, 1970), the damping value in the horizontal direction for this slab size is about 34%. The analytically-computed total damping based on logarithmic decay was found to be 27%. Based on the estimated soil strains, the test-compatible material damping values are expected to be in the range of 3% to 4%. The analytical material damping used was 2%. The test radiation damping is estimated as:

$$D_{radiation} = D_{total} - D_{material} = 29\% - (3\% \text{ to } 4\%) = 25\% \text{ to } 26\%$$

The velocity response of the slab from SASSI analysis and the best fit logarithmic decay are shown in Figure 4. The analytical radiation damping is calculated to be 25% (=27%–2%). A summary of the results are provided in Table 2.

Table 2. Comparison of test and analytical results

	Test results	Analytical (SASSI) results
Frequency	20 Hz	20.5 Hz
Total damping	29%	27%
Material damping	3-4%,	2%
Radiation damping	25-26%	25%

This simple test versus analysis confirms that soil radiation damping can be accurately modelled using finite element programs. More importantly, this validation example confirms that high soil radiation damping does in fact occur in large foundations.

REFERENCES

Seed, H.B., Idriss, I.M. 1970. Soil moduli and damping factors for dynamic response analysis. Report No. EERC 70-10. University of California, Berkeley.
Woodward-McNeill Associates, 1974. Development of soil-structure interaction parameters, proposed units 2 and 3, San Onofre Nuclear Generating Station. Los Angeles.
Clough, R. and Penzien, J. 1975. Dynamics of Structures. University of California, Berkeley: McGraw-Hill.
Newmark, N. and Rosenbleuth, E. 1971. Fundamentals of Earthquake Engineering. Englewood Cliffs, N.J.: Prentice-Hall, Inc.
Lysmer, J., Tabatabaie, M., Tajirian, F., Vahdani, S ., Ostadan, F. 1981. SASSI-A System for Analysis of Soil-Structure Interaction. Report No. UCB/GT/81-02, Geotechnical Engineering, University of California, Berkeley.
Richart, F.E., Jr., Hall, J.R., Jr., Woods, R.D. 1970. Vibrations of Soils and Foundations. Englewood Cliffs, N.J.: Prentice-Hall, Inc.

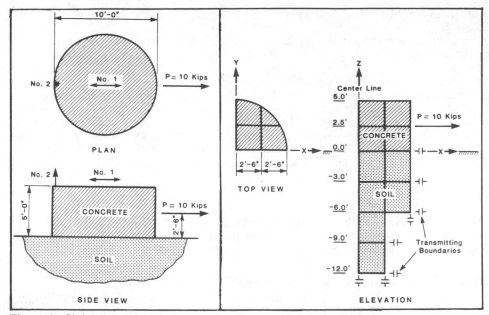

Figure 1. Slab geometry
and instrument locations

Figure 3. One-quarter finite element model

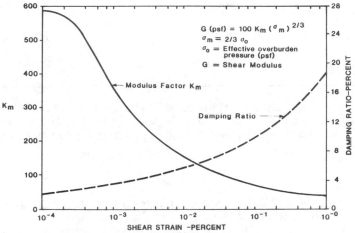

Figure 2. Modulus factor K_m and damping
ratio vs. shear strain (San Mateo sand)

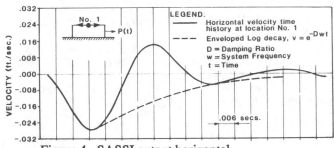

Figure 4. SASSI output horizontal
velocity response at slab center

L Inelastic behaviour of metals and constitutive equations

A constitutive model for the inelastic cyclic behaviour of beam sections

F.Brancaleoni & C.Valente

Department of Structural and Geotechnical Engineering, Rome, Italy

1 ABSTRACT

A constitutive model devoted to the prediction of the inelastic response of beam sections subjected to complex histories of imposed deformation has been developed.

The analytical formulation is given within the framework
of nonconventional plasticity of two surfaces type, restated
in terms of stress resultants and sectional deformation.

Features of the proposed model are: the allowance for a
smooth elastic-plastic transition by means of a plastic
stiffness assumed to be a nonlinear function of the distance
between the surfaces at the loading point and the capability
to reproduce the evolving elastic range via the hypothesis
of overlapping surfaces.

The intended application is herein specialized to the case
of a plane metal section subjected to coupled effect of
axial load and bending moment.

Test comparisons have been obtained using a different
higher level numerical model of "sublayer" type.

2 INTRODUCTION

Wide attention is given in the literature to the analysis of
nonlinear structures and topic in this field is the modeling
of behaviour of beam sections. The adopted models can be
divided into three classes (Meyer 1981): those based on
relations of global type (e.g. Kaldjian 1967, Baber 1979);
the so called "sublayer" or "semi-finite element" models
relative to a discretization of the section (e.g. Baron
1969, Toma 1983, Cook 1985) and those derived from the concept of interaction surfaces. Models of the first type are
practicable as long as a single stress resultant component
is involved, while the seconds give in any case excellent
results but at cost of large computational effort. A com-
promise can be achieved by means of model of the third kind.
From a classical plasticity point of view the interaction
surfaces are given the rôle of yielding surfaces in the
generalized stress space and if only monotonic proportional

loading paths are considered, acceptable results are obtained. Little attention has apparently been devoted to using more sophisticated descriptions of the behaviour of beam sections where more complex, nonproportional and cyclic, loading paths are involved. In these cases a gradual (rather than an abrupt) transition from elastic to plastic states should be modeled together with a more realistic and complex treatment of the interaction (e.g.Brancaleoni 1984).

It is the aim of the present work to propose a simple model to predict the response of a metal beam section under complex histories of imposed deformations. In particular reference is made to a formulation of nonconventional plasticity of the two surfaces type restated in terms of stress resultants and associated sectional deformations.

The model description is specialized and fully developed in the case of a plane section subjected to coupled effect of axial load and bending moment. Due to the lack of experimental results an alternative solution procedure has been further developed. This was produced using a different higher level numerical model based on the discretization of the section. The results so obtained are herein referred as the "exact" solution and are used to supply the basic features to build in the proposed model and to constitute the necessary test comparisons.

3 NUMERICAL EXPERIMENTS

The numerical investigation has been performed via a "sublayer" model where each layer has been given a stress-strain relation of elastic-perfectly plastic type; within this assumption a 20 layer partition of the section has been shown to be sufficiently accurate for the present purposes. Histories of imposed deformations have been considered so that stress resultants follow by integration over the section. Different shaped sections have been studied with explicit reference to the couple of axial force N and bending moment M. Fig.1 shows typical interaction cuves referred to initial yielding (D_y) and limit behavior (D_l) halved by symmetry; two extreme cases ($\mathbb{I}$, $\blacksquare$) and the one adopted in the application ($\bigcirc$) are drawn normalized with respect to the proper yielding (N_y,M_y) or limit (N_l,M_l) values.

As obvious, irrespective of the shape of the section, the virgin elastic nucleus is always bounded by the straight lines $|M| = |1-a|\ N_y(2\varrho^2/h)$, where $a=N/N_y$ varies between 0 and 1; ϱ is the inertia gyrator and h the height of the section. On the contrary, apart the increment in pure bending strenght, the shape of D_l depends on the shape of the section and expands until a parabola in the ideal case of a rectangular one. D_l is fixed in the generalized stress space and bounds all the stress points attainable by the section that can have either elastic or elastic-plastic nature depending on the evolution of D_y. To this end one must note, fig.2, that different loading paths ending at a same point give rise to the same D_y; it will therefore suffice to discuss only radial loading paths located in $(N/N_y)^+$ and

$(M/M_y)^+$. For self containedness and in view of the applications the results are limited to a tubular section but they
apply as well in any other case.

Indicating by ß the ε-normalized slope in the plane ε,
$\chi(h/2)$ (note ß equals the $N-M[h/(2\varrho^2)]$ slope in the elastic
range) five loading paths are of interest:

1. ß = 90°, fig.3: the section always remains elastic until is istantaneously and totally plastified $N_y=N_l$ (note
also that this is the only accessible point of D_l); D_y
remains unchanged.

2. ß = 0°, fig.3: the stress point stably tends towards M_l
(reachable only in the limit),with the transition sharpness
(M_l/M_y ratio) depending on the section. D_y results both
strecthed and translated towards M_l; a progressive symmetric
reduction in size is observed along the N direction.

3.-4. 40°~ß~90° or 0°~ß~30°, fig.4: in any case, after the
initial radial elastic path, the stress point deviates
towards N_l or M_l according to the prevailing stress component (note that this happens whatever the loading path,
not only for the radial one). Depending on the ductility
value, through ß, the stress point continues in this fashion
until N_l is reached or M_l is approached. In the first case
D_y moves along N and flattens according to the load direction (continued loading until N_y resembles the virgin D_y);
in the second case D_y behaves as for ß=0° but a flattening
in the load direction is observed.

5. 30°~ß~40°, fig.4: this range corresponds to more or
less accentuated transition phenomena resulting in an initial loss in the M component quickly followed by a recover
to return to the features observed in the case 0°~ß~30°.

Stabilized cycles of imposed deformation result in closed
cycles of stress resultants, figs.2,4 and therefore D_y is
always recovered so that shakedown of the section must be
expected, while the effects of a monotonic axial load in the
shape and size of D_y are clearly depicted in fig.3.

4 QUALITATIVE DESCRIPTION OF THE MODEL

The above numerical investigaton emphasizes how, in order to
model properly the behaviour of a section, two main topics
must be considered: a smooth transition in the elastic-
plastic range and an elastic domain of evolutionary type.

As a first step towards this direction a model is proposed
in which the above features are achieved by means of: the
presence of two surfaces, one of which can be recognised as
the classical yield surface, and the use of a plastic
stiffness taken as an appropriate function of the distance
between the surfaces at the loading point.

One of the two surfaces (D_m: inner and movable) coincides,
in its starting position, with the elastic nucleus of the
section in its virgin state (D_y); once the stress point contacts D_m this acts as a classical loading surface translat-
ing according to some specific kinematic rule. The other
surface (D_f: outer and fixed) always coincides with D_l; once
the stress point contacts D_f this acts as a classical yield

surface in decribing perfectly plastic behaviour. Note the important fact that in the present formulation D_f is considered everywhere attainable. A second important point is the description of the evolution of the loading surface subjected to the constraint of remaining inside the limit surface; this feature is herein very simply simulated by means of a mechanism of rigid translation associated with the hypothesis of overlapping surfaces; consistently the elastic range results in an evolutionary domain bounded partly by D_m and in the remainder by D_f; a typical situation is shown in fig.5. Finally the elastic-plastic coupling effect between stress resultants is obtained via the consistency condition and the normality rule associated with the loading surface as above defined, while the inelastic behaviour is described by a plastic stiffness taken to be a function of the ratio between the current and initial (start of yielding) distance of the stress between D_m and D_f.

5 ANALYTICAL DESCRIPTION OF THE MODEL

The two surfaces stress space theory (Krieg 1975, Dafalias 1976) is restated in terms of the generalized section quantities $\underline{s}=[N, M(h/2\varrho^2)]$ and $\underline{e}=[\varepsilon, \chi(h/2)]$ normalized with respect to N and ε; the following surfaces are introduced:

$$F_1 = (N/N_l)^{r1} + (M/M_l)^{r2} - 1$$

$$F_2 = |(N-N_c)/N_y| + |(M-M_c)/M_y| - 1$$

F_1 is fixed in the generalized stress space and bounds all the admissible stress states; F_2 is, in turn, the bound of the virgin elastic nucleus when $\underline{s}_c = [N_c, M_c(h/2\varrho^2)] = 0$ and hardens kinematically according to the motion of its center $\underline{s}_c$. As a new feature F_1 and F_2 are allowed to overlap and the region istantaneously occupied by their intersection identifies the evolving elastic range whose contour is given the rôle of loading surface.
   The said assumptions togheter with the additive strain decomposition $\underline{\dot{e}} = \underline{\dot{e}}^e + \underline{\dot{e}}^p$ lead to the following constitutive relations which define the model:

$$\underline{\dot{e}}^e = (1/K^e) \cdot \underline{\dot{s}} \qquad ; \qquad \underline{\dot{e}}^p = \dot{L}_i(1/K^p) \cdot \underline{n}_i$$

and which apply accordingly to the simultaneous satisfaction of the following conditions (Ortiz 1985):

$$F_i \leq 0 \qquad ; \qquad \dot{L}_i \geq 0 \qquad ; \qquad F_i \cdot \dot{L}_i = 0$$

$K^e=E \cdot A$ and $K^p=H \cdot A$ are respectively the elastic and plastic modulus times the cross area A; $\underline{n}_i$ is the unit exterior normal to F_i; $\dot{L}_i=\underline{\dot{s}}_i:\underline{n}_i$ is the loading index where (:) stands for scalar product and the subscript i implies (without summation convention) F_1 or F_2 depending where the point $\underline{s}$ is on. It comes down that F_1 and F_2 harden independently and as a consequence of the overlapping hypothesis a discontinuity

point is admitted (intersection point of F_1 and F_2, fig.5)
equipping the model by an additional constraint: cases in
which both $F_1=F_2=0$ and $\dot{L}_1>0$, $\dot{L}_2>0$ are solved assuming i=1 in
the above equations; roughly speaking the dominant rôle is
given to F_1 the limit "strong" surface and F_2 is simply updated providing $F_2=0$.

The set of constitutive relations is completed by a further equation defining the motion of F_2:

$$\dot{\underline{s}}_c = K_c \cdot (\dot{e}^P : \dot{e}^P)^{\frac{1}{2}} \cdot [\underline{u}/(\underline{u}:\underline{n}_2)]$$

where $K_c=K^P$, as easy verified by $\dot{F}_2=0$, and $\underline{u}$ the unit normal
in the $(\underline{s}-\underline{s}_c)$ direction and refers to Ziegler's rule.
Standard algebraic manipulations lead to:

$$\dot{\underline{s}} = \underline{K}:\dot{\underline{e}} \qquad \text{where} \quad \underline{K} = K^e \begin{bmatrix} 1-n_1^2 K^* & -n_1 n_2 K^* \\ -n_1 n_2 K^* & 1-n_2^2 K^* \end{bmatrix}$$

and $K^* = [E/(H+E)] \cdot L^*$; $L^*=L^*(\dot{L}_i)$ being the Heavyside step
function with $L^*(0)=0$. The hardening modulus H is determined
by the initial δ_{in} and current δ distances, measured along
the $(\underline{s}-\underline{s}_c)$ direction, between the loading and limit surface
(Dafalias 1979) according to the formula:

$$H = C \cdot [d/(1-d)]$$

where C is a shape parameter to be identified and $d=\delta/\delta_{in}$,
ranging from 1 to 0, a percentage measure of the actual
plastic load δ with respect to the more recent past history
δ_{in}. Following (Dafalias 1979) d is updated at the initiation of each new loading process or during the same loading
path if an increase over unity is encountered.

6 MATERIAL PARAMETERS AND COMPARISON WITH EXPERIMENTS

1. Parameters of the proposed model are the following four
couple of quantities: the yielding (N_y,M_y) and limit (N_l,M_l)
values of N and M; the shape of the limit surface (r1,r2)
and the elastic and plastic moduli (E,H).

The constants N_y,M_y and respectively N_l,M_l are easily
evaluated by means of the stress value at conventional yield
point (σ_y) of the uniaxial $\sigma-\varepsilon$ curve; while the exponent
r1 and r2 must be fitted on the shape of the limit surface
of the examined section. Due to the uncoupled elastic
behaviour, E comes down again from the $\sigma-\varepsilon$ curve. H is
automatically evaluated providing a value for the shape
constant; referring to a pure bending test the nonlinear
differential equation of moment-plastic curvature gives:
$(1/\varrho^2 A) dM/d\chi_p = H = C \cdot [\Delta M1/\Delta M2]$ so that a curve fitting
procedure can be used to evaluate C via the integrated form:
$C = -\Delta M2 - \Delta Mlog(\Delta M2/\Delta M)$ where the Δ, against a given M,
are clearly shown in fig.6. The correction factor δ_{in}/s_r
(s_r being the radius of D_f in the $\underline{s}$ direction) is then applied to C to account for different loading paths.

2. The intended application refers to a mild steel tubular section given the following characteristcs: $E= 206 \cdot 10^3$ MN/m^2 and $\sigma_y = 0.36 \cdot 10^3$ MN/m^2. The limit surface has been choosen as a parabola ($r_1=2$ and $r_2=1$) and therefore small deviations in the response must be expected due to the noncoincidence of the real and adopted surface, figs. 7a,8a. The value $C = 1.75 \cdot 10^3$ MN/m^2 has been finally calculated averaging between first and repeated plastic load branches, fig.6.

Two very different loading histories are choosen to check the model capabilities. Case 1 ,fig.7, refers to cyclic radial path with similar rates of growth for ε and χ ; case 2 ,fig.8, refers to cycles of increasing χ with costant applied ε . In every case the "exact-E" solution and the model "response-R" are drawn in the N-M, N-ε, M-χ plane normalized by the conventional yield values.

A satisfactory behaviour can be pointed out even if a very simple form has been choosen for the shape parameter, figs. 7b,c and 8b,c but only a qualitative prediction of the elastic range can be achieved by the present model, figs. 7a,8a; as a consequence poor transition phenomena can be predicted in the elastic-plastic range if the load path is strongly varying, anyway the limit behaviour is always recovered by the presence of the fixed surface.

7 CONCLUSION

An initial study in modeling the response of metal beam sections is hehrein presented. In order to get an experimental basis a first model of "sublayer" type related to a higher numerical level has been adopted; subsequently a second model that generalizes the nonconventional plasticity of two surfaces type has been developed. The good results already obtained by this second simpler model encourage towards further work in this field.

ACKNOWLEDGMENTS

The authors would like to thank Prof. Ciampi for help and useful suggestions during the course of this research work.

REFERENCES

Meyer, C.1981. Dynamic finite element analysis of reinforced structures. IABSE Colloquium Delft. 33:65-84.
Kaldjian, M.J. 1967. Moment-curvature of beams as Ramberg Osgood functions. J. Eng. Struct. Division 93:53-65.
Baber, T.T. & Y.K. Wen. 1979. Stochastic equivalent linearization for hysteretic, degrading multisory structures. Civil Engineering Studies no.471, University of Illinois, Urbana-Champaign, Illinois.
Brancaleoni, F., V. Ciampi & R. Di Antonio. 1984. A finite element of plane beam with a nonconventional formulation of plasticity. Trans. of AIMETA 7 Trieste. 5:451-462.

Toma, S. & W.F. Chen. 1983. Cyclic inelastic analysis of
    tubular column sections. Comp. & Struct. 16:707-716.
Cook, N.E. & K.H. Gerstle. 1985. Load history effects on
    structural members. J. Eng. Struct. Division 111:628-640.
Baron, F. & M.S. Venkatesan. 1969. Inelastic response for
    arbitrary histories of loads. J. Eng. Struct. Division
    95:763-786.
Krieg, R.D. 1975. A practical two surface plasticity theory.
    J. Appl. Mech. 42:641-646.
Dafalias, Y.F. & E.P. Popov. 1976. Plastic internal variable
    formalism of cyclic plasticity. J. Appl. Mech. 98:645-651.
Ortiz, M. & E.P. Popov. 1985. Accuracy and stabilty of
    integration algorithms for elastoplastic constitutive
    relations. Int. J. Num. Meth. Eng. 21:1561-1576.

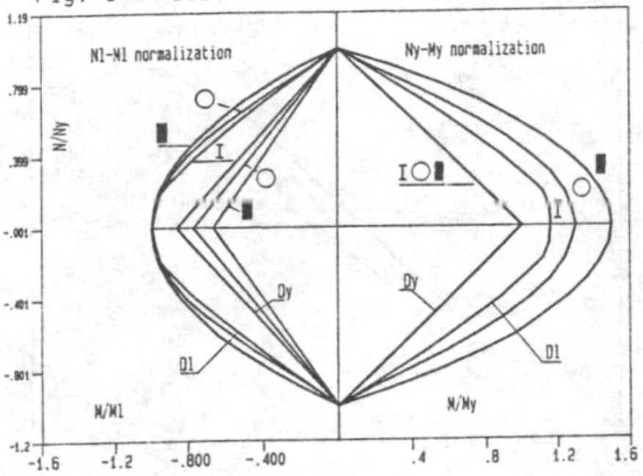

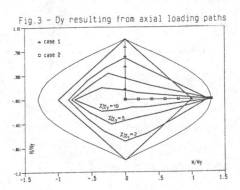

Fig. 1 - Yield and limit interaction curves

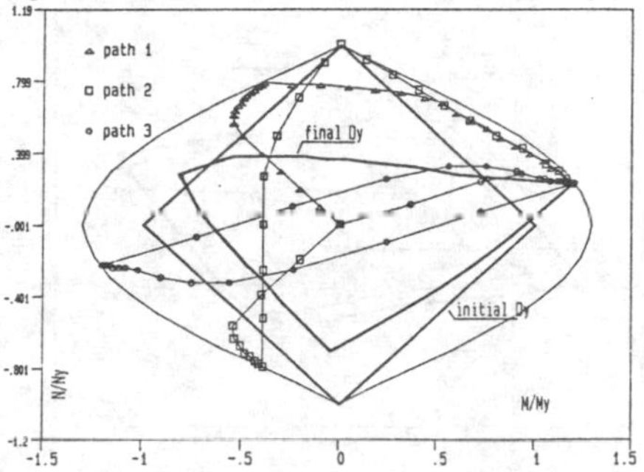

Fig. 2 - Yield surface for different loading paths

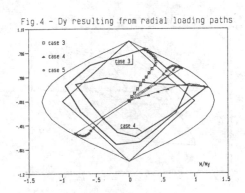

Fig.3 - Dy resulting from axial loading paths

Fig.4 - Dy resulting from radial loading paths

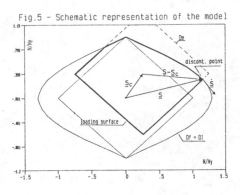

Fig.5 - Schematic representation of the model

Fig.6 - Moment-Pl.Curvature: cyclic response

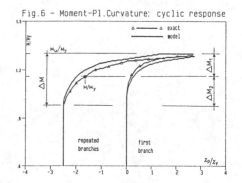

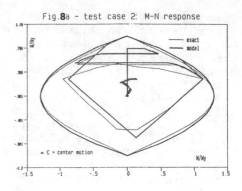

Fig.**7**a – test case 1: M–N response

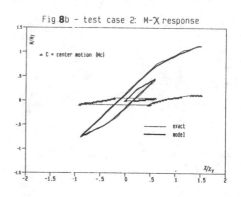

Fig.**8**a – test case 2: M–N response

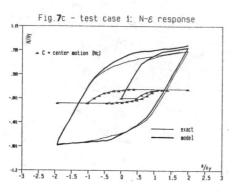

Fig.**7**b – test case 1: M–Χ response

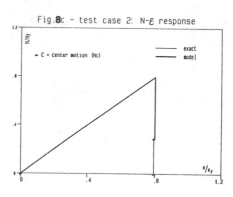

Fig.**8**b – test case 2: M–Χ response

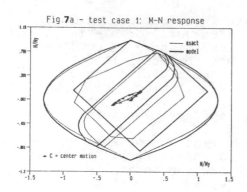

Fig.**7**c – test case 1: N–ε response

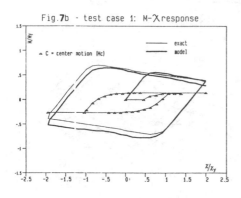

Fig.**8**c – test case 2: N–ε response

Creep-fatigue interaction of type X6CrNi1811 stainless steel in unirradiated and irradiated condition

H.-J.Lehmann

Rheinisch-Westfälischer Technischer Überwachungs-Verein e.V., Essen, FR Germany

1 OBJECTIVE OF THE INVESTIGATION

In the framework of the licensing procedure for the German fast broeder reactor SNR-300 an evaluation was conducted to determine the minimum safety margin between failure of material and the bilinear creep-fatigue envelope of ASME CODE N47 /1/. To obtain a comparable response with regard to the creep-fatigue damage, an attempt was made to analyse uniaxially loaded specimens using the ORNL model /2/ applying the same constitutive equations and average material data as is used for inelastic lifetime analyses.

For this purpose 59 creep-fatigue tests of structural steel X6CrNi1811 (similar to AISI 304ss) were evaluated.

2 SPECIMEN GEOMETRY AND PERFORMANCE OF TESTS

The usual small-size "hour-glass"-shaped specimens were used for the creep-fatigue tests. Within the measure length of the specimens the diameter is variable (R = 100 mm) and has a minimum value of 8.8 mm in mid-position (Fig. 1). The advantage of this geometry is the location of creep and fatigue at the same locus. On the other hand this geometry demands a correction of total strain range by the application of a suitable strain concentration factor.

The tests were carried out under strain-controlled conditions with a strain rate of $\dot{\varepsilon} = 3 \cdot 10^{-3}$ s^{-1}.

The hold times were inserted at maximum stress in tension. All tests were conducted at a temperature of T = 550°C. The dosis of the irradiated specimens amounts to 10^{19} n/cm^2 and $1.7 \cdot 10^{21}$ n/cm^2 (E > 0,1 MeV).

3 METHOD OF EVALUATION

The evaluation of the creep-fatigue tests was separated into following parts:
a. Calculation of the strain concentration factor
b. Calculation of the starting stress of relaxation
c. Extrapolation of short hold times to longer ones corresponding to the stationary phases of the service

d. Calculation of stress relaxation and corresponding creep damage
e. Calculation of the fatigue damage

The ORNL model is not suitable for calculation strain concentration factors for a membrane stress state governing in a test specimen.

The application of a bilinear stress-strain relationship leads to an excessive overestimation of strain if the bilinear yield strength is exceeded.

Thus the strain concentration factor was calculated with elastic-plastic analyses by application of nonlinear average stress-strain-relationships of the 10th cycle.

The strain concentration factor (SCF) depends on the total strain range and has the following values for "hour-glass"-shaped specimens.

$\Delta\varepsilon_t$ /%/	SCF
0.4	1.10
0.6	1.14
1.0	1.21
1.2	1.25
1.5	1.31

The calculation of the starting stress for the relaxation was performed, with

$$\sigma_{ss} = SY1 \cdot (1 - E_m /E) - 0.5 \cdot \Delta\varepsilon_t$$

σ_{ss} = starting stress for relaxation
SY1 = bilinear yield strength of 10th cycle
E_m = Modulus of plasticity
E = Young's modulus
$\Delta\varepsilon_t$ = total strain range

This does not give adequate agreement with the measured values.

The difference does not only depend on the bilinear hardening law chosen, but is also due to the step of 10th cycle. A step to the 10th cycle is valid provided that significant softening occurs during longer hold time periods. Because of the short hold times of the experiments there is an underestimation of the starting stress. This means that the ORNL model is not suitable for calculating the starting stress of creep-fatigue specimens if hold-times are short. Thus the measured values are taken as basis for the further evaluation steps.

To compensate for this inconvenience and to provide a relation with the ORNL model, the creep damage was calculated uniaxially in relation to the starting stress. For this purpose a reference cycle similar to a service cycle was considered. A total strain range of 0,4 %, a temperature of 550 $^\circ$C and a maximum hold time of 100 h was selected.

For comparison purposes the calculations were conducted with an average nonlinear and an average bilinear hardening law of the 10th cycle of X6CrNi1811 stainless steel.

The results are shown in Table 1.

270

Table 1

Time-independent hardening law	Hold time /h/	Cyclic stress of 10th cycle /MPa/	Creep damage D_c per load cycle
nonlinear	0,5	145,2	6.4×10^{-5}
bilinear	0,5	125,5	2.2×10^{-5}
nonlinear	10	145,2	3.5×10^{-4}
bilinear	10	125,5	1.9×10^{-4}
nonlinear	100	145,2	6.2×10^{-4}
bilinear	100	125,5	4.4×10^{-4}

The difference in creep damage dependent on the starting stress
caused by the different hardening laws increases with increasing hold
time. To determine the creep damage in relation to the scattering of
the time-independent material data, an upper and a lower limit of the
cyclic hardening for the 10th cycle was assessed from material data
for X6 CrNi 1811.

The assessment led to a value of about $\pm$ 40 % in relation to the
average behaviour. The corresponding variation of the bilinear
hardening law provides, for the reference cycle with 100 hours hold
time, the following values for creep damage per load cycle.

$$\hat{D}_c = 1 \times 10^{-3}$$

$$\overline{D}_c = 4.4 \times 10^{-4}$$

$$\overset{\vee}{D}_c = 3.1 \times 10^{-4}$$

The maximum creep damage D_c caused by the scattering of cyclic
material behaviour is about 2.3 times higher than the creep damage D_c
calculated within the inelastic lifetime analyses as based on average
material behaviour.

This uncertainty in creep damage has to be covered by the safety
margin between the failure of specimens and the creep-fatigue envelope.
An extrapolation is necessary to provide a statement for long hold
time periods, comparable with those in service. Most of the test
series were performed with maximum hold times of 30 minutes and 60
minutes.

From test series with hold times up to six hours in unirradiated
and 48 hours in irradiated condition, the cycles to failure in
relation to hold time were evaluated (Fig. 2).

The gradient of the curves can be separated into three parts. The
first one, up to a hold time of about one minute, shows a slight
decrease of the cycles to failure. This part can be allocated to
plasticity. The deformation of the material during this time interval
of relaxation is caused by dislocation movement in the grains.

In the second part, up to hold times of about 100 minutes, additional creep takes place, leading to a strong decrease in the cycles
to failure. This is the region of the creep-fatigue interaction.

271

The third part, above 100 minutes hold time, is due to pure creep. Calculations have shown that the values of the strain rates in this region are comparable with those of secondary creep. The deformation mechanism of the material is grain boundary gliding. If this deformation mechanism is dominant, the number of cycles to failure tends to limiting values, the saturation points (Fig. 2). At these points the fatigue damage remains constant.

The calculated strain rates at the saturation points are

$$\dot{\varepsilon} = 1 \times 10^{-8} \; s^{-1} \; \text{for unirradiated condition}$$

and

$$\dot{\varepsilon} = 8.6 \times 10^{-10} s^{-1} \; \text{for irradiated condition.}$$

These values correspond with hold times of 3.5 h and 30 h for unirradiated and irradiated condition respectively. For extrapolation purposes it is essential that these hold times provide a minimum creep-fatigue damage. A further increase in hold times leads to greater creep damage and an increasing safety margin, since the fatigue damage remains constant. In relation to the strain rate and deformation modus the total elongation reveals a comparable behaviour /3/ (Fig. 3). At low strain rates the elongation tends also to a minimum limit. The strain rates of this region correspond very well with those calculated for the minimum number of cycles to failure.

The calculation of stress relaxation has been based on the biexponential creep equation according to Blackburn /4/.

$$\varepsilon_c = \varepsilon_x \left[1 - \exp(-st)\right] + \varepsilon_t \left[1 - \exp(-rt)\right] + \dot{\varepsilon}_m \cdot t$$

The stress-dependent and temperature-dependent coefficients are taken from AISI 304ss /4/; they also accord with the German steel X6 CrNi 1811.

The strain-hardening rule, based on total creep-strain, in combination with the creep law has been used to calculate the variable stresses.

The calculation of stress relaxation has been conducted numerically.

The results were compared with measured relaxation curves of two heats of steel X6 CrNi 1811.

A good agreement between calculation and experiment was found (Fig. 4).

The creep damage was determined using the number of cycles to failure resulting from the experiments and average creep-rupture strength values (with multiplier 0.72 according to ASME CODE N47) of steel X6 CrNi 1811 /5/. For calculating the fatigue damage, the effective strain range was determined by multiplying the strain range with the strain concentration factor (SCF). The allowable number of cycles was determined using the fatigue curve for inelastic analysis of the ASME CODE N47, which is also in accordance with the fatigue behaviour of steel X6 CrNi 1811.

4 RESULTS

A total of 59 creep fatigue and fatigue tests were evaluated. The results show that the creep-fatigue damage rule according to ASME CODE N47 is suitable for estimating the creep-fatigue usage of structures. The minimum safety margin between the failure of the specimen and the creep-fatigue envelope amounts to about 2.5 (Fig. 5).

With arrangement of the results according to strain range, the following curvatures of creep-fatigue damage as a function of hold time can be derived (Fig. 6).

For short hold times the total damage strives towards a maximum (1-3 minutes hold time) which corresponds to plasticity. A decrease of total damage ist observable in the region of creep-fatigue inter-action with longer hold times. The minimum is reached for hold times corresponding with the above saturation points. A further increase in hold time leads to increasing creep damage with constant fatigue da-mage. This leads to increasing safety margins. The minimum safety margins correspond to following parameters:

- low strain ranges and hold times corresponding to
  saturation points
- irradiated material condition
- welded joints.

With respect to the uncertanty of creep damage (factor 2.3) in relation to the scattering of time-independent material data, further tests with welded large-scale specimens in the low strain range regime are necessary for a safe design.

REFERENCES

/1/ ASME CODE CASE N47
    Class 1 Components in Elevated Temperature Service
    Sect. III, Div. 1,
    ASME, New York 1984

/2/ Interim Guidelines for Detailed Inelastic
    Analysis of High-Temperature Reactor System Components
    ORNL-5014, Dec. 1974

/3/ B. van der Schaaf; M.I de Vries; J.D. Elen
    Effect of Irradiation on Creep-Fatigue Interaction
    of DIN 1.4948 Stainless Steel Plate and Welds
    at 823 K
    IAEA Specialist's Meeting on "Properties of Primary
    Circuit Structural Materials including Enviromental
    Effects
    Bensberg, Germany, Oct. 17/21, 1977

/4/ L.D. Blackburn
    Isochronous Stress-Strain Curves for Austenitic
    Stainless Steel
    Am. Soc. of Mech. Eng., New York, 1972

/5/ VdTÜV-Werkstoffblatt 313
    Hochwarmfester austenitischer Walz- und Schmiedestahl
    X6CrNi1811 (W.-Nr. 1.4948)
    Febr. 1982

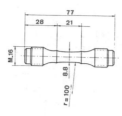

Fig. 1 Geometry of the creep-
fatigue specimens

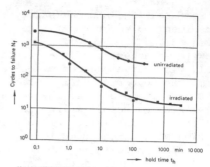

Material: X6 CrNi 1811 stainless steel, heat 3
Temperature = 550°C
■ $\Delta\varepsilon_t$ = 1,5%, Fluence: 1,7 x 10^{21} n/cm^2 (E > 0.1 MeV)
● $\Delta\varepsilon_t$ = 1,0%

Fig. 2 Cycles to failure in
relation to holdtime

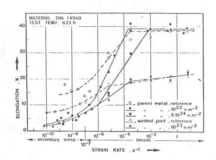

Fig. 3 Influence of strain rate
on creep elongation and
tensile elongation /3/

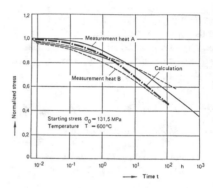

Fig. 4 Comparison of measured
and calculated
relaxation behaviour

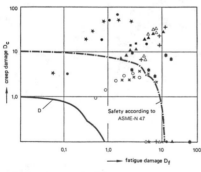

X base material, heat 1, $\Delta\varepsilon_t$ – 0.6%
● base material, heat 1, $\Delta\varepsilon_t$ – 0.6%, irradiated
○ welded joint, heat 1, $\Delta\varepsilon_t$ – 0.6%
★ welded joint, heat 1, $\Delta\varepsilon_t$ – 0.6%, irradiated
+ base material, heat 2, $\Delta\varepsilon_t$ – 0.6%
■ base material, heat 2, $\Delta\varepsilon_t$ – 0.4%
△ base material, heat 2, $\Delta\varepsilon_t$ – 1.0%
■ base material, heat 2, $\Delta\varepsilon_t$ – 0.4%
▲ base material, heat 3, $\Delta\varepsilon_t$ – 1.0%

Fig. 5 Results of the creep-
fatigue tests

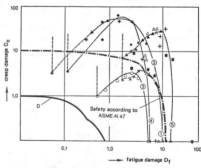

① X base material, heat 1, $\Delta\varepsilon_t$ – 0.6%
② ● base material, heat 1, $\Delta\varepsilon_t$ – 0.6%, irradiated
③ ○ welded joint, heat 1, $\Delta\varepsilon_t$ – 0.6%
④ ★ welded joint, heat 1, $\Delta\varepsilon_t$ – 0.6%, irradiated
+ base material, heat 2, $\Delta\varepsilon_t$ – 0.6%
■ base material, heat 2, $\Delta\varepsilon_t$ – 0.4%
△ base material, heat 2, $\Delta\varepsilon_t$ – 1.0%
■ base material, heat 2, $\Delta\varepsilon_t$ – 0.4%
⑤▲ base material, heat 3, $\Delta\varepsilon_t$ – 1.0%

Fig. 6 Creep damage in relation to fatigue
damage with total strain range as
parameter

274

On the strain accumulation at thermal cycling with moving level of heat-transfer liquid

D.A.Gokhfeld & O.F.Cherniavsky
Cheliabinsk Polytechnical Institute, USSR

1 INTRODUCTION

The problem of strain accumulation phenomenon under thermal cycling conditions has received much attention in nuclear energy industry. The possibility of this phenomenon in thin-walled tubes in particular has been taken into account in the Design Code [1] as far as in the end of the sixties. As the operating temperature and temperature rates in the new types of energetical plants having liquid-metallic heat-transfer agent increase, the problem requires more general and stringent analysis. It is essential to remember that strain accumulation is possible even at negligible mechanical loads [2]. And as it was shown in the reference the quasi-stationary temperature fields with respect to the coordinate system moving along with the heat source have extremum properties in the mentioned sense. The actual temperature fields in some fast reactor core components under certain conditions prove to be close to the quasi-stationary ones. This increases the probability of their geometry changes which may clear ly endanger the serviceability of a structure and lead to its collapse. Some results of theoretical analysis in this field and corresponding experimental data are discussed in the paper.

2 APPROACHES TO THE PROBLEM

There are two main approaches to the deformation state

analysis of structures subjected to temperature (and load as well) fluctuations. The most classical approach consists in detailed step-by-step study of stress, strain and displacement evolution due to the given loading history. As it was shown [3] the proposed recently rational version of the structural model based on the well known Masing scheme has many essential advantages. In spite of its relative simplicity it permits one to reflect various deformation processes – isothermal and non-isothermal, proportional and non-proportional, plastic and creep in rather good agreement with the experimental data. But if the step-by-step procedure is employed the calculations remain rather time consuming even when modern computers are used. At creep conditions cyclic deformation processes stabilize slowly and there are some difficulties when employing the extrapolation procedure [4].

3 THE SHAKEDOWN THEORY

Another approach to the problem in question consists in application of limit analysis methods [2]. As the limit analysis is based on the more simple elastic-perfectly plastic model of medium the corresponding predictions do not pretend to be exact. But the qualitative definiteness of the results obtained and simplicity of the calculation procedures make this approach more convenient for engineers especially at the first stages of the design work.

The approximate methods of elastic shakedown theory are the most simple and clear. They permit one to determine the lower and upper bounds for the limit conditions concerning the beginning of strain accumulation processes. In particular, the corresponding lower bound for axisimmetric plates and shells can be obtained using the modificated Melan theorem relationships [2]:

$$\int_{-h}^{h} \min_{\tau}(\sigma_S - \sigma_i^{(e)})dz \geqslant 0, \quad \int_{-h}^{h} \max_{\tau}(-\sigma_S - \sigma_i^{(e)})dz \leqslant 0,$$

$$\int_{0}^{h} \min_{\tau}[(\sigma_S - \sigma_i^{(e)})z]dz + \int_{-h}^{0} \min_{\tau}[(-\sigma_S - \sigma_i^{(e)})z]dz \geqslant 0,$$

$$\int_0^h \max_\tau [(-\sigma_s - \sigma_i^{(e)}) z dz + \int_{-h}^0 \max_\tau [(\sigma_s - \sigma_i^{(e)}) z] dz \leqslant 0,$$

where $\sigma_i^{(e)}$ $(i=1,2)$ is the circumferential or meridio-
nal elastic stress due to the given loading history, σ_s
is the yield stress depending on duration of hold-time
periods in loading cycle and current temperature, $2h$ is
the shell thickness, τ is the current time.

The upper bound condition can be obtained making use of
the approximate kinematical method [2].

The shakedown analysis reflects the main qualitative
regularities of the ratchetting processes such as non-
simultaneous (i.e. not isochronous) inelastic deformation
in different parts of the body. Under conditions realiza-
tion of the kinematically admissible mechanism can include
both plastic and creep deformation processes.

It is worth noting that the truth of the calculations
depends essentially on the adopted value of yield stress
correspondingly to the perfectly plastic schematization
of stress-strain diagram. Thus the properties of the material must be investigated in loading conditions close
to the real ones.

4 REACTOR CORE SHELL ANALYSIS

Let us consider the shakedown methods application for a
shell which is an important core component of the fast
reactor BN-600. The temperature fields of the shell operating in the zone of liquid natrium level are given in
Figure 1. When the fast reactor is scrammed by the safe-
guard system the natrium temperature decreases rapidly.
At the same time lowering of its level takes place due
to alteration of the liquid volume. The distribution of
the average temperatures in the shell cross-sections in
successive time moments after reactor has been screamed
is illustrated by Figure 1b. Figure 1c is showing varying of the differencies between the middle and outer

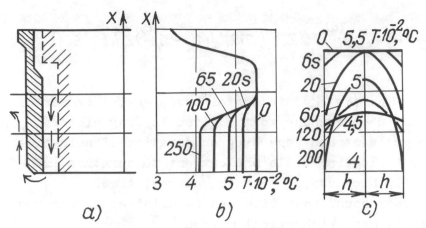

Figure 1. Temperature distribution in the shell; a —sketch
of reactor core shell; b - average temperatures; c - temperature distribution in a section.

surfaces temperatures of the shell. It is distinctly seen
that the zone of maximum axial temperature gradient is
moving along the shell axis. The maximum thermal stress
distribution in the shell is not isochronous one and this
is the necessary condition for stress accumulation at
thermal cycling. The calculations have been made to determine the value of safety factor. It turns to be equal to
1.3-1.5. If the thermal gradient will surpass the value
then in the zone of the shell within the limits of the
natrium level fluctuations a corrugation will be formed
 5 .

5 EXPERIMENTAL VERIFICATION
The shakedown analysis made above can be verified by special tests. Thin-walled tubular specimens were subjected
to repeated actions of quasi-stationary axisymmetric temperature field. They were made of low   carbon
steels and chromium manganese-silicon steel. The specimen
outer diameter was 69 mm and the wall thickness of 1 or
in some cases 2 mm.
  A specially designed experimental set (Figure 2) has a
high-frequency induction plant of 25 kw power. The set
is motor-operated (1) to rotate the specimen (2) slowly

and to realize its reciprocating motion along the axis.
The automatic equipment controls the testing prescribed
regimes (4), speed of specimen's reciprocating motion (5)
its maximum temperature (6) and variation of its diameter (7) during a heating-cooling cycle. Instantaneous
specimen temperature fields at various heating intensity
are illustrated schematically by Figure 3. Each cycle
consisted of the downward passage of the tubular specimen
with a suitable velocity enabling it to reach the prescribed maximum temperature and of rapid upward passage
during which the induction heater is de-energized. When
moving downward the specimen's temperature distribution
becomes quasi-static one (vigorously, in its part which
is remouted from the edges). The form of the ensuring
thermal wave can be varying by changing the velocity and
the distance between the heater and the level of cooling
liquid. The temperature distribution along specimen's
thickness was close to the uniform one.

The experimental data obtained are close to results of
the theoretical analysis. They show that the type of realizing ratchetting mechanism depends on asymmetry of
the thermoelastic stress cycle which is defined by the
form of the generated thermal wave. The character of the
yield stress temperature dependence plays an essential
role as well. For example, at corresponding values of
temperature gradients the temperature fields of the type
shown in Figure 3 lead to progressive reduction of specimen's diameter (Figure 4). Under another conditions such
type of temperature field can be formed (Figure 5) that
each passage of specimen will lead to increase its diameter (see Figure 6, the curves obtained for specimen's
of low-carbon steels). But in the case of the considered
alloyed steel due to the character of its yield stress-
temperature dependence the latter direction of the deformation process have not been realized.

It is worth mentioning that at relatively low temperature level the cyclic strain increments ceases after certain

279

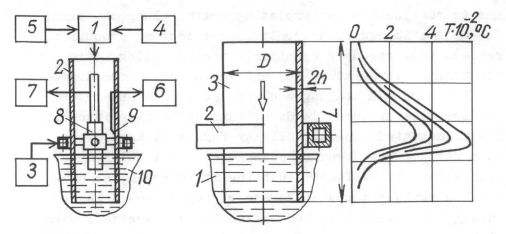

Figure 2. Block di-
agram of the experi-
mental set.

Figure 3. Instantaneous shell temperatures at different intensity of
heating.

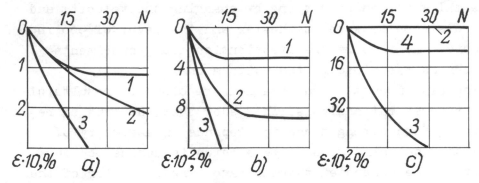

Figure 4. Strain accumulation in shells. a,b – low-carbon steels, c – low-alloyed steel (1-t_{max}=350°C, 2-400°,
3-600°, 4-500°).

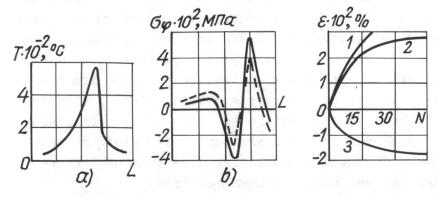

Figure 5. Temperature field (a) and
circumferential stresses distribution
(b), 1-inner, 2-outer surface.

Figure 6. Strain ac-
cumulation. 1,2 –
low-carbon steels,
3-low-alloyed steel.

number of passages. It is due to formation of corresponding residual stress field and strain hardening effect.
At higher temperature levels the ratchetting stabilizes
and finite strain increments become independent of the
cyclic number. It should be noted that coinsidence between experimental data and the theoretical results is
not only qualitative but quantitative as well.

6 CONCLUSION REMARKS
The analysis shows that the cyclic temperature stresses
if not local ones have to be attributed to the general
stress category as they can lead to essential changes in
geometry of a structure. Now the stress classification
recommended in [1] does not take into account this possibility.

REFERENCES

1. Rules for Construction of Nuclear Vessels. ASME Boiler
   and Pressure Vessel Code, Section III, New York, 1968.
2. Gokhfeld D.A. and Cherniavsky O.F. Limit analysis of
   structures at thermal cycling. Sijthoff and Noordhoff
   Int. Publ. Alphen aan den Rijn – The Nethelands, Rock-
   ville – USA, 1980, 536 p.
3. Gokhfeld D.A. and Sadakov O.S. Plasticity and creep of
   structural elements under repeated loading, in Russian.
   Mashinostrojenije, Moscow, 1984, 256 pp.
4. Gokhfeld D.A., Madudin V.N., Sadakov O.S., Cherniavsky
   O.F. Strain accumulation analysis under cyclic thermal
   actions at creep conditions, in Russian. Problemy proch
   nosti (Journal of Strength Problems, Acad. of Sciences
   of Ukr.SSR), Nr.9, 1980, pp.8-12.
5. Gokhfeld D.A., Cherniavsky O.F., Zhukov V.V., Tatarsky
   Yu.N. On conditions for incremental collapse realizing
   in reactor core components, in Russian. Prikladnyje
   problemy prochnosti i plastichnosti (Applied problems
   of strength and plasticity), Gorky State University,
   Gorky, 1983, pp.20-26.

M Structural reliability – Probabilistic safety assessment

Design of an expert system to assess structural damage in nuclear primary piping

S.F.Garribba
Istituto di Ingegneria Nucleare, Politecnico di Milano, Italy
A.A.Lucia
Joint Research Centre, European Community, Ispra, Italy

1 INTRODUCTION

The research efforts on expert systems and their scopes have rapidly increased during the last few years. The interest in these systems is due to their success in dealing with real-world problems by properly incorporating all the knowledge which may become available (Buchanan 1983, Hayes-Roth 1983). Through this process prediction capability may improve and specialize in time. Particularly, use of expert systems to assess damage and residual life of structural components would permit the continued use of plants that might otherwise be retired from service unnecessarily (Ogawa, 1985). Given these prospects, a project has been initiated to develop an expert system which is able to model the damage of structural components from the primary piping of LWRs during their operational life. Corrective measures could then be devised for timely, safe and economic repair and/or for improving the knowledge about the damaging mechanisms.

In principle, the expert system which is now in its planning stage would have a threefold goal. It could help nuclear power plant operators in the interpretation of the state of damage of primary piping and in the selection and planning of the best corrective or repair action. Next, the expert system could provide a tool for audits, since it would offer a coherent and comprehensive frame to deal with the several sources of approximation which are encountered in structural diagnosis. A special importance indeed, is attached to the interpretation of imprecise and/or uncertain knowledge and the optimum combination of varying degrees of evidence. Finally, the system may be adopted as a means for collecting and organizing information, so as to allow the validation of the theoretical models that are adopted for understanding structural behavior. Although the expert system would be designed to fit nuclear applications, it is apparent that the underlying concepts might also find use to assess performance and remaining life for a larger class of structural components.

The expert system roughly consists of a procedure for inferring intermediate or definitive conclusions on structural damage, basing on the domain of knowledge and the accumulating input data (like load history and non-destructive inspection). In this procedure flexibility and effectiveness are the two major requirements. For sake of flexibility, the expert system must allow updates of the knowledge base without demanding its complete restructuring. The requirement leads to a distributed architecture, where several expert modules or blocks are connected through a communication and control network. The expert modules collect and process information, allow access to the knowledge base in order to make its contents understandable to the user. For sake of effectiveness, a criterion of optimization exists to focus the attention of the expert system on a particular portion of knowledge and to exploit the selected pieces of knowledge in an efficient way. An overall co-ordinating function is foreseen to insure coherence in the degree of approximation and detail which can be adopted by the different expert modules.

In the effort towards designing and constructing the expert system four main steps are envisaged. First, are the design of the architecture of the expert system and the identification of the main functions which must be assigned to the different expert modules or blocks. Second, is the definition of the knowledge mechanisms and their representation in terms of rules and metarules. Third, is the preparation of prototypes for the expert modules and their testing. As the fourth step, are the design and enactment of a capability for controlling the interconnections among the expert modules to make information flows compatible and to optimize the operation of the whole system.

2 ARCHITECTURE OF THE EXPERT SYSTEM

EXULT (Expert System for Structural Ultimate Life-Time prediction and management) is an expert system that relies upon the generic information concerned with the design of the structural component, its manufacturing and preliminary testing process. Furthermore, use is made of the specific information which is continuously accumulated during component operation and maintenance. All information is retrieved and evaluated. This practice offers the opportunity to enhance and finalize the forecasts about the damage of the structural component and its remaining life, while time progresses. Forecasts would be therefore more reliable as their meaning appears to be more critical.

As in any other expert system, the development environment of EXULT can be thought of as an empty shell into which the knowledge engineer and the domain expert insert their own rules to define the knowledge base. They choose a reasoning method to select the inference mechanisms that will apply the rules (Hayes-Roth 1983). The consultation environment includes a set of fundamental inference processing functions, semantic analyzer and temporary archives (Figure 1).

The inference mechanisms of EXULT underlie an organization made of a number of modules or blocks that are linked as in the scheme of Figure 2. User or analyst can exploit interactively the functions performed by the modules of EXULT (and by their combinations) for the tasks of diagnosis and planning. In each module the rules and decision criteria can be modified under a set of metarules. Since rules and decision criteria are made manifest, comparisons become possible with lifetime predictions made by other groups of analysts.

The modular array of the expert system seems to allow an easier representation of the base of knowledge, and to permit an incremental construction of the system. More specifically, (i) the number of expert functions introduced in the system can be increased (or decreased) without requiring modifications in the overall archite--

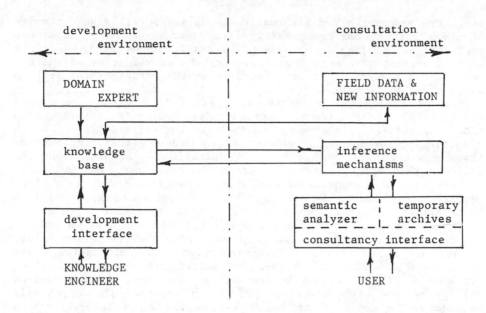

Figure 1. Main roles in the design and operation of EXULT. The obvious interactions between knowledge engineer, domain expert and potential user are not shown.

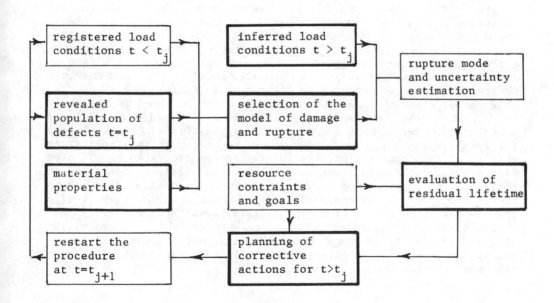

Figure 2. Architecture of EXULT displaying the main information links among the different expert modules and their dynamics. Blocks meaning expert modules are distinguished by heavy lines.

cture and procedures for information management; (ii) single modules of the expert system can possibly find separate usage in practical applications; (iii) the system can be easily inspected to make its content understandable to the user; (iv) the knowledge paradigms can be varied and adapted locally according to the nature of the problem which must be afforded.

The requirement of effectiveness refers to the sequential and hierarchical organization of functions performed by the expert system and its modules. Tools are devised to focus the attention of EXULT on a particular portion of knowledge and to tune the analysis. Metarules, semantic nets and frames are established for combining the evidence that may come from different sources. Knowledge is continuously updated. Computations can be performed iteratively at various levels of refinement. Both heuristic and formal methods are used to characterize the relationships existing among the different rules.

The aim of EXULT is to exploit the information the user enters, in conjunction with the knowledge base containing the production rules, to arrive at a decision or recommendation. The information links established in EXULT and between its expert modules may work along two models. If information is provided by the user, the information will be matched with antecedents of the rules. The rule search will commence at a point where match is found. If no information is provided, questions are asked, beginning with the antecedents in the first rule, and continuing until all antecedents of a rule are found to be valid.

On the other hand, control links are organized according to a hierarchical structure. In each expert module the highest level rules would be contained in the first control flow, lower level rules in the subsequent levels of control. Within each level, sequential control would be maintained, with branching occurring only if all antecedents in a rule are found to hold. The co-ordinating module would operate at a metalevel, in the sense that the system utilizes the metaknowledge contained in the strategy-rule base to perform a suitable control on the expert modules and to evaluate the effect of its own co-ordinating activity. The organization that ensues would reduce the number of rules which are required (Guida 1981).

Both bayesian and non-bayesian approach are proposed as the basis for updating the various bodies of evidence. Specifically, fuzzy measures of uncertainty seem to provide a powerful mathematical tool for establishing a common framework and representing the implication aspects of the reasoning process. Since the expert system should also find use for controlling routine procedures and ordinary estimates of residual life, knowledge paradigms must be suitable (i.e. deep enough) to recover approximations and defaults that characterize common engineering practice. A further advantage of fuzzy set theory is the provision for fuzzy linguistic variables. These variables indeed seem to be the instrument which deals in the most appropriate way with the adjectives employed by human experts in many circumstances (Goodman 1985).

3 THE EXPERT MODULES

In its first version EXULT would comprehend six expert modules. Nothing prevents to add further expert functions, provided that adequate resources are available to the project. In the meanwhile, the non- expert functions must rely upon conventional computer codes and procedures. The expert modules contain metarules and are allowed to establish rules, and to collect or interpret data. In the following a brief description is given of the basic functions performed by these knowledge-base modules. The description also contains hints at the main areas of approximation which ought to be dealt with and some solutions which can be proposed.

3.1 Definition of material properties

The properties of the material consist of constants and parameters that are contained in the models of damage and are used for the computation of rupture characteristics. In principle, properties come from a variety of sources and refer to testing conditions which hardly reflect the effective environment surrounding the structural component. For instance, piping material tests focus on understanding quantitatively how cracks form and propagate in piping material, leading to coolant leakage. Moreover, piping systems may be tested to failure under simulated earthquake, hydrodynamic and water hammer loads. Recurrent approximations are due to (i) inadequate weighting of data sources, or careless pooling of etherogeneous information; (ii) treatment of uncertainties effected under too stringent assumptions; (iii) disregard towards some characteristics of the material, although they are relevant in determining the physical performance of the structural component.

To contain the approximations that may arise in the proces of data combination, use could be made of the calculus of fuzzy restrictions. In this frame, data first are classified according to attributes specifying type of material, type of loading, environment and so on. Second, attributes are assigned a weight vector. This vector allows to consider the different meaning that attributes may have from the viewpoint of damage model. Then, the structural component has the material defined in terms of a degree of membership for each class of attributes. Next, a correspondence index can be defined to compare different materials by the help of a combination of weights and degrees of membership (Garribba, 1987a).

3.2 Revealed population of defects

During the useful life of the component the distribution of cracks and other imperfections is determined by the help of a non-destructive testing examination performed at a number of discrete times t_j (j = 1,2,3...). The degree of accuracy of the inspection, the distribution of space locations examined, and the concentration of time istants t_j can be decided basing upon some criteria.

A most critical issue in the validation of non-destructive inspection techniques for critical structural components is that of unrealistic criteria. A validation set may be defined which contains the minimum number of parameters sufficient to define the inspection technique.

Needless to recall that there still is a large gap between the capabilities of non-destructive examination methods in detecting discontinuities and in assessing their size and location. A first source of error connects with the technique adopted (i.e. ultrasonic testing, gamma-ray examination, etc.), the measurement chain and the standard adopted. A second type of error relates to the process of

289

signal interpretation, a process which is heavily affected by human errors. Furthermore, it must be noted that cracks and imperfections may entail complex shape and orientation. This complexity is often disregarded when data are registered in terms of few and simple geometric parameters (like crack length, crack diameter and depth). Given the limits in the available information, it can be thought of representing probabilities in terms of intervals. Alternatively, one can base upon fuzzy measures of uncertainty which seem to provide an effective tool to interpret the inspection process of the piping system, and to manage the related imprecision and uncertainty (Garribba, 1987b).

3.3 Inferred load conditions

Loading conditions which affect the structural component for time $t > t_j$ must be estimated by means of their magnitude (load size, shape, location, etc.) and frequency of occurrence. Loads include mechanical stress, temperature, chemical attack and neutron irradiation. In a generic pressure vessel, for example, operational loading may be due to vibration during normal service and pressure fluctuations. It represents a high-cycle low-intensity (internal) loading. Whereas in nuclear reactor components, low-cycle high-intensity (internal) loading may be due to accidental events like loss of feedwater, scram, loss of offsite power, failure of coolant pumps, shutdown, and so on. Moreover, among the major external hazards deserving attention are earthquakes. This is a type of loading for which extensive macro-and microseismic models should be developed for the particular site under investigation.

Potential sources of error are of several kinds. Particularly, (i) some loads are disregarded, even if they have an impact on structural behavior; (ii) computation of local stress signifies numerous approximations; (iii) representation of multiple combinations of load transients requires a suitable analytical apparatus; (iv) treatment of uncertainties may go beyond reasonable assumptions. In the inference of loading conditions, models are needed to compute the frequencies of occurrence. A rather general representation can be effected in terms of fuzzy marked random-point processes. Proper marks allow to express attributes that identify the transients. Depending on the problem, the fuzzy random process could be compressed into a simpler representation, like a marked point process having a denumerable (crisp) mark space.

3.4 Selection of the model of damage

A variety of damage patterns coexist in determining the degree of damage of the structural component and its mode of rupture. Intriguing problems are encountered because experimental evidence may be lacking, or models do not interpret experimental evidence. It is hard to provide adequate consideration for the interference among different defects and to recognize possible coupling between stress and strength. Difficulties also arise with regard to the expression of multidimensional stochastic processes that develop in space and in time.

Clearly, there are interactions among inspection, loading conditions and models for calculation of damage. Standards allow for differences in acceptance of crack and/or defect indication based upon its size and location in the cross section. Standards also allow the acceptance of cracks and/or defects greater than the permitted sizes. Credit can be taken for lower stresses around the flaw, for its shape and location, and actual material properties. The challenge is to develop the base of knowledge to perform this type of analysis.

The fundamental damage patterns which are referred to assume failure of the structural component due to propagation of cracks or defects. In principle, several crack and defect growth models are of use. Owing to the availability of (non-equivalent) crack growth models, computation may end with different results (Dufresne, 1982).

In the selection of the growth model, weight is given to data regarding material properties, and past history (i.e. registered load conditions and defects). Both imprecision and uncertainty influence defect distributions, load spectrum and material data. Therefore, it can be supposed that crack-growth may follow a hybrid treatment. The aggregation of information allows for at least two different criteria of implementation that might be called consensory aggregation and competitive aggregation (R.R. Yager 1985). If the value assigned to one constituent element of a (failure) model permits freedom of choice of the values assigned to constituent elements of other models, then a consensory aggregation is appropriate. Conversely, if the acceptance as valid of a value for one constituent element of a model affects the options open to the constituent elements of other models, the aggregation should be competitive. In such a situation the choice of a crack growth model affects the options open for the acceptance of other pieces of information. Whichever aggregation process is adopted, however, it must allow for evolution, because imprecision and uncertainty would be progressively reduced during the useful life of the component. As a consequence, the lifetime prediction based on this expert module could take advantage of new information, and the estimation capability would improve.

3.5 Evaluation of residual lifetime

An integrated life assessment methodology must consider all the different modes (and corresponding models) to estimate the remaining life of the structural component reliably and accurately. Several failure modes can indeed coexist and contribute. Erosion/corrosion usually cause primary piping weakening by processes such as thinning, pitting, and hydrogen embrittlement. Fatigue and corrosion-assisted fatigue failures and corrosion-assisted or neutron-radiation-assisted fatigue failures are detected by non-destructive examination methods. Past cyclic operation from the historical data review together with a crack-propagation-type fracture model and an anticipated future operating practice, is then used to determine the remaining life of the structural component susceptible to these failure modes (Dufresne 1982).

The lifetime of the structural component should be computed by assigning a subjective measure of uncertainty to the (up-)crossing of some safety level (or surface), for any time period of interest in t > t_j. In practice, it is assumed that each failure mode defines a division of some space in two parts separated by a smooth surface called the limit state. The two parts are the safe set and the failure set relative to the failure mode. If relevant physical variables are taken to be random variables, the survivability of the

291

structural component would be simply defined as the probability of obtaining an outcome in the safe set. A common approach consists in relying upon the theory of cumulative stochastic processes and to find for the residual lifetime values that can be more easily evaluated against the assumed goals.

Nevertheless, it may happen that limit state and physical variables cannot be easily assigned by means of precise terms. The concept of failure and the corresponding verbal descriptors are somewhat arbitrary. And probability distribution assignments to the relevant physical variables may perhaps become inconsistent because they are simply taken from analytical distributions with mathematically nice properties. In these circumstances, the crossing problem would imply the intercomparison of fuzzy random variable rather than orthodox (i.e. probabilistic) stochastic processes. Due to the complexity of the problem, ways should be looked for reducing it to more manageable, although approximate form.

3.6 Planning of corrective actions

A number of corrective actions can be devised and launched either to increase the degree of safety during a given period of time, or to prolong the life of the structural component, or to attain the two goals. If the predicted remaining life of the piping system or of its component parts is less than the user's goal (i.e. utility's target), various recommendations can be developed to extend the life of the primary piping to that goal. Groups of feasible actions consist of: repair; derating (conversely, uprating) of operational requirements; incremental (conversely, decremental) changes in the procedures of monitoring and non-destructive examination. Recommendations could also be made for the periodic monitoring of new areas and for specific record keeping, for refinements in the models adopted and further iterations in the computation. The recommended monitoring could also address to obtain temperatures, loads and other data at selected locations to improve life expectancy predictions at a later date.

Clearly, the effects of each recommendation should be quantified. Available actions are resource-constrained and ought to be conceived as an integral part of the nucelar power plant safety and reliability program. Cost effectiveness could be the criterion for ranking available options. On this basis a cost-to-benefit evaluation of competing alternatives would be performed to assist the user (i.e. the plant operator) in selecting recommendations for implementation.

4. SUMMARY AND CONCLUSIONS

The design, construction and testing of an expert system with the capabilities which have been described is an enormous task demanding the convergent and continuous effort of a large work force. This remark is one outcome of the planning activity that has been conducted so far. The final goal is at predicting the residual lifetime of primary piping of LWRs and at selecting the optimum corrective actions. Primary piping integrity and nuclear power plant availability would be enhanced if an expert system is used for continuous on-line information processing to validate measurements obtained from non-destructive examination, and to provide primary piping safety verification as well as fault identification. This

292

expert system should base upon the generic information concerned with mechanical design, manufacturing process, and the specific information accumulated during operation, maintenance and inspection.

In the present phase of the project, focus is on the definition of expert system architecture and on the simulation of some cognitive procedures that can be followed to deal with recurrent areas of error and/or approximation.

As far as the architecture is concerned, it has been decided to rely upon a distributed network of expert functions where several modules co-operate. Conversely, the means for representation of knowledge reward further investigation. A comprehensive framework is offered by fuzzy measures of uncertainty which signify a substantial departure from classical probability theory. In this respect, it is very important to examine the descriptive capability of inference models consisting of fuzzy variables, fuzzy implication and fuzzy reasoning. It is also needed to establish connections with formal and heuristic methods and procedures which are well established and widely used. There are quite a few recent theoretical studies and efforts in this direction. The key question remains open, however, of whether the increment of power offered by the new non-orthodox techniques for representation of uncertain knowedge is worthy the price.

ACKNOWLEDGMENT

Work partially performed under Study Contract no. 2827-85-11 ED awarded by the Commission of European Communities to Politecnico di Milano.

KEY WORDS

Knowledge elicitation; Structural damage and survivability; Baysian and non-bayesian updating of evidence.

REFERENCES

Buchanan, B.G. and R.O. Duda 1983. Principles of rule-based expert systems. In M.C. Yovits (ed.), Advances in computers 22, p. 163-216. New York: Academic Press.

Dufresne J., A.C. Lucia, J. Grandemange and A. Pellissier-Tanon 1982. Etude probabiliste de la rupture de cuve des réacteurs à eau sous pression. JRC European Communitites, Ispra: Rapport EUR 8682.

Garribba, S., A.C. Lucia, A. Servida and G. Volta 1987a. Fuzzy measures of uncertainty for evaluating non-destructive crack inspection. To be published in Structural Safety.

Garribba, S., E. Guagnini and P. Mussio 1987b. Uncertainty reduction techniques in an expert system for fault-tree construction. In B. Bouchon and R.R. Yager (eds.), Uncertainty in knowledge-based systems. To be published. Dordrecht: Reidel Publishing.

Goodman, I.R. and H.T. Nguyen 1985. Uncertainty models for knowledge-based systems: a unified approach to the measurement of uncertainty. Amsterdam: North-Holland.

Guida, G. and M. Somalvico 1981. Multi-problem solving: knowledge representation and system architecture. Information Processing Letters 13: 204-214.

Hayes-Roth, F., D.A. Waterman and D.B. Lenat (eds.) 1983. Building expert systems. Reading: Addison-Wesley.

Ogawa, H., K.S. Fu and J.T.P. Yao 1985. SPERIL-II. An expert system for damage assessment of existing structure. In M.M. Gupta, A. Kandel, W. Bandles, J.B. Kiszka (eds.), Approximate reasoning in expert systems, p. 731-744. Amsterdam: Elsevier Science.

Yager, R.R. 1985. Aggregating evidence using quantified statements. Information Science 36: 179-206.

Application of sampling to acceptance reinspection of nuclear plant components

M. Amin & S.L.Chu
Sargent & Lundy, Chicago, Ill., USA

1 INTRODUCTION

In recent years, the acceptance sampling technique has been extensively used in nuclear plant sites to provide an assurance (a quantifiable confidence level) that the installed items (components, supports, structures, equipment, etc.) are acceptable. This demonstration of assurance of the adequacy of installed items has often become necessary because of concerns about the potential existence of discrepancies or lack of proper documentation.

The technique of acceptance sampling has been effectively used since the 1940s to evaluate the acceptability of manufactured products (American National Standard, 1981; Duncan, 1974). When acceptance sampling is used for quality verification of installed work by reinspection, an engineering evaluation of discrepancies must be integrated into the acceptance criteria to avoid an unreasonable rejection of the work based on very conservative acceptance criteria. Other provisions must also be added at the implementation phase of sampling so that conclusions derived can be logically defended. This is particularly true because of the limited experience with the use of sampling to assess the adequacy of constructed components.

Unless the aforementioned factors are directly addressed when formulating a sampling plan for reinspection, the economy and efficiency sought by using sampling might be lost.

Application of acceptance sampling to resolution of nuclear plant construction issues has received attention in the power industry. A recent document prepared by Nuclear Construction Issues Group (NCIG) deals with the use of sampling for reinspection of welds. The document NCIG-02 (Nuclear Construction Issues Group, 1987) was reviewed by the U.S. Nuclear Regulatory Commission staff and was found to be acceptable. The use of sampling for first-line quality control inspection of installed work is being discussed in several committees, but definition of parameters, defining populations suitable for inspection, has not yet received a comprehensive review.

This paper describes the considerations necessary for selecting sampling plans to assess the quality of installed work by reinspection. Recommendations for major decisions in plan formulation and evaluation of sampling results to provide additional assurance in the implementation phase are discussed.

2 MAJOR DECISIONS IN PLAN FORMULATION

When sampling is being considered to evaluate an installed work, several basic choices must be made: (1) definition of population, (2) selection of acceptance criteria, and (3) selection of sampling plan from several commonly available types. Specific recommendations for these items are discussed in this section.

2.1 Definition of population

The definition of population is the most important step in obtaining meaningful results from a sampling inspection. Two factors are significant when defining a population: the specification of the inspection item, and supporting documentation for homogeneity of the population. The selection of the inspection item determines how population and sample sizes are to be counted. Items are usually inspected using checklists, and each entry in the checklist is an attribute with its corresponding acceptance criteria. These inspection acceptance criteria are invariably determined by using conservative assumptions. If the item being inspected has attributes found to be discrepant relative to inspection acceptance criteria, it does not necessarily mean that the item is inadequate for its intended function. This observation leads to the requirement for defining the inspection item, which must be a unit for which a margin can be logically evaluated by using conventional engineering analysis and, if necessary, testing. For purposes of clarity, the margin m is defined herein for load-bearing components. Margin can be similarly defined for other devices.
The margin m is

$$m \quad = \frac{\text{Capacity}}{\text{Load}} = \frac{CAP}{L} \qquad\qquad (\text{Eq. } 1)$$

where

CAP = credible capacity of inspection item for failure type under consideration

L   = load effect on the inspection item for the load combination of interest and measured in the same unit as CAP

For example, suppose it is necessary to use visual weld inspection to ascertain the adequacy of the welding on certain HVAC duct supports of a project. Each support contains a number of welded connections, and each connection may have several welds. For each weld several attributes must be inspected. In the engineering analysis or design, the effects of external load on the support or its connections are usually determined, and margin values for the support or individual connections of a support are reviewed. Therefore, the item being inspected in this instance is logically taken to be an HVAC duct support or support connection rather than the individual welds or weld attributes.
Population homogeneity must be documented to show that sampling results remain applicable to the population. Based upon the nature and scope of concerns raised, it must be decided whether a single or multiple population should be evaluated. For example, if the concern is limited to a particular contractor's work, e.g., structural steel erector, a single population consisting of all the structural steel

296

erected by that particular contractor can be selected as the population. On the other hand, if the concern is related to more than one
contractor, e.g., both the structural steel and the piping contractor,
more than one population should be considered.

Similarly, if the concern is limited to the work done by a few
specific craftsmen or to the installation done using a particular work
procedure, the components installed by those specific craftsmen or
installed using that particular work procedure can be included under one
single population. However, if the concerns are numerous and
nonspecific and affect the entire plant construction, the total plant
must be divided properly into many populations. In such cases, the
history of procurement, manufacture, fabrication and installation of
components should be reviewed. Major changes in organizations, specifications, work procedures, training programs, etc., provide checkpoints
that should be considered for deciding whether single or multiple
populations should be evaluated.

When formulating a sampling evaluation, it is also necessary to review
the sampling results to verify that the assumed homogeneity remains
valid. For example, if shop and field welds are considered in a single
population, sampling results should show that the two types of weld have
a similar trend of discrepancies.

2.2 Acceptance criteria

To realistically assess in-place quality, any sampling plan used for
this purpose should contain an acceptance criterion in terms of margin
of the items being inspected. A discrepant inspection item not meeting
the requirements of inspection acceptance criteria might still be adequate when actual conditions of the item and its function are
considered.

Based on strength parameters used to calculate CAP in Equation 1, two
types of margin-based acceptance criteria can be specified: safety-
significant and design-significant. For a design-significant criterion,
applicable design allowable stresses are used to calculate CAP in Equation 1. The corresponding margin is

$$m_D = \frac{CAP\ (D)}{L} \qquad \text{(Eq. 2)}$$

where

CAP (D) = capacity of inspection item based on applicable allowable
stresses

The design significant acceptance criterion for the discrepant item is
stated as

$$m_D \geq 1.0 \qquad \text{(Eq. 3)}$$

The safety-significant criterion considers a discrepant item to be
acceptable, even if design allowables are exceeded, as long as the item
can still be expected to perform its intended function. In this case,
CAP in Equation 1 is calculated by considering reasonable lower-bound
limits for excessive deformation, ductility, or capacities obtained
during destructive testing. The safety-significant margin is defined as

$$m_S = \frac{CAP\ (S)}{L} \qquad \text{(Eq. 4)}$$

297

a. Illustration of capacities and margins

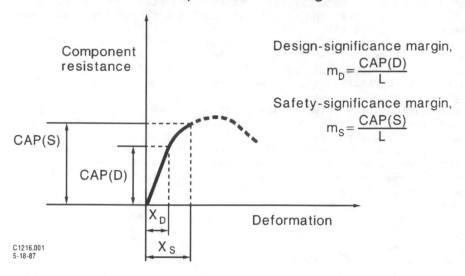

C1216.001
5-18-87

b. Illustration of acceptance criteria and low-margin condition

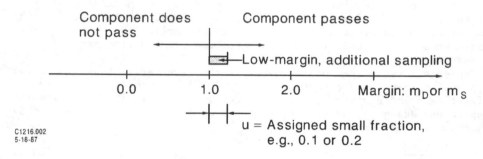

C1216.002
5-18-87

Figure 1. Margins, margin-based acceptance criteria and low-margin conditions.

where

CAP(S) =   capacity of inspection item based on reasonable lower limits of deformation or failure load

The safety-significant acceptance criterion in terms of margin is stated as

$$m_S \geq 1.0 \qquad\qquad (Eq.\ 5)$$

The margin-based criteria to be used should be specified when the sampling plan is being formulated considering the aims of the evaluation. Figure 1 shows the margins and acceptance criteria discussed above.

298

2.3 Sampling plan type

Four different types of sampling plans are commonly available for inspection by attributes: single, double, multiple, and item-by-item
sequential plans (Duncan, 1974). Table 1 lists examples of different
types of plans for a specific population size and a specific choice of
four basic sampling parameters. Plans 2 through 5 in Table 1 provide an
equivalent degree of protection as judged by their operating characteristic (OC) curves shown in Figure 2. The zero-defective acceptable
plan identified as Plan 1 in Table 1 (single sampling plan with acceptance number, $a_n = 0$, and rejection number, $r_n = 1$) is more conservative
than the other sampling plan types. See Figure 2.

Although the zero-defective acceptable sampling plan is least likely
to be used to determine acceptability of manufactured products, this
plan is considered most suitable for assessing the adequacy of a population of installed items. There are three reasons for this choice.

First, the zero-defective acceptable plan has zero for its acceptance
number; other plans have acceptance numbers greater than zero. Since
the acceptance criterion of discrepant items is to be margin-based
(safety- or design-significant as selected based on circumstances), it
is not defensible to use a plan with an acceptance number that is not
zero. Such a nonzero choice would imply that a design-significant
discrepancy, for example, may be observed in the sample and that the
population is still considered acceptable. It may be argued that not
finding a defective item in the sample does not imply that there are no
defectives in the population. The acceptable quality limits being used
in the sampling evaluations at this time are not values derived from
safety goals using system-oriented studies. Therefore, the practice of
not admitting a potential failure in the sample is prudent.

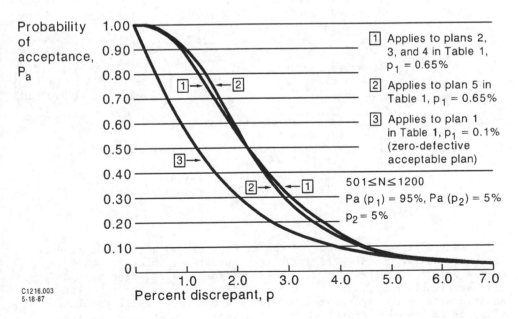

C1216.003
5-18-87

Figure 2. OC curves of sampling plans shown in Table 1.

299

Table 1. Examples for types of acceptance sampling plans: $501 \leq N \leq 1200$.

Reference No. in Discussion	Sampling Plan S	Cumulative		
		Sample Size, n	Acceptance No., a_n	Rejection No., r_n
1	Single: zero-defective acceptable	56	0	1
	$p_1 = 0.1\%$ (Duncan, 1974)			
2	Single:			
	$p_1 = 0.65\%$ (American National Standard, 1981)	125	2	3
3	Double:	80	0	3
	$p_1 = 0.65\%$ (American National Standard, 1981)	160	3	4
4	Multiple:	32	*	2
	$p_1 = 0.65\%$ (American National Standard, 1981)	64	0	3
		128	1	4
		160	2	4
		192	3	5
		224	4	5
5	Item-by-item sequential $p_1 = 0.65\%$ (Duncan, 1974)	$n > 0$	$\left(\begin{array}{c} -1.4107 \\ + \\ 0.0216n \end{array} \right)$ **	$\left(\begin{array}{c} 1.4107 \\ + \\ 0.0216n \end{array} \right)$

*Acceptance not permitted at this sample size.
**Acceptance not permitted for $n < 66$.
The four parameters: p_1 = acceptable quality level for each plan;
$Pa(p_1) = 95\%$;
p_2 = limiting quality = 5%; and $Pa(p_2) = 5\%$
(see Section 2.3 for definitions).

The second reason for choosing the zero-defective acceptable plan is
its conservative OC curve relative to other plans, which is shown in
Figure 2. The third reason is related to sample size. For zero-defective acceptable plans, sample size is determined from the familiar
equations based on hypergeometric distribution for a small population
size, or on binomial distribution for a large population size; i.e.,

$$C \quad = 1 - P_a (p_2) \qquad\qquad\qquad\qquad\qquad (Eq.\ 6)$$

$$R \quad = 1 - p_2 \qquad\qquad\qquad\qquad\qquad\qquad (Eq.\ 7)$$

$$P_a(p_2) \quad = \begin{cases} \dbinom{RN}{n} \Big/ \dbinom{N}{n} & ,\ \text{for small populations} \qquad (Eq.\ 8a) \\[2ex] R^n & ,\ \text{for large populations} \qquad (Eq.\ 8b) \end{cases}$$

where

N =   population size

n =   sample size

p_2 =   limiting quality, a maximum value of proportion discrepant in the population that is considered tolerable

R =   reliability

$P_a (p_2)$ =   probability of accepting population using sample n when proportion defective in the population is p_2

C =   confidence that is the probability of at least one discrepant item identified in the sample n

$\dbinom{y}{x}$ =   combination symbol

The relative advantage is that for the same reliability and confidence levels, the sample size associated with a zero-defective acceptable plan is smaller than those required by other plans. This is also shown by the examples in Table 1. The sample sizes for confidence of 95% and reliability of 95% are given in Table 2 as a function of the population

Table 2. Sample size for zero-defective acceptable sampling plan.

Confidence = 95%, Reliability = 95%

Population	Sample
31 or less	All
32 - 40	31
41 - 60	38
61 - 80	42
81 - 100	44
101 - 120	47
121 - 140	48
141 - 160	50
161 - 225	52
226 - 440	54
441 - 1100	56
1101 or more	58

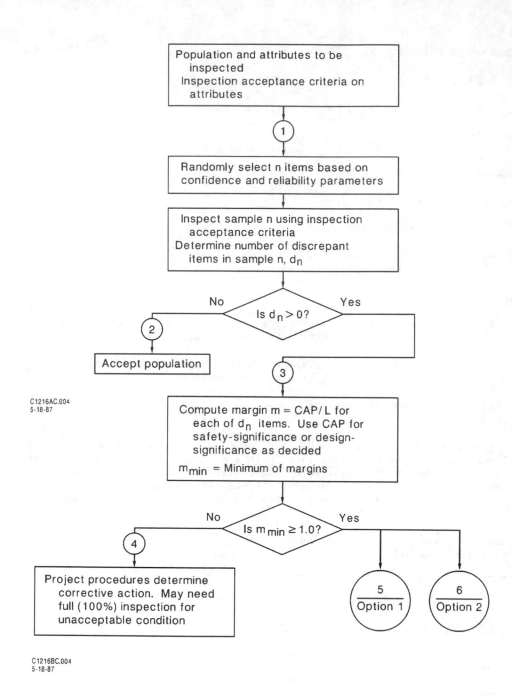

Figure 3. Flowchart for implementation of sampling to assess the adequacy of in-place quality.

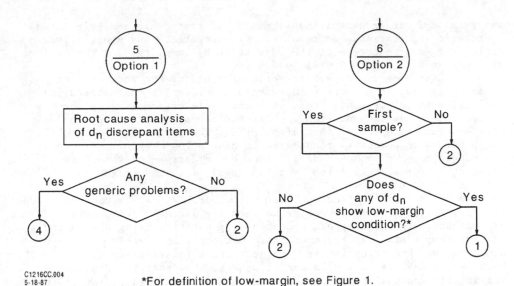

C1216CC.004
5-18-87

*For definition of low-margin, see Figure 1.

Figure 3. (Cont.)

size. These values of confidence and reliability are typically used at nuclear sites to resolve concerns by reinspection sampling.

3 IMPLEMENTATION

The flowchart for assessing the adequacy of in-place quality by sampling based on the discussion of previous sections is given in Figure 3. In addition to using the conservative zero-defective acceptable sampling plan, the procedure in Figure 3 provides an added assurance of quality, which is contained in Options 1 or 2 (locations 5 and 6) of Figure 3. The flowchart and its options are briefly described below.

If no discrepancies in the random sample n are found when using the inspection acceptance criteria ($d_n = 0$), the population is accepted at location 2 of the flowchart and sampling stops. If discrepant inspection items are found (d_n is greater than zero), a margin evaluation of these items is made; see location 3 of the flowchart. Here the capacity used for the margin evaluation depends on whether a safety-significant or a design-significant acceptance criterion is adopted. To minimize the evaluation effort, for each inspection item, the evaluation will be made with available loads and nominal resistance values. If necessary, the loading and resistance values will be refined to use actual loads and in-place resistance parameters.

If any of the discrepant items fails to meet the margin-based accep-tance criterion (location 4 in Figure 3), sampling cannot be used to as-sess quality. Project procedures will determine corrective action. This may include full (100%) inspection for the unacceptable discrepant condition and appropriate disposition of the resulting nonconform-ances. If the sample passes the margin-based acceptance criterion, one of the two options specified (location 5 or 6 in Figure 3) must be followed to further evaluate the sampling results before the population is accepted. In Option 1 (location 5 in Figure 3), a root cause analysis of d_n discrepant items is made. If root cause analysis shows no generic implication, the population is accepted. Any generic implication suggests rejection of the population by sampling, and the action of location 4 in Figure 3 will be followed.

Option 2 recommends a specific action instead of the general root cause analysis. If a certain low-margin condition is encountered in evaluating the discrepant items, one more cycle of sampling is performed. If no discrepant item in this additional sample fails the margin-based criterion of initial sampling, the population is accepted. A component is considered to have a low-margin condition if its margin m (m_D or m_S depending on whether a design- or safety-significant acceptance criterion is being used) is in the range

$$1.0 \leq m \leq (1 + u)$$

where u is an assigned fraction such as 0.1 or 0.2. While the selection of u is arbitrary, its use serves to identify narrowly acceptable discrepancies in the sample. Accepting a population using double the initial sample size, as recommended in Option 2, is clearly more conservative than accepting it with just the initial sample size. Moreover, a Bayesian analysis can be used to show that for the same reliability and confidence levels, the size of the expanded sample for a zero-defective acceptable plan, which depends on the number of low-margin conditions observed in the initial sample, is less than or equal to the initial sample size. Therefore, doubling the initial sample size is conservative.

It is noted that an arbitrary expansion of sampling beyond the original sample size, in order to search for a final decision, is known to lower the confidence associated with the selected sampling plan. The concept of sampling expansion, when a low-margin condition is observed, incorporated into Option 2, has been suitably defined to avoid this difficulty.

4. SUMMARY AND CONCLUSIONS

This paper discusses the use of acceptance sampling to ensure the adequacy of an installed engineering work when a reinspection of the in-place quality is required. The recommendations are summarized as follows:

1. While inspections are performed using checklists with many entries (attributes), the unit of inspection should be an item for which it is possible to evaluate a margin using conventional engineering analysis and testing.

2. The main reasons for considering the defined population as homogeneous should be documented before sampling. Sampling results and information should be reviewed in order to confirm the assumed population homogeneity.

3. Although items being inspected can be accepted during inspection using inspection acceptance criteria, the sampling plan must also contain a margin-based acceptance criterion to evaluate the acceptability of items that are found to be discrepant during the inspection. This avoids the potential rejection of a population on the basis of usually conservative inspection acceptance criteria. Depending on circumstance, the margin-based criterion may specify safety-significance or design-significance as the limit of acceptability of discrepant items.

4. For assessment of in-place quality, the use of zero-defective acceptable sampling plans is recommended. When the acceptance number is not zero, the defense of assurance provided by the sampling evaluation

is not currently practical because of the unavailability of values for the acceptable and the limiting quality levels derived from system-oriented studies.

5. Because of the unavailability stated in the item above, the implementation of sampling must also contain additional steps for evaluating the impact of found discrepancies on the acceptability of the population. Two alternatives are suggested for providing this additional assurance: root cause study of discrepancies, and expanded sampling when certain low-margin conditions are encountered.

REFERENCES

American National Standard, "Sampling Procedures and Tables for Inspection by Attributes," American Society for Quality Control, ANSI/ASQC Z1.4-1981 (Also known as Military Standard 105D).
Duncan, A. J., Quality Control and Industrial Statistics, Richard D. Irwin, Inc., Homewood, Illinois, 1974.
Nuclear Construction Issues Group, "Sampling Plan for Visual Reinspection of Welds," NCIG-02, Revision 2, March 24, 1987.

A design-oriented stochastic finite element method

A.Haldar & S.Mahadevan
Georgia Institute of Technology, Atlanta, USA

1 INTRODUCTION

Conventional procedures for structural analysis and design treat the parameters of a structural system consisting of loads, geometry, material properties and boundary conditions as deterministic or known constants. However, in reality, all these parameters are random; their precise values are not known. To address the uncertainty in such a problem, three sets of analyses of a system are required, at least in the nuclear industry, considering the best estimate as well as the upper and lower bound estimates of the parameters. A systematic finite element-based approach to probabilistic structural analysis leads to the stochastic finite-element method (SFEM).

The stochastic finite element method can provide additional information about the performance of the structure, such as response statistics or reliability; however, it would be more useful if such information could be used in designing the structure. Further, it is well known that a structure designed using a conventional deterministic procedure exhibits non-uniformity of risk at the element level. This basic deficiency can be removed in a reliability-based design which uses the stochastic finite element method.

2 A RELIABILITY-BASED DESIGN PROCEDURE

2.1 Trial structure and design variables

A structure designed according to a simplified deterministic procedure is used as the starting point in the reliability-based design method proposed here. In the proposed method, a probabilistic description of the variables related to loads, structural geometry, material properties and boundary conditions is necessary. Such a description can be given in terms of first and second moments and the corresponding probability density function. Usually, a complete probabilistic description of a variable is very difficult to obtain. Sometimes, a known probability distribution is assumed based on experience to represent a random variable, as a reasonable approximation.

2.2 Design criteria

In conventional deterministic procedures, the final design has to satisfy all the strength and serviceability requirements with a factor of safety. In reliability-based design, the structure is required to satisfy all the limit states with a desired level of reliability.

Let $X = \{X_1, X_2, \ldots, X_n\}^T$ be the vector of random variables of the structure. A well-known measure of structural reliability is the reliability index β, which is the minimum distance from the origin to the limit state surface $G(Y)=0$, where Y is the vector of random variables of the structure transformed to the standard normal space. The probability of failure is then given by $p_f=1-\phi(\beta)$ where ϕ is the cumulative distribution function for a standard normal variable.

2.3 Design iteration

In the proposed reliability-based design procedure, the reliability indices corresponding to every limit state are computed at each iteration. The improvement in the design is achieved by iterating with the design variables, thus ensuring that all the reliability indices are in a small narrow desired range. However, for complicated structures, the relationships between the design variables and the response variables are not available in closed form. Hence, an optimization procedure is necessary which would ensure that the design criteria be satisfied within a few systematic iterations.

3 COMPUTATION OF RELIABILITY INDICES

The reliability index β can be estimated as $\beta = \sqrt{y*^T y*}$, where y* is the point of minimum distance from the origin on a particular limit state surface. Rackwitz and Fiessler (1978) proposed a numerical iterative scheme to evaluate y* by establishing a sequence of linearization points y_i, $y_{i+1}\ldots$, according to the rule

$$y_{i+1} = [y_i^T \alpha_i + \frac{G(y_i)}{\nabla G(y_i)}] \alpha_i \tag{1}$$

where $\nabla G(y) = \{\partial G(y)/\partial y_1, \ldots \ldots \partial G(y)/\partial y_n\}^T$ is the gradient vector of the performance function and $\alpha_i = -\nabla G(y_i)/ \nabla G(y_i)$ is the unit vector normal to the limit state surface away from the origin. In many cases, the performance function G(y) may not be available in analytical form. Der Kiureghian and Ke (1985) used a linear transformation from X to Y and obtained an expression for $\nabla G(y)$ using the chain rule of differentiation. Further, the differentiation of the finite element equations after an appropriate transformation from the displacement to a general response makes it possible to compute $\nabla G(y)$ numerically.

4 OPTIMIZATION ALGORITHM FOR RELIABILITY-BASED DESIGN

Any model for structural optimization has to address the following three aspects: (i) the objective function that should be minimized or maximized; (ii) the constraints which limit the region in which optimum design can be achieved and (iii) the algorithm for finding the optimum solution. These are discussed in the following sections.

4.1 Constraints

All the constrains to be considered here are related to the reliability of the structure. In terms of reliability indices, these constraints can be expressed as

$$\beta_i \geq \beta_i^o \quad ; \quad i = 1,2,\ldots,m \tag{2}$$

where β_i^o is a preassigned minimum value for the reliability index β_i corresponding to the ith limit state, and m is the number of limit states. Alternatively, the constraints can also be expressed in terms of probability of failure, p_f . Note that, in this formulation, it is possible to specify different desired reliabilities for different limit states, since the limit states may not all have equal importance.

4.2 Objective function

In the context of the present paper, we are concerned only with a design that satisfies all the reliability constraints. Hence, the optimization problem is reduced to merely satisfying the constraints without any objective function to minimize or maximize.

4.3 Optimization algorithm

A penalty-function formulation is used here, which changes the problem under consideration into one of unconstrained optimization. Using this formulation, the objective function can be written in terms of reliability indices as

$$f(X_d) = \sum_{i=1}^{m} \alpha[\max\{(\beta_i^o - \beta_i(X_d)), 0\}]^2 \tag{3}$$

where X_d is the vector of design variables that can be manipulated and α is the penalty parameter. As the value of α is increased, greater satisfaction of the constraints is achieved. This form of the objective function ensures that when β_i is greater than β_i^o, no penalty is incurred. Hence, the minimum value of the objective function is zero when all the constraints are satisfied. The β_i's are computed through an iterative numerical procedure as described in Section 3.

For iterative design improvement, a derivative-free optimization algorithm that appears to be well-suited is the method of Hooke and Jeeves with discrete steps (Bazaraa and Shetty 1979). This method performs two types of search: local exploratory search and pattern search. Let $d_1 \ldots d_n$ be the coordinate directions and x_k be the vector of design variables at the kth iteration. Given a step size Δ, the method evaluates the objective function for a positive movement in the first coordinate direction. The move is termed a success if the move produces a lowering of the value of the objective function for this problem, and that move is accepted. If the move produces a failure, then a step in the negative coordinate direction is tried and accepted if successful. If neither a positive nor a negative step along a coordinate direction is successful, then there is no further move along that coordinate direction. This exploratory search is repeated along every coordinate direction and the (k+1)th iteration point is determined.

The method uses pattern search to accelerate convergence to the solution. The direction of movement towards the solution is indicated by $x_{k+1} - x_k$; this is multiplied by an acceleration factor to amplify the step size and to obtain the intermediate point z_{k+1}. At this point, exploratory search is performed once again to determine the next iteration point x_{k+2}. If the exploratory search at z_{k+1} results in failure in every coordinate direction, then that pattern search is discarded and exploratory search is performed at x_{k+1} to determine the next iteration point x_{k+2}.

The choice of the step size Δ is governed by the required accuracy ε. Initially, Δ is larger than ε; when the procedure stops, replace Δ by $\Delta/2$ and continue the search. Finally when $\Delta \leq \varepsilon$ and there is no further improvement, the procedure stops and the solution is reached. However, for the case of the objective function as formulated above, the procedure can be stopped as soon as the objective function reaches the value zero, at any evaluation. The acceleration factor may be taken as 1.0.

5 NUMERICAL EXAMPLE - DESIGN OF A PLANE FRAME

A frame to be designed using the proposed methodology is shown in Fig. 1. The frame is subjected to a lateral wind load P and a uniform distributed load W. The two columns are required to have identical cross-sections. There are nine random variables to be considered: P, W, I_1, A_1, I_2, A_2, R_2 and R_3, where R_1, R_2 and R_3 are the random capacities of the members. Out of these nine basic variables, the design variables are I_1, A_1, I_2, and A_2. Statistical descriptions of all the random variables are given in Table 1. The correlation coefficients between the random variables, following Der Kiureghian and Ke (1985), are assumed as follows: $\rho_{A_iA_j} = \rho_{I_iI_j} = \rho_{A_iI_j} = 0.13$ for $i \neq j$, $\rho_{I_iA_i} = 0.95$, $\rho_{A_3R_i} = \rho_{I_3R_i} = 0.5$, $\rho_{R_1R_2} = 0.8$, $\rho_{R_2R_3} = 0.7$, $\rho_{R_1R_3} = 0.9$. All other correlation coefficients are assumed to be zero.

The performance function that each member is required to satisfy is of the form

$$g_i(R,S) = 1 - \frac{F_3}{I_iR_2(1 - \frac{F_2}{R_3A_i})} \tag{4}$$

where F_2 is the axial force in member i and F_3 is the maximum bending moment in member i. This performance function, adapted from Der Kiureghian and Ke (1985), is used only for the purpose of illustration.

The steps of the reliability-based design procedure (using reliability indices) are shown in Table 2. A discrete step size $\Delta = 0.3$ and an acceleration factor of 1.0 have been used. The desired minimum reliability index β_i is assumed to be 3.0 for all i. During the iterative improvement, the variables I and A for each member are grouped together and changed simultaneously, since both would change if the section is changed.

For comparison, the same structure is designed according to standard design practices satisfying the same performance function, starting

310

from the same trial structure. The two designs are compared in Table 3. It can be observed that for the conventional design the risk at the element-level is non-uniform. Two out of the three members have reliability indices less than 3.0, indicating that they have higher risk of failure. On the other hand, the proposed methodology produced an optimum design with desired reliability for all the members.

6 CONCLUSIONS

A stochastic finite element-based structural design procedure is proposed here. The method is explained with the help of an example. It is apparent that the procedure is not only successful but also improves the design in an optimum way, making the structure more reliable. The proposed method is rational and easy to implement. It can also be extended to other structural optimization problems with different objective functions, including minimum weight or minimum cost design.

7 ACKNOWLEDGEMENT

This paper is based upon work partly supported by the National Science Foundation under Grants No. MSM-8352396, MSM-8544166, and MSM-8644348. Financial support received from the American Institute of Steel Construction, Inc., Chicago, is also appreciated. Any opinions, findings and conclusions or recommendations expressed in this publication are those of the writers and do not necessarily reflect the views of the sponsors. The writers also would like to thank Professor C.M. Shetty of Georgia Institute of Technology for a technical discussion which helped to formulate the optimization algorithm used in this paper.

REFERENCES

Bazaraa, M.S. & C.M. Shetty 1979. Nonlinear programming: theory and algorithms. New York. John Wiley & Sons.
Der Kiureghian, A. & J.-B. Ke 1985. Finite-element based reliability analysis of frame structures. Proc. 4th International Conference on Structural Safety and Reliability. p. I-395-I-404. Kobe: IASSAR.
Rackwitz, R. & B. Fiessler 1978. Structural reliability under combined load sequences. Computers and Structures. 9:489-494.

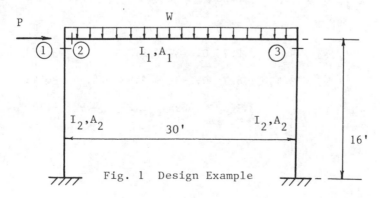

Fig. 1 Design Example

Table 1. Description of Basic Random Variables

No.	Symbol	Units	Mean	Coefficient of Variation	Description
1	P	kips	8.00	0.40	Type I
2	W	kips/ft	16.00	0.18	Lognormal
3	I_1	ft^4	1.63	0.24	Normal
4	A_1	ft^2	3.13	0.33	Normal
5	I_2	ft^4	1.33	0.12	Normal
6	A_2	ft^2	4.00	0.18	Normal
7	R_1	$kips/ft^2$	700.00	0.14	Lognormal
8	R_2	$kips/ft^2$	500.00	0.10	Lognormal
9	R_3	$kips/ft^2$	1400.00	0.11	Lognormal

Table 2. Reliability-based Design of the Plane Frame in Fig. 1

Trial No.	I_1	A_1	I_2	A_2	Reliablity Indices at points 1, 2 and 3	Objective function
1	1.63	3.13	1.33	4.0	$\beta_1=1.527$ $\beta_2=1.525$ $\beta_3=2.104$	5.148
2	1.93	3.43	1.63	4.3	$\beta_1=2.207$ $\beta_2=1.827$ $\beta_3=3.105$	2.169
Intermediate	2.53	4.03	2.23	4.9		
3	2.83	4.33	2.23	4.9	$\beta_1=3.333$ $\beta_2=3.242$ $\beta_3=4.161$	0

Table 3. Comparison of conventional and proposed design results

Member No.	Conventional method				Proposed method			
	I	A	β	Pf	I	A	β	Pf
1	1.63	4.3	2.207	0.0137	2.23	4.9	3.333	0.0004
2	1.93	3.43	1.827	0.034	2.83	4.33	3.242	0.0006
3	1.63	4.3	3.105	0.009	2.23	4.9	4.161	0.00015

Using component test data to develop failure probabilities and improve seismic performance

G.S.Holman & C.K.Chou
Lawrence Livermore National Laboratory, Calif., USA

1 INTRODUCTION

Over the past decade methods have been developed to probabilistically
assess how large earthquakes would affect nuclear power plants,
particularly the associated risk to public health and safety. These
probabilistic risk assessment (PRA) techniques combine "event trees",
which describe the postulated accident scenarios (or "initiating
events") capable of causing core melt, with "fault trees" describing
the likelihood of equipment failures leading to a reduction in or loss
of the ability of certain plant systems to perform their designated
safety functions given that an initiating event occurs. A key element
in the fault tree analysis is the "fragility" -- or likelihood of
failure -- of various components under postulated accident conditions.
    Application of these analysis techniques, both in NRC-sponsored
research such as the Seismic Safety Margins Research Program (Bohn
1983) and in commercial PRA studies, has indicated that potential
accidents initiated by large earthquakes are one of the major
contributors to public risk. However, component fragilities used in
these analyses are for the most part based on limited data --
primarily design information and results of component "qualification"
tests -and engineering judgement. The seismic design of components,
in turn, is based on code limits and NRC requirements that do not
reflect the actual capacity of a component to resist failure;
therefore, the real "seismic margin" between design conditions and
conditions actually causing failure may be quite large. These
elements combine to produce fragilities that are not only highly
uncertain but also, in the view of many experts, are overly
pessimistic descriptions of the likelihood of failure for many
components. The observed performance of mechanical and electrical
equipment in non-nuclear industrial facilities that have experienced
strong-motion earthquakes tends to support this view. However, this
same experience has also indicated that although a component may
itself perform well in an earthquake, poor or inadequate support
conditions may increase the likelihood of its "failure" in a safety
sense. This also holds true in certain cases for aging or
environmental effects, which may require attention if an "adequate"
description of fragility is to be achieved.
    In order to improve the present fragility data base and develop
realistic input for probabilistic risk assessments and seismic margin
studies, the NRC commissioned a Component Fragility Research Program

313

(CFRP). The CFRP is being conducted in two phases. Phase I comprised parallel efforts to (1) develop and demonstrate procedures for testing components to obtain new fragilities data, (2) identify by systematic ranking those components that most influence plant safety and are therefore candidates for NRC testing, and (3) compile and evaluate existing fragilities data obtained from various sources. The results of these three activities form the basis for a comprehensive evaluation of component behavior, based both on available data and on new testing, to be performed in Phase II of the program.

The CFRP supports the need for realistic inputs for probabilistic risk assessments and margin studies. This research seeks to test the hypothesis that electrical and mechanical components have greater seismic capacity that is presently assumed in seismic risk assessments, and, as a consequence, that the significance of the earthquake threat might be diminished in licensing decision making. This paper presents an approach to fragility testing which can provide the following while remaining within practical constraints on time and resources:

1. More realistic component failure probabilities for PRA applications. Improved descriptions of component fragility, based on actual failure data, will reduce the uncertainty inherent in subjective fragilities drawn from design information and results of equipment qualification testing.

2. Better understanding of component failure modes, of how various individual factors affect failure, and of the real "margin" between design or qualification requirements and conditions that might actually cause failure. This understanding also provides a basis for enhancing the seismic capacity of critical equipment through specific hardware modifications.

3. Guidance for development of seismic review procedures for existing plants, for interpretation of existing qualification or fragility data, and for specification of test procedures of equipment for which in-depth testing to "failure" is warranted.

Although testing alone cannot remove entirely the subjectivity inherent in current descriptions of component fragility, it can, if structured properly, improve our ability to more realistically assess seismic risk while at the same time contributing to elimination of unnecessary licensing delays to respond to seismic issues.

2 DEFINITION OF COMPONENT "FRAGILITY"

"Fragility" is a term commonly used to describe the conditions under which a component (or, in general, a structure, a piping system, or piece of equipment) would be expected to fail. In this paper we address seismic fragility, in other words, what levels of seismically induced input motion would be required to cause component failure; it is important to keep in mind, however, that fragility can in principle be defined for any input condition affecting component performance. Failure can be characterized as either functional (e.g., erratic behavior, failure to perform intended function) or structural, or as the exceedance of some predetermined performance criteria (such as a limit given in a design code).

One interpretation of component fragility -- which we will refer to as the "fragility level" -- evolves from qualification testing. In seismic qualification testing, a component is subjected to input motion characterized by a specified waveform describing input level

314

(seismic acceleration) as a function of frequency. The component is "qualified" if it continues to perform its intended function when its response to this input motion -- the "test response spectrum," or TRS -- meets or exceeds pre-determined acceptance limits (the "required response spectrum," or RRS). In qualification testing, the TRS is usually measured at the component support points.

Although it may establish the adequacy of a component for a particular seismic environment, a successful qualification test does not directly provide data on what input motion levels actually result in component failure. This can be (and sometimes is) done by retaining the original input spectrum and then increasing the input level until "failure" (however it is defined) occurs. The TRS at failure represents the "fragility level" of the component; the difference between the fragility level and the qualification level thus represents the seismic margin or "ruggedness" of the component.

Fragility is described differently when used for PRA purposes or for other types of probabilistic analysis. In this case, the fragility of a component represents the probability of its failure -- or more rigorously speaking, the probability of attaining a defined "limit state" -- conditioned upon the occurrence of some level of forcing or response function. It may be expressed in terms of a local response parameter (for example, input motion at the component mounting location) or can be tied to a more global forcing function such as free field peak ground acceleration (PGA). Note however that when fragility is anchored to a forcing function, the further removed the component is from that forcing function, the more factors there are (such as structural response and soil-structure interaction) that must be considered in the fragility description.

The probability of failure is typically described by a family of "fragility curves" plotted at various levels of statistical confidence (see Fig. 1). The central, or "median" function represents the fragility analyst's best estimate of the "true" fragility of the component taking into account all significant factors which, in the analyst's judgement, might contribute to failure. The central point (50% probability of failure) on this curve represents the "median capacity" of the component; ideally, this probabilistic value would correspond with the deterministic "fragility level" of the component. The fragility function is a cumulative distribution usually characterized by a log-normal function with this median value and a logarithmic standard deviation β_R which describes the "random" variation in the parameters affecting fragility. In a description of seismic fragility, for example, this parameter might represent the differences in real earthquake ground motion compared to the input motion that a component is subjected to in qualification or fragilities testing. Note that the random uncertainty controls the slope of the fragility function; the less random uncertainty, the steeper the fragility function becomes. As random uncertainty is reduced towards zero, the fragility "curve" approaches a step function with its break point at the fragility level of the component.

The 5% function and 95% function in Fig. 1 represent the "modeling uncertainty" in the median fragility function. These bounds, which may also be referred to as 5% and 95% confidence limits, are based on the assumption that there is uncertainty in the median capacity; this uncertainty is characterized by a logarithmic standard deviation β_U. Simply stated, the 95% confidence limit implies the following:

1. There is a 95% subjective probability ("confidence") that the "true" fragility function for the component would be equal to or

greater than the 95% function.

2. There is only a 5% subjective probability that the actual fragility level is less than the median capacity indicated for the 95% fragility function.

Modeling uncertainty, often described as "lack of knowledge" about the component in question, reflects the adequacy (or inadequacy) of information -- component damping values, for example -- used by the fragility analyst to form his judgements about component capacity. Thus, modeling uncertainty in fragility descriptions has a subjective rather than a "random" basis as is true in the statistical sense.

For any given component, empirically developing a statistically meaningful seismic fragility would require that a large population of identical components (e.g., several hundred or several thousand) be subjected to successively higher levels of acceleration and the distribution of failures (however "failure" is defined) be recorded as a function of acceleration level. Practical constraints on time and resources clearly make this infeasible for a single component under well-defined load conditions, let alone for the effectively infinite combinations and permutations of component type, manufacturer, mounting, and loading conditions that could be identified for actual nuclear power plants. Therefore, an alternative approach is necessary to experimentally gain an insight into fragility.

Our approach to assessing fragility takes advantage of the fact that for practical PRA applications, a limited or "lower bound" fragility description may be adequate. In a probabilistic analysis, failure occurs only when the probability distributions of response and fragility overlap; therefore, only the lower tail end of the fragility curve may be of interest from a PRA standpoint. For components having a high seismic capacity (high "ruggedness"), the overlap of the response and fragility distributions could conceivably be so small under all credible loading conditions as to imply that the probability of failure is negligibly low.

One method of developing a "lower bound" fragility is to estimate the so-called "HCLPF" (for High Confidence, Low Probability of Failure) capacity for the component. The HCLPF capacity considers both the random and modeling uncertainty in the median capacity. The definition of HCLPF used in the CFRP is that adopted by the LLNL Seismic Design Margin Program (Budnitz 1985), namely that value of the forcing or response function (such as seismic acceleration) for which we have "95% confidence" that the probability of "failure" is less than 5 percent. According to this definition, if the median capacity of a component is defined by a peak acceleration with value A, the corresponding HCLPF capacity (i.e., HCLPF acceleration) is obtained from the numerical relationship:

$$A_{HCLPF} = A \exp{[-1.65(\beta_R + \beta_U)]} \tag{1}$$

where β_R and β_U represent the random and modeling uncertainties, respectively. The median capacity A can be determined by component tests, either to actual failure or to some threshold or "cut-off" limit. The cut-off might be applied, for example, in testing certain components whose actual median capacities were significantly above any response levels of regulatory interest.

The HCLPF capacity provides a practical means of addressing variations that inevitably arise between actual plant conditions and test conditions, variations that might otherwise be difficult to parametrically quantify by testing alone. For example, the random uncertainty β_R allows for variations in real earthquake motion compared to test input motion, variations in building floor response, or (e.g., for cabinet-mounted electrical devices) random variations in cabinet response. The modeling uncertainty β_U can account for variations in real damping values, or in component mounting conditions, or in the response of functionally similar components of different size or supplied by different manufacturers. These uncertainties can be quantified by systematically structuring test conditions in the form of "sensitivity studies" to investigating the effect of various parameters on the measured median capacity of the device tested. This was the basic approach taken in our Phase I demonstration tests.

The HCLPF approach has the added advantage that, in the absence of complete fragility data, a "lower bound" fragility can still be defined for a seismically qualified component by assuming its qualification level also represents its HCLPF capacity. Engineering judgement can then be applied to estimate the uncertainty parameters and thus make inferences about the median capacity.

Note that because the HCLPF capacity by definition presumes a five percent probability of failure, while "qualification" implies no failure, this approach tends to be conservative. It may in fact be overly conservative if qualification levels are low, as would be the case for many plants in the eastern United States. However, HCLPF capacities based on "high level" qualification data -- from plants in the western United States, for example -- can provide useful lower bound fragilities for plants having relatively low design basis earthquakes. The CFRP prioritization report (Holman 1985a) describes how we used this approach to infer the actual capacity of selected electrical equipment.

In itself, the HCLPF capacity is a useful parameter on which to base regulatory decisions concerning seismic performance. However, extreme care must be exercised in selecting "reasonable" values of β_R and β_U when using a HCLPF capacity derived from qualification data to infer the actual capacity or "fragility level" of a component. The reasons for this are two-fold:

1. As shown in Fig. 2, the slope of the fragility curve becomes more shallow as random uncertainty (β_R) increases. Therefore, the resultant median capacity on the 95% curve (and, for constant β_U, the inferred fragility level) also increases with increasing random uncertainty. As shown in Fig. 3, however, if the fragility level of the component is known (e.g., from actual failure tests), then the HCLPF capacity derived from the median capacity decreases with increasing random uncertainty.

2. Similarly, as modeling uncertainty (β_U) increases, the offset between the 95% fragility function and the 50% function also increases, implying an increase in the inferred fragility level. If, on the other hand, the fragility level is known, an increase in modeling uncertainty drives the HCLPF capacity towards lower (i.e., more conservative) values.

The above exercise illustrates how a "bottom-up" approach towards estimating median capacity (i.e., inferred from HCLPF capacity) can imply that fragility level increases with uncertainty, which is clearly non-conservative. This observation suggests that the reverse approach -- basing HCLPF capacities on measured fragility levels -- is preferable for assessing seismic performance. For a given fragility level, such a "top-down" approach yields lower (i.e., more conservative) HCLPF capacities as uncertainty increases.

3 GENERAL APPROACH TO FRAGILITIES TESTING

As discussed earlier, resource and time constraints make it
impractical to develop meaningful fragility descriptions by empirical
means alone. Even for nominally identical components, variations among
models and manufacturers, and in mounting and loading conditions imply
that any fragility estimate will be based in large part on engineering
judgement. Meaningful fragilities testing can, however, be conducted
within practical constraints if it focuses on understanding how various factors influence component behavior rather than on developing
fragilities explicitly. Such testing would be conducted according to
the following steps:

   1. Identify a representative component or sample of components for
testing. Such a sample might not necessarily be "generic" in the
purest statistical sense, but should attempt to include significant
variations within a given component type (armature- vs reed-type
relays, for example).

   2. Identify failure modes and the relevant forcing or response
functions. Characterize "failure" (either functional or structural)
in terms of a suitable parameter that can be measured experimentally.

   3. Identify factors or "technical issues" affecting component
failure; such factors might include, but not necessarily be limited
to, variations in mounting, input motion, and component damping.
Design experimental program to parametrically vary those factors
judged to be important.

   4. Perform tests to identify when failure occurs, i.e., component
median capacity. Such tests might, for example, follow qualification
procedures (e.g., use of same input spectra) but at elevated input
motion levels. Alternatively (particularly for "high capacity"
components), test to a pre-determined level to establish a "lower
bound" failure threshold.

   5. Use test results to empirically estimate median capacity; note
that this result is equivalent to the deterministic "fragility level"
generated as a part of some qualification tests. Based on component
behavior over all conditions considered, develop estimates of random
variability and modeling uncertainty in the empirically derived median
capacity.

   Although this approach does not remove subjectivity from the process
of describing component fragility, it does improve the basis on which
these judgements are made. Furthermore, it can provide an improved
basis for interpreting data from other sources or for defining test
conditions if more rigorous testing of a specific component is necessary.

4 FRAGILITY TESTING OF A MOTOR CONTROL CENTER

To demonstrate our approach to fragilities testing, we performed
fragility tests on a three-column Westinghouse Five-Star motor control
center containing 8 Westinghouse motor controllers of various types
and sizes as well as 14 relays of different types and manufacturers.
The Five-Star is the current basic model marketed by Westinghouse for
industrial and power system applications; it is essentially identical
to the Type W motor control center manufactured by Westinghouse from
1965 to 1975, various configurations of which are found in many
nuclear power plants of this vintage. The particular electrical
devices selected represented a sample of standard MCC devices typical

in function of those found in actual plants, but are not necessarily generic for all similar devices.

To investigate the effect of base flexibility on the structural behavior of the MCC and on the functional behavior of the electrical devices, we conducted multiple tests on each of the following four mounting configurations:

1. Four bolts per column with top bracing.
2. Four bolts per column with no top bracing.
3. Four bolts per column with internal diagonal bracing.
4. Two bolts per column with no top or internal bracing. This was the "standard" mounting configuration recommended by the MCC supplier.

We performed a total of 56 test runs, including 43 biaxial random motion tests (vertical plus one horizontal axis). Table input motions in the random motion tests ranged up to 2.5 g zero period acceleration (ZPA), which yielded in-cabinet spectral accelerations up to 20 g and higher at the device locations. The desired response spectrum applied in each test was characteristic of qualification spectra (i.e., 250% of ZPA between 4 Hz and 15 Hz), each test run was about 45 sec in duration of which approximately 30 sec was strong motion.

In these tests we investigated both the functional behavior of the individual electrical devices -- relays and starters -- and the structural response of the MCC cabinet for various levels of table input motion and for four different cabinet mounting configurations. Device "fragility" was characterized by contact chatter correlated to local in-cabinet response measured at the device location, with functional "failure" being defined as the first sign of chatter. Among the topics investigated were (1) the relative susceptibility of normally-open and normally-closed contacts, (2) the chatter susceptibility of energized vs deenergized contacts, and (3) the ability of the devices to respond to commanded changes of state during strong seismic excitation. Details of the MCC test program are provided by our Phase I demonstration test report (Holman 1986b).

We later applied these experimental results to develop probabilistic fragility descriptions for each type of electrical device in the MCC, referencing fragility to local ZPA at the device location. In addition to "best estimate" descriptions of device fragility, we estimated the random variability and modeling uncertainty to arrive at a "high confidence, low probability of failure" (HCLPF) capacity for each type of device.

In these tests we also observed the structural behavior of the MCC cabinet for various levels of table input motion and for the different mounting configurations. Although substantial damage (cracked and broken welds) was observed in later tests, the cabinet nevertheless withstood some 20 strong motion tests at ZPA levels up to 1.9g before any significant damage was observed. In general, the results of our demonstration tests suggested the following:

1. In general, two distinct response modes can be identified for the MCC cabinet. The first of these, which we refer to as the "frame" response, reflects global motion of the MCC structure. Results of low-level (0.2g sine-sweep) transmissibility tests indicated that the cabinet frame resonance lies between about 3.5 Hz and 12 Hz, depending on cabinet mounting configuration. The second mode, typically lying between about 14 Hz and 26 Hz, reflects the local response of the individual draw-out units (or "buckets") which house the relays and starters. We refer to this as the the "bucket" response; the resonant frequency measured for each draw-out unit is referred to as its "bucket" resonance.

2. Contact "chatter" appears to be influenced more by spectral acceleration than by ZPA, although the likelihood of chatter will, on the whole, increase with ZPA. Device fragilities based on spectral accelerations should in principle be "more appropriate" but at the same time would be more difficult to apply in a PRA.

3. Chatter appears to be influenced more by low-frequency input motion (i.e., less than about 10 Hz) than by high-frequency motion. Consequently, the lower-frequency "frame" resonances will affect device performance more than higher-frequency local "rattling" of the cabinet structure. This observation was further supported by the results of supplementary single-axis sinusoidal tests on one armature-type relay and one motor starter which showed these devices to be most sensitive to input motions in the 2.5 Hz to 8 Hz range.

4. Normally-closed contacts were more prone to chatter than normally-open contacts. At no time, however, was chatter observed in energized contacts regardless of their normal (i.e., deenergized) state; consequently, deenergized normally-closed contacts were found to be most susceptible to seismically-induced chatter. Furthermore, at no time did device chatter affect the ability of the devices to respond normally to commanded changes of state; in other words, the devices performed as intended even during strong motion.

5. Reed-type relays, which we did not observe to chatter at any time during our tests, appear to be more resistant to seismic motion than the more conventional armature-type relays.

6. Top bracing of the motor control center can increase the seismic capacity of both the MCC structure (by limiting cabinet motion) and the internal electrical devices (by increasing the resonant frequency of the cabinet frame).

5 DEVELOPMENT OF DEVICE FRAGILITIES

When developing a probabilistic description of fragility, one is faced with selecting the best reference parameter to define the fragility as well as with quantifying the uncertainty in the chosen parameter. Based our data, it appears that the devices tested are most sensitive to spectral accelerations in the 2.5 to 8 Hz range. However, most fragilities are expressed in terms of ZPA, as in the Fragility Handbook (Cover 1983) developed as part of the Seismic Safety Margins Research Program, and in commercial PRAs. Zero period acceleration is also a convenient parameter commonly computed by equipment qualification engineers. For this reason, the initial focus in our MCC tests was on ZPA and less on frequency-dependent effects; we therefore selected local ZPA as the parameter on which to base "single-parameter" fragilities that would be workable in PRA applications. Note as well that fragility descriptions incorporating frequency effects would be more complex, would require detailed information on device structural characteristics and input motion frequency content, and would be accordingly more difficult to apply in probabilistic risk assessments.

Of the devices tested, sufficient data was collected to develop fragility descriptions for the following six:

1. Westinghouse Size 2 full-voltage non-reversing motor starters (auxiliary contacts only).

2. Westinghouse Size 2 full-voltage reversing motor starters (auxiliary contacts only).

3. Westinghouse Size 3 full-voltage non-reversing motor starter (auxiliary contacts only).
   4. General Electric Type CR relays.
   5. Westinghouse Type AR relays.
   6. Square-D Type X relays.

We based device fragilities on the first sign of chatter in normally-closed contacts in their deenergized state. In some runs contact chatter was observed at a lower ZPA level that associated with no chatter in other runs. In these cases, the lowest "chatter" level and the highest "no chatter" level were equally weighted and used in a regression analysis of the data. All fragilities discussed here are described as lognormal distributions, each with a median capacity expressed in terms of local ZPA and its random variability and modeling uncertainty (β_R and β_U, respectively) expressed as logarithmic standard deviations developed from the raw data.

The threshold of chatter for each device at each location and for each cabinet mounting configuration tested was used as a data point. Most devices were tested at two locations for four cabinet mounting configurations, resulting in eight data points. However, some test conditions produced no chatter; in these cases, no failure data point was defined.

The method chosen to best fit the data to a lognormal distribution was to perform a linear regression analysis on the logarithm of ZPA versus probability. The best fit of the data and the one standard deviation bounds on the best fit were plotted on probability paper for each device with the data points ordered in ascending ZPA from 1 to N data points. The probability of failure for the first data point was taken as 1/N+1, the second as 2/N+1, and so forth. The data points thus represent a cumulative distribution function from 1/N+1 to N/N+1. Figure 4 shows a typical set of fragility curves so developed, in this case for the Westinghouse Type AR relays tested. In general, the median capacity ("best estimate" curve, 50 percent failure probability) ranged from 5.2 to 6.1 g local ZPA with HCLPF capacities ranging from 3.2g to 4.3g local ZPA. When considered as a single group, these relays have a median estimated capacity of 5.6g and a HCLPF capacity of 3.6g. Keep in mind these are local in-cabinet levels, not the motion at the cabinet base. Corresponding capacities for the starters tested were similar or higher depending on starter size. Note that these results assume that input motion is oriented with the direction of contact motion; the results of our tests indicated that virtually no chatter occurs when the direction of input motion is perpendicular to that of contact motion.

6 IMPLICATIONS FOR FUTURE FRAGILITIES TESTING

As noted in the introduction to this paper, resource and time constraints make it impractical to explicitly develop probabilistic fragility descriptions by empirical means alone. Meaningful yet cost-effective fragilities testing can, however, be conducted within these constraints if it seeks not to explicitly develop "generic fragilities" broadly applicable to wide ranges of components, but rather to enhance understanding of how certain components fail ("failure modes"), what the important factors are that affect component performance, and what the relative influence of these factors is.

Testing in the form of "sensitivity studies" provides one method of gaining this understanding. In our Phase I demonstration tests, for example, we investigated MCC behavior (primarily functionability of electrical devices) through a carefully structured series of parametric sensitivity tests. The test results provided actual seismic capacities of the specific components tested, as well as a basis for estimating "single-parameter" fragility descriptions (i.e., referenced to local ZPA) including confidence limits and practical "lower bound" (i.e., HCLPF) seismic capacities. The tests also suggested possible hardware modifications to increase seismic capacity, such as top bracing of the cabinet or use of reed- rather than armature-type relays. More importantly perhaps, the test results suggest that other descriptions of fragility -- incorporating frequency effects, for example -- might be "more appropriate" for characterizing component behavior.

The "sensitivity study" concept applies to the interpretation of existing (e.g., qualification) data as well. For example, as part of our Phase I component prioritization activities, we developed fragility descriptions for five components based on "high-level" seismic qualification data (Holman 1986a). Although not true "fragility" data, these test results provided useful information on component behavior under conditions exceeding any anticipated change in peak ground acceleration for plant sites in the eastern United States. Here we assumed that the qualification test results represented the HCLPF capacity for each. Besides providing a basis for developing workable probabilistic fragility descriptions, these tests yielded insight into the influence of such parameters as support arrangement, cabinet rigidity and mass distribution on seismic capacity. In some cases the results of these tests identified practical -- and often relatively minor -- hardware modifications which substantially improved the seismic performance of the equipment tested.

It is important to recognize, however, that "sensitivity studies" in qualification testing often arise out of necessity as equipment is modified to meet requirements. The fragility analyst must therefore pay careful attention to the specific modifications made, particularly when seeking to apply data to similar components.

Clearly, even for functionally identical components, variations among manufacturers and models, in size and type, and in mounting and loading conditions imply that any fragility estimate -- or, for that matter, other methods of assessing component performance -- will be based to a certain extent on engineering judgement. It is important that this judgement be supported by as firm a technical basis as possible within practical constraints. Testing for "understanding" rather than for explicit fragilities would provide the following:

1. Guidance to the fragility analyst as to what should be considered in developing a specific fragility description for a specific component.

2. An improved basis for interpreting and applying test data obtained from other sources. This is particularly valuable, for example, using qualification data to assess actual component capacity.

3. An improved basis for defining test conditions if more rigorous testing of a specific component is required.

4. Guidance for developing screening techniques for reviewing actual plant equipment ("walkdown" techniques) and suggesting modifications for enhancing the seismic capacity of critical components.

We demonstrated through our Phase I tests how these objectives can be achieved for a motor control center and its internal devices. Testing following this basic approach could be similarly applied to develop failure probabilities and improve the seismic performance of other types of components.

REFERENCES

Bohn, M.P., et al. 1983. Seismic Safety Margins Program Phase I Final Report. Lawrence Livermore National Laboratory, Report NUREG/CR-2015.
Budnitz, R.J. 1985. An Approach to the Quantification of Seismic Margins in Nuclear Power Plants. Lawrence Livermore National Laboratory, Report NUREG/CR-4334.
Cover, L.E. 1983. Seismic Safety Margins Research Program: Equipment Fragility Data Base. Lawrence Livermore National Laboratory, Report NUREG/CR-2680.
Holman, G.S. and C.K. Chou 1986a. Component Fragility Research Program: Phase I Component Prioritization. Lawrence Livermore National Laboratory, Report NUREG/CR-4899.
Holman, G.S. and C.K. Chou 1986b. Component Fragility Research Program: Phase I Demonstration Tests. Lawrence Livermore National Laboratory, Report NUREG/CR-4900.

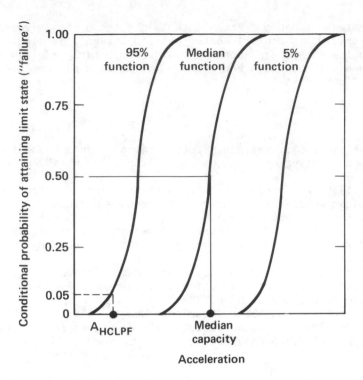

Figure 1.   Typical curve set representing component fragility.

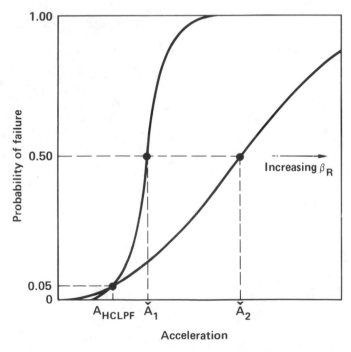

Figure 2.   Typical 95% fragility function showing how increasing
random uncertainty affects median capacity derived from a HCLPF value.

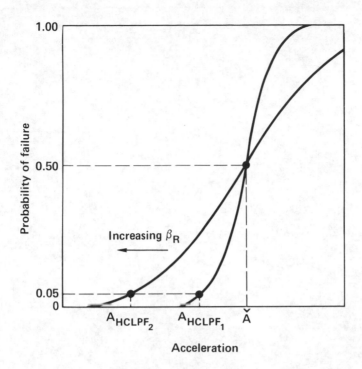

Figure 3. Typical 95% fragility function showing how increasing random uncertainty affects the HCLPF value derived from a constant median capacity.

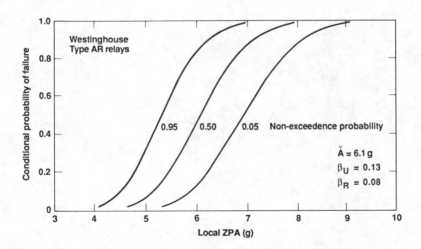

Figure 4. Fragility curves developed for Westinghouse Type AR relay.

Engineering application of stochastic earthquake motion models with non-linear soil amplification*

H.Kameda

Urban Earthquake Hazard Research Center, Disaster Prevention Research Institute, Kyoto University, Japan

ABSTRACT

Stochastic earthquake motion models are developed that are a combination of nonstationary random process model for the basement rock level scaled for earthquake magnitude and distance and the soil amplification model taking into account the nonlinear response of overlying soil layers. The usefulness of the model is demonstrated by two example engineering applications: (1) a closed-form random vibration solution for the site-dependent response spectra, and (2) simulation of risk-consistent earthquake motions.

1. INTRODUCTION

Stochastic ground motion models are a powerful tool for probabilistic assessment of seismic hazard and structural response. Pioneering works in this area used stationary process models (Housner and Jennings 1964). They were extended to models with nonstationary variation of rms intensity (Shinozuka and Sato 1967, Amin and Ang 1968, Goto and Toki 1968, Iyengar and Iyengar 1968, Jennings and Housner 1969, Ruiz and Penzien 1971, etc.). Because of their nature with nonstationary variation of intensity and stationary frequency content, such models will be referred to as "amplitude modulated process models."

The amplitude-modulated process models apply successfully to a broad class of random variation of structural response. However, there are cases where they fail to be valid, particularly when inelastic response is involved (Kameda 1977). This requires "evolutionary process models," with evolutionary power spectra (Priestley 1965, Kameda 1975).

As earthquake motions to take place "in the future" is of our major engineering concern, the stochastic ground motion models should be composed of physically justifiable model parameters and should be scaled for source-site parameters that can be estimated for future earthquakes, typically, magnitude and epicentral distance. An evolutionary process model was developed for simulation of ground motions scaled for magnitude, epicentral distance, and local soil conditions (Kameda,

---

*Basic concepts of this study were presented in the RILEM symposium on Stochastic Methods in Materials and structural Engineering, Los Angeles, April 1986.

327

Sugito and Asamura 1980). An alternative evolutionary process model was developed to deal with ground motions on rock surface (Kameda and Sugito 1984), which will be a basis for the rock surface motion model developed in this study.

There is a significant nonlinearity in the soil amplification of earthquake motions by surface layers overlying the bedrock. This aspect was evaluated by using a simple conversion model (Kameda and Sugito 1985) for peak ground motions. This concept was extended to the intensity parameter of the evolutionary process model (Sugito, Kameda, et.al. 1986). It will be implemented in the analysis in the following chapters.

It is important to formulate the stochastic earthquake motion models in a way that is convenient for engineering application. The models developed herein have been designed in such a direction, and yet without losing its physical basis and data-based nature. On this basis, various useful engineering applications can be developed for probabilistic assessment of seismic hazard and structural response.
Their typical examples are shown in 4. and 5.

2. EVOLUTIONARY PROCESS MODEL OF STOCHASTIC EARTHQUAKE MOTIONS ON ROCK SURFACE

A prediction model for free rock surface motions is developed. The free rock surface motion herein means the earthquake motion that should take place on the surface of the bed rock when overlying soil deposits are hypothetically removed. The definition of bed rock herein is that with shear wave velocities of 600--700 m/sec.

Kameda and Sugito (1984), and Sugito and Kameda (1985) proposed a stochastic ground motion model for rock surface. The model was developed to provide an evolutionary spectrum of the rock surface motion where the peak rms intensity, the starting time, and the duration, which are frequency dependent, were defined as functions of earthquake magnitude and epicentral distance. As those models were formulated empirically by using best-fitting functions that are not necessarily convenient for analytical treatments, they have been remodeled (Kameda, Ueda and Nojima 1986) by using the spectral functions represented in terms of rational functions for the convenience of engineering applications. This modified model is explained in the following, and will be used throughout this study.

The nonstationary earthquake acceleration is represented by

$$x(t) = \sum_{k=1}^{m} \sqrt{2G(t,\omega_k)\Delta\omega} \cdot \cos(\omega_k t + \varphi_k) \tag{1}$$

in which $G(\omega,t)$ = nonstationary power spectrum for time t and angular frequency ω, ϕ = independent random phase angles distributed uniformly over $0 \sim 2\pi$, and m = number of superposed harmonic component.

2.1 Evolutionary process model

The evolutionary spectrum $G_r(t,2\pi f)$ for the rock surface motion model is represented by

$$G_r(t,2\pi f) = \gamma^2 \alpha_f(t) G_{nr}(t,2\pi f) \tag{2}$$

where

$$G_{nr}(t,2\pi f) - \frac{2\beta_g(t)}{\pi^2 f_p(t)} \cdot \frac{\{f/f_p(t)\}^2}{[1-\{f/f_p(t)\}^2]^2+4\beta_g^2(t)\{f/f_p(t)\}^2} \qquad (3)$$

in which γ = peak RMS intensity, $\alpha_f(t)$ = normalized mean-square intensity variation, $G_{nr}(t,2\pi f)$ = normalized nonstationary frequency content, $f_p(t)$ = time-dependent predominant frequency, and $\beta_g(t)$ = time-dependent spectral shape parameter. The time-dependent parameters $\alpha_f(t)$, $f_p(t)$ and $\beta_g(t)$ have been modeled by

$$\alpha_f(t) - (t/t_m)^2 \exp\{2(1-t/t_m)\} \qquad (4)$$

$$f_p(t) - f_{po}+A_1(t-t_m), \quad \beta_g(t) - \beta_{go}+B_1(t-t_m)$$

where t_m = value of t at which $\alpha_f(t) = 1$, $f_{po} = f_p(t_m)$, $\beta_{go} = \beta_g(t_m)$, and A_1 and B_1 are time-independent constants. Empirical formulas for these model parameters have been developed, and are shown in Table 1.

Table 1. Empirical formulas for determining model parameters for
rock surface motions

$\log \gamma - 1.950+0.5371 \cdot M-1.991 \cdot \log(\Delta+30.0)$	(5)
$t_m - 19.77-7.35 \cdot M+0.7196 \cdot M^2+0.0023(M-1) \cdot \Delta$	(6)
$f_{po} - 4.124+(0.0115-0.0048 \cdot M+0.000272 \cdot M^2) \cdot \Delta$	
$\quad +(-0.7959+0.2577 \cdot M-0.01743 \cdot M^2) \cdot 10^{-4} \cdot \Delta$	(7)
$\beta_{go} - -0.2306+0.2967 \cdot M-0.0174 \cdot M^2$	
$\quad +(-0.0193+0.0049 \cdot M-0.0003 \cdot M^2) \cdot \Delta$	(8)
$A_1 - \begin{cases} -11.76+3.187 \cdot M-0.2158 \cdot M^2 & ; M\leq 7.5 \\ 0 & ; M>7.5 \end{cases}$	(9)
$\log B_1 - -0.02160-0.5713(\Delta+30.0)$	(10)
$f_{pmin} - 6.78 \cdot 10^{-0.1M}$	(11)
$\Delta_o(M) - 1.06 \cdot 10^{0.242M} -30.0$	(12)

The limiting epicentral distance $\Delta_o(M)$ in Eq.(12) represents the boundary of the epicentral region, within which it is assumed as first order approximation that the ground motion is uniform.

Kanai-Tajimi type spectral function (Tajimi 1960) has been used widely to represent ground motion characteristics, as it permits straightfoward interpretation as absolute acceleration response of a simple oscillator to white noise input. Although the Kanai-Tajimi spectrum is a good model to represent over-all characteristics of earthquake motions, it fails to be realistic in low frequency regions as it does not vanish at zero frequency. It has been found out that the velocity response type spectrum function matches much better with the rock surface motion model proposed by Kameda and Sugito (1984). For this reason, the spectrum function given by Eq.(3) is employed throughout this study.

2.2 Amplitude-modulated process model (AMPM)

If the time-dependent model parameters $f_p(t)$ and $\beta_g(t)$ are replaced by their values at t = t$_m$, then we have an amplitude-modulated process model with

$$G_r(t,2\pi f) - \gamma^2 \alpha_f(t) G_{nr}(t_m,2\pi f) \qquad (13)$$

with

$$\begin{aligned} f_p(t) &= f_p(t_m) - f_{po} \\ \beta_g(t) &= \beta_g(t_m) - \beta_{go} \end{aligned} \qquad (14)$$

In this case, the intensity function $\alpha_f(t)$ is the only time-varying model parameter.

From Eq.(9), it is pointed out that as the earthquake magnitude M increases the predominant frequency $f_p(t)$ becomes less time dependent in which case the amplitude-modulated process model is justified.

2.3 Stationary process model

Further simplification leads to a stationary process model:

$$G_r(t,2\pi f) = G_r(2\pi f) - \gamma^2 G_{nr}(t_m,2\pi f) \qquad (15)$$

Using a stationary process model for earthquake motion causes a disadvantage, as the ground motion duration is involved as an arbitrary parameter. However, one can enjoy in return its analytical simplicity, as in 4.

3. IMPLEMENTATION OF NONLINEAR SOIL AMPLIFICATION

3.1 Formulation of the conversion factor

On the basis of the stochastic earthquake motion model for rock sur-face, the earthquake motions on the surface of the overlying ground is evaluated by using a simple conversion factor $\beta_R(f)$; i.e., the evolutionary spectrum of the ground surface motion is represented by

$$G_s(t,2\pi f) - \{\beta_R(f)\}^2 G_r(t,2\pi f) \qquad (16)$$

Herein the conversion factor is modeled by the following rational function of frequency f.

$$\beta_R(f) - \frac{1+2a^2(f/f_s)^2}{\{1-(f/f_s)^2\}^2+4h_s^2(f/f_s)^2} \qquad (17)$$

Eq.(17) involves three parameters a, f_s, and h_s which will be determined from the site parameters and the spectral intensity of the input rock surface motion by taking account of the nonlinear soil amplification effect.

3.2 Site parameters

The following two site parameters are used (Kameda and Sugito, 1985).

(i) soil softness S_n defined by Kameda, Sugito and Goto (1982):

$$S_n - 0.264 \int_0^{d_s} \exp\{-0.04N(x)\} \exp(-0.14x)dx - 0.885 \qquad (18)$$

330

in which $N(x)$ = SPT blow count at depth x (in meters), and d_s = total depth of the blow count profile, and

(ii) depth d_p (in meters) of the upper boundary of the bed rock.

The soil softness parameter S_n characterizes the soil properties within a depth of 15~20m from the ground surface, with S_n = 1 for $N(x) \cong 0$, S_n = 0 for $N(x) \cong 17$, and S_n = -0.4 for $N(x) \cong 50$. The base rock depth d_p will explain the effect of deeper part of the soil layers where detailed soil data are difficult to obtain. The combined use of S_n and d_p will maximize the utilization of soil data that are commonly available from engineering construction sites.

3.3 Evaluation of the model parameters

Nonlinear soil amplification was implemented by Kameda and Sugito (1985) in the conversion factor for peak ground motions in a simple function of S_n, d_p and the input rock motion. The same concept was extended, and the conversion factor β_α for the square-root of the peak value of the evolutionary spectrum along the time for each frequency has been formulated by Sugito, Kameda et.al (1986). The results were given in a tabular form for individual frequencies. By applying them to the functional form of Eq.(16), the following procedure for determining the model parameters a, f_s and h_s is developed.

$$f_s = \left[\frac{f_1^2 f_2^2 f_3^2 \{f_1^2 \beta_1 (\beta_2-\beta_3)+f_2^2\beta_2(\beta_3-\beta_1)+f_3^2\beta_3(\beta_1-\beta_2)}{f_1^2 f_2^2 (1-\beta_3)(\beta_1-\beta_2)+f_2^2 f_3^2(1-\beta_1)(\beta_2-\beta_3)+f_3^2 f_1^2(1-\beta_2)(\beta_3-\beta_1)} \right]^{1/4} \quad (19)$$

$$h_s = \sqrt{\frac{\beta_1}{2(\beta_1-\beta_2)} \left(X_1-\frac{\beta_2}{\beta_1}X_2\right)} \quad, \quad a = \sqrt{\frac{\beta_1\beta_2}{\beta_1-\beta_2}(X_1-X_2)} \quad (20)$$

where

$$X_i = 1-\frac{1}{2}(1-\beta_i)\frac{f_s^2}{f_i^2} - \frac{f_i^2}{2f_s^2} \quad; \quad i=1,2 \quad (21)$$

in which β_j ; $j=1,2,3$, is the value of $\beta_\alpha(f)$ at a specified frequency $f = f_j$.

The value of $\beta_\alpha(f)$ is determined from

$$\beta_\alpha(f) = \begin{cases} 10^{r_{c\alpha}}\{\sqrt{G_r(t_m,2\pi f)}\}^{r_{1\alpha}} & ; \quad \sqrt{G_r(t_m,2\pi f)} \geq \alpha_{re}(f) \\ 10^{r_{c\alpha}}\{\alpha_{re}(f)^{r_1}\} & ; \quad \sqrt{G_r(t_m,2\pi f)} < \alpha_{re}(f) \end{cases} \quad (22)$$

in which $r_{c\alpha}$ = effect of linear amplification, and $r_{1\alpha}$ = effect of nonlinear amplification, which are given by

$$\begin{cases} r_{o\alpha} = r_{o\alpha}(f) = u_{oo}(f)+u_{o1}(f)S_n+u_{o2}(f)\log d_p \\ r_{1\alpha} = r_{1\alpha}(f) = u_{1o}(f)+u_{11}(f)S_n+u_{12}(f)\log d_p \end{cases} \quad (23)$$

The coefficients on the right-hand side of Eqs.(23) are listed in Table 2. Note that u_{10}, u_{11}, u_{12} vanish for frequencies below 1 Hz; i.e., the nonlinearity of soil amplification is accounted for only in the higher frequency region above 1 Hz. This does not preclude the possibility of nonlinear amplification at low frequencies, but its evaluation will require additional site data to include even deeper

331

f (Hz)	μ_{00}	μ_{01}	μ_{02}	μ_{10}	μ_{11}	μ_{12}
0.13	0.0	0.006	0.0	0.0	0.0	0.0
0.19	−0.073	0.007	0.071	0.0	0.0	0.0
0.25	−0.169	0.008	0.156	0.0	0.0	0.0
0.31	−0.201	0.040	0.198	0.0	0.0	0.0
0.37	−0.212	0.069	0.222	0.0	0.0	0.0
0.43	−0.202	0.083	0.231	0.0	0.0	0.0
0.49	−0.192	0.098	0.236	0.0	0.0	0.0
0.55	−0.168	0.110	0.230	0.0	0.0	0.0
0.61	−0.151	0.121	0.227	0.0	0.0	0.0
0.67	−0.130	0.135	0.218	0.0	0.0	0.0
0.73	−0.115	0.139	0.216	0.0	0.0	0.0
0.79	−0.104	0.149	0.212	0.0	0.0	0.0
0.85	−0.083	0.156	0.203	0.0	0.0	0.0
0.91	−0.061	0.162	0.194	0.0	0.0	0.0
1.03	−0.035	0.178	0.184	0.0	0.0	0.0
1.21	0.0	0.195	0.171	−0.002	−0.010	−0.002
1.45	0.032	0.256	0.160	−0.004	−0.103	−0.004
1.75	0.087	0.322	0.148	−0.010	−0.206	−0.019
2.11	0.158	0.348	0.130	−0.015	−0.262	−0.048
2.53	0.236	0.330	0.118	−0.023	−0.283	−0.086
3.01	0.318	0.254	0.104	−0.035	−0.266	−0.125
3.55	0.389	0.150	0.087	−0.055	−0.228	−0.155
4.15	0.452	0.0	0.067	−0.089	−0.152	−0.176
4.81	0.482	−0.128	0.054	−0.155	−0.114	−0.164
5.53	0.507	−0.218	0.031	−0.239	−0.072	−0.144
6.25	0.527	−0.284	0.002	−0.295	−0.053	−0.130
7.03	0.540	−0.342	−0.026	−0.350	−0.040	−0.119
7.87	0.559	−0.401	−0.054	−0.401	−0.030	−0.105
8.77	0.560	−0.468	−0.075	−0.441	−0.024	−0.095
10.03	0.552	−0.555	−0.100	−0.500	−0.020	−0.072

part of the grounds.

The rock motion intensity $\alpha_{re}(f)$ separating the regions of linear amplification and nonlinear amplification is determined from

$$\alpha_{re}(f) = 10^{(l_0 + l_1 S_n)} \qquad (24)$$

where

$$\begin{cases} l_0 = 1.135 - 0.643 \cdot \log f + 2.256(\log f)^2 - 2.913(\log f)^3 \\ l_1 = -0.350 + 0.286 \cdot \log f - 4.960(\log f)^2 + 4.888(\log f)^3 \end{cases} \qquad (25)$$

Fig.1 shows the conversion factor $\beta_R(f)$ determined from Eq.(17) for different combinations of the site parameters S_n and d_p and different combinations of the earthquake magnitude M and the epicentral distance Δ. Observe the nonlinear soil amplification effects in the different values of $\beta_R(f)$ for same S_n, d_p but different M, Δ values.

4. APPLICATION-I: RANDOM VIBRATION SOLUTION FOR ATTENUATION OF SITE-DEPENDENT RESPONSE SPECTRUM

The stochastic earthquake motion model developed in the previous chapters can be easily applied to random vibration analysis, and the spectral moments of linear structural response are obtained analytically. They can be combined with the response spectrum method developed by Der-Kiureghian (1980).

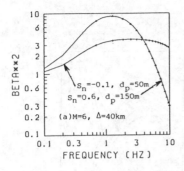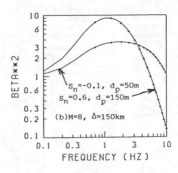

Fig.1 Spectral conversion factor

In this manner, a closed-form random vibration solution for the attenuation of response spectra has been obtained. The solution provides a site dependent response spectra with nonlinear soil amplification that was incorporated in the ground motion model.

Thus the random vibration solution has been obtained (Kimura and Kameda, 1987). Its result is a lengthy formula, but can ben evaluated exactly and esily, once its computer code is implemented.

Fig.2 shows a numerical result of mean response spectra for given values of M, Δ, S_n and d_p. They are compared with the empirical results developed by Sugito, Kameda, Goto and Hirose (1986). Fig.3 shows theoretical results of the uncertainties of the response spectra in terms of the coefficient of variation. They are compared with the results derived from statistical analysis of strong motion data. Observe that the theoretical uncertainty with a deterministic excitation intensity is much smaller than the data-based uncertainty from the strong motion data, whereas the total uncertainty by considering an uncertain excitation intensity (the value δ_γ =0.427 has been determined on the basis of Kameda and Sugito (1985)) agrees fairly well, particularly in the period range below 3 seconds.

5. APPLICATION-II: SIMULATION OF RISK-CONSISTENT EARTHQUAKE GROUND MOTIONS

In engineering application of the stochastic ground motion model, determination of its model parameters is an important question. Probabilistic seismic hazard analysis is an established powerful tool for assessment of the characteristics of earthquake motions in future. However, the conventional probabilistic seismic hazard analysis deals only with a single intensity parameter, and does not apply to simultaneous determination of several parameters characterizing the ground motion.

In order to solve this problem, Kameda, Ueda and Nojima has proposed a method to determine all ground motion parameters needed for simulation on a consistent probabilistic basis. It enables one to simulate risk-consistent ground motions. The details of the method are presented elsewhere (Kameda, Ueda and Nojima 1986). Its outline is as follows.

The earthquake ground motion model developed in 2. requires one to determine its peak rms intensity γ, duration t_m, predominant frequency f_p, and spectral shape parameter β_g, which are all defined as functions

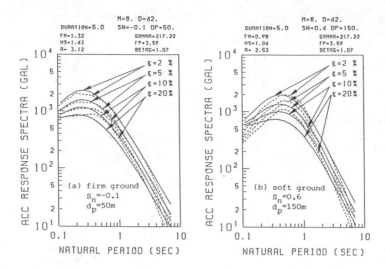

Fig.2 Mean response spectra (solid line: random vibration solution,
dashed line: empirical formula; M=8, Δ =62km)

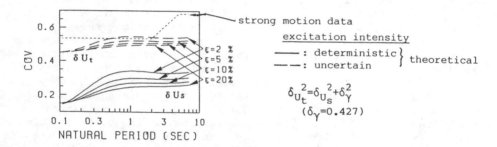

Fig.3 Uncertainties of response spectra

of M and Δ by Eqs. (4) (6), or Eqs. (5) (8). As M and Δ are treated as
random variables in probabilistic seismic risk analysis, all these
parameters should also be treated as random variables.

With this notion, the intensity parameter γ is first determined from
the conventional seismic hazard analysis; i.e., the value of γ is
determined as $\gamma_o(p_o)$, a function of the annual probability of exceedence
p_o. Then other parameters, t_m, f_p and β_g are determined as conditional
mean values given that $\gamma > \gamma_o(p_o)$. If these parameters are commonly
denoted by G, their estimates are defined by

$$\bar{G}(p_o) - E[G|\gamma > \gamma_o(p_o)] \tag{26}$$

In this manner, one can avoid the excessive complexity of dealing with
the joint probability function of these parameters, and determine all of
them as functions of p_o only. Using the parameters determined in this
way, risk-consistent ground motions can be generated for an aribtrary
value of p_o. Fig.3 shows simulated risk-consistent ground motions for
Tokyo, for p_o=0.005 (or return period of 200 years), and for various soil
conditions.

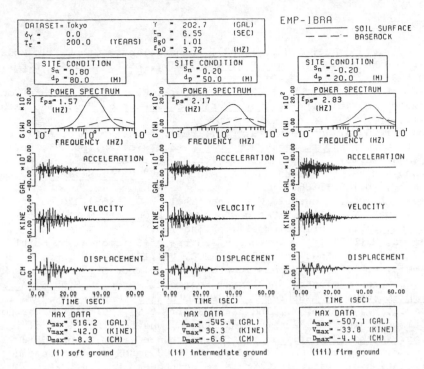

Fig.4 Simulated risk-consistent earthquake ground motions
   (p_0=0.005; Tokyo)

6. CONCLUSIONS

Following conclusions may be derived from this study.
1) Stochastic earthquake motion models for rock surface motions have been developed. The basic model incorporates evolutionary spectrum.
An equivalent amplitude-modulated process model and an equivalent stationary process model has also been proposed. All spectral functions are represented by rational functions that are convenient for random vibration application.
2) A conversion factor incorporating nonlinear soil amplification has been proposed to transform the spectral function of the rock surface motion to that of the ground suface motion. It is also defined in terms of a rational function of the frequency.
3) Two example applications of the models developed herein have been presented. One is a closed-form random vibration solution for site-dependent response spectra, and the other is simulation of risk-consistent earthquake ground motions.

7. ACKNOWLEDGMENT

The author wishes to express his appreciation to his co-worker Dr. M. Sugito of Kyoto University for his efforts to develop the concepts presented herein. Contributions by Mr. J. Kimura, former graduate student and Mr. N. Nojima, graduate student at Kyoto University in the development of the results presented in 4. and 5. are gratefully acknowledged.

REFERENCES

1. Amin, M., and Ang, A.H-S. (1968), "Nonstationary Stochastic Model of Earthquake Motion," Jour. Eng. Mech. Div., ASCE, Vol.94, No.EM2, pp.559-583.
2. Goto, H., and Toki, K.(1969), "Structural Response to Nonstationary Random Excitation," Proc. 4th World Conf. Earthq. Eng., Santiago, Vol.I, pp.39-54.
3. Housner, G.W., and Jennings, P.C.(1964), "Generation of Artificial Earthquakes," Jour. Eng. Mech. Div., ASCE, Vol.90, No.EM1, pp.113-150.
4. Iyengar, R.M., and Iyengar, K.T.S.R.(1969), "Nonstationary Random Process Model for Earthquake Accelerogram," Bull. Seism. Soc. Amr., Vol.59, pp.1163-1188.
5. Jennings, P.C., Housner, G.W., and Tsai, N.C.(1969), "Simulated Earthquake Motions for Design Purposes" Proc. 4th World Conf. Earthq. Eng., Santiago, Vol.I, pp.145-160.
6. Kameda, H.(1975), "Evolutionary Spectra of Seismogram by Multi-filter," Jour. Eng. Mech. Div., ASCE, Vol,101, No.EM6, pp.787-801.
7. Kameda, H.(1977), "Stochastic Process Models of Strong Earthquake Motions for Inelastic Structural Response," Proc. U.S.-S.E. Asia Symp. on Eng. for Natural Hazards Protection, Manila, pp.71-85.
8. Kameda, H., Sugito, M., and Asamura, T.(1980), "Simulated Earthquake Motions Scaled for Magnitude, Distance, and Local Soil Conditions," Proc. 7th World Conf. Earthq. Eng., Istanbul, Vol.2, pp.295-302.
9. Kameda, H., Sugito, M., and Goto, H.(1982), "Microzonation and Simulation of Spatially Correlated Earthquake Motions," Proc. Int'l. Earthquake Microzonation Conf., Seattle, Vol.III, pp.1463-1474.
10. Kameda, H., and Sugito, M.(1984), "Prediction of Strong Earthquake Motions on Rock Surface Using Evolutionary Process Models," Proc. Conf. Struc. Analysis & Design of Nuclear Power Plants, Porto Alegre, pp.161-186.
11. Kameda, H., and Sugito, M.(1985), "Earthquake Motion Uncertainties as Compared between Ground Surface Motion and Bedrock Input Motion--Characterization Using Evolutionary Process Models," 8th SMiRT Conf., Brussels, M1K1/3, pp.297-302.
12. Kameda, H., Ueda, K., and Nojima, N.(1986), "Simulated Earthquake Motions Coinsistent with Seismic Hazard Analysis," Proc. 7th Japan Earthq. Eng. Symp., Tokyo, pp.181-186.
13. Priestley, M.B.(1965), "Evolutionary Spectra and Nonstationary Processes," Jour. Royal Statistical Society, Ser. B, Vol.27, pp.204-237.
14. Ruiz, P., and Penzien, J.(1971), "Stochastic Seismic Response of Structures," Jour. Eng. Mech. Div., ASCE, Vol.97, No.EM2, pp.441-457.
15. Shinozuka, M., and Sato, Y.(1967), "On the Numerical Simulation of Nonstationary Random Processes," Jour. Eng. Mech. Div., ASCE, Vol.93, No.EM1, pp.11-40.
16. Sugito, M., and Kameda, H.(1985), "Prediction of Nonstationary Earthquake Motions on Rock Surface," Proc. JSCE, Struc. Eng./Earthq. Eng., Vol.2, No.2, pp.149-159.
17. Tajimi, H.(1960), "A Statistical Method of Determining the Maximum Response of a Building Structure during an Earthquake," Proc. 2nd World Conf. Earthq. Eng., Tokyo/Kyoto, Vol.2, pp.781-797.

Seismic risk assessment of a BWR*

J.E.Wells, D.L.Bernreuter, J.C.Chen, D.A.Lappa, T.Y.Chuang & R.C.Murray
University of California, Lawrence Livermore National Laboratory, USA
J.J.Johnson
EQE, Inc., Sam Ramon, Calif., USA

1. BACKGROUND

The simplified seismic risk methodology (Shieh, 1985) developed in
the United States NRC Seismic Safety Margins Research Program
(SSMRP) (Smith, 1981) was demonstrated by its application to the
Zion nuclear power plant (Bohn, 1984) a pressurized water reactor
(PWR). The simplified seismic risk methodology was developed to
reduce the costs associated with a seismic risk analysis while
providing adequate results. A detailed model of Zion, including
systems analysis models (initiating events, event trees, and fault
trees), SSI and structure models, and piping models, was developed
and used in assessing the seismic risk of the Zion nuclear power
plant (FSAR) a PWR. The simplified seismic risk methodology was
applied to the LaSalle County Station nuclear power plant, a
boiling water reactor (BWR), to further demonstrate its
applicability, and if possible, to provide a basis for comparing
the seismic risk from PWRs and BWRs.

2. METHODOLOGY OVERVIEW

Seismic risk analysis can be considered in five steps: seismic
hazard characterization (seismic hazard curve, frequency
characteristics of the motion); seismic response of structures and
components; structure and component failure descriptions; plant
logic models (fault trees and event trees); and probabilistic risk
quantification. Key elements of the LaSalle simplified seismic
risk analysis are to:

1. develop the seismic hazard at the LaSalle site including
    the effect of local site conditions,

*This work was supported by the United States Nuclear Regulatory
Commission under a Memorandum of Understanding with the United
States Department of Energy

2.   compare best estimate seismic response of structures, components, and piping systems with design values for the purposes of specifying median responses in the seismic risk calculations,

3.   develop building and component fragilities for important structures and components,

4.   investigate the effects of hydrodynamic loads on seismic risk,

5.   develop the systems models (e.g., event and fault trees), and

6.   estimate the seismically induced core damage frequency.

3. SEISMIC INPUT

The methodology used to develop the hazard curves (the probability of exceeding any given level of peak ground acceleration, PGA) is mostly based on the methodology and data given in the NRC sponsored Eastern United States Seismic Hazard Characterization study (Bernreuter, 1985) Descriptions of the zonation, seismicity, and choice of ground motion model are based on the opinions of eleven seismicity experts and five ground motion experts. These expert opinions were used to supplement the available data. One of the important features of the methodology is to identify and assess the uncertainty in all the parameters of the analysis, and propagate that uncertainty using a Monte Carlo simulation approach. The parameters used in the simulations are described by probability distributions. The zonation maps and the ground motion models have discrete probability distributions. All other variables have continuous probability distributions.

4. SEISMIC RESPONSE OF STRUCTURES AND COMPONENTS

For each level of earthquake described by the seismic hazard curve, three aspects of seismic response are necessary to perform the seismic risk analysis: median level (or best estimate) response, variability of response, and correlation of responses. Seismic responses are required for all structures and components contained in the plant logic models. These responses, together with fragilities, allow for the calculation of seismically induced failure probabilities.

Several approaches to developing median level responses are possible. For this analysis a limited amount of re-calculation of the response using best-estimate methods and parameters, and scale factors applied to the design responses were used. The basic strategy for developing median level responses was to perform selected probabilistic response analyses of the LaSalle structures for two ranges of earthquakes -- a lower level earthquake and a higher level earthquake, henceforth called acceleration range 1 and 2, respectively. Two acceleration levels were considered in order to permit interpolation for other earthquakes of different peak accelerations. Results of the analyses gave probability distributions for two types of response: in-structure forces and moments to be used in the fragility evaluation of the structures themselves and in-structure response spectra at equipment and component locations for their fragility evaluation.

5. STRUCTURE AND COMPONENT FRAGILITIES

The development of structure fragilities proceeded as follows. A review of the seismic design analysis results and development of a preliminary set of structure element capacities initiated the task. Simultaneously, a preliminary SMACS analysis (Johnson, 1983) was performed for a single earthquake simulation at near the SSE level to provide a basis of comparison with the design results. Having reviewed the design analysis results and structure model, changes in the structure model, which better captured the expected behavior of the structure, were recommended and incorporated into the SMACS analysis. Additional preliminary SMACS analyses were performed, loads generated, and an assessment of the model modification made.

A comparison of the capacities of the structural elements with the SMACS generated median loads showed very large ratios for acceleration ranges 1 and 2 which make it highly unlikely that structural failures will occur for any excitation level considered in the hazard curve.

Component fragilities were developed for major LaSalle components identified as important in terms of systems behavior and risk. LaSalle specific design reports and equipment qualification data were used as the principal basis for fragility assessment. Median level responses were used in the fragility assessment as generated from the SMACS analyses. A comparison of median response for acceleration ranges 1 and 2 with the capacities of components considered shows large margins. Thus the failure probabilities of these components due to seismic events are quite small.

6. HYDRODYNAMIC LOADS

In a seismic risk analysis of a BWR, it is necessary to consider the combined effects of seismic loads and one or more hydrodynamic loads since seismic events can potentially induce hydrodynamic events (Safety Relief Valve, SRV) discharge, pool swell, condensate oscillation, and chugging). For LaSalle, we evaluated which of these loads were likely to occur simultaneously with an earthquake and developed an approach for combination of the seismic and hydrodynamic responses. The approach was applied to the combination of seismic and SRV discharge responses. Based on

339

the results of several case studies, simple combination rules for obtaining the median and the logarithmic standard deviation of the combined responses from the median and logarithmic standard deviation of the individual responses were developed. The seismic risk analysis is performed for the combined seismic and SRV events, however, the contribution to risk due to hydrodynamic loads was small.

7. PLANT LOGIC MODELS

The accident sequences in the LaSalle analysis were identified using event trees. These event trees were developed by the Risk Methodology Integration Evaluation Program (RMIEP). The event trees included the large, medium, and small loss-of-coolant accidents (LOCAs) and loss-of-offsite power (LOSP). The LOSP was assumed to occur as a result of an earthquake in the range being studied.

Fault tree analysis was used to define the failure paths in the accident sequences. These fault trees, as with the event trees, were developed by the RMIEP and modified by LLNL for seismic initiators. The RMIEP developed fault trees for all the major safety systems at LaSalle and included the support systems, such as electric power and component cooling water. Consequently, the trees are quite large, involving thousands of gates and events. The trees became even larger upon modification for seismic events. To make the problem manageable, the fault trees were probabilistically culled (Wells, 1983) to obtain the minimal cut sets.

The system fault tree minimal cut sets were combined according to the event tree logic to give Boolean expressions for each accident sequence. Each accident sequence was culled so that it contained only the most probable cut sets.

8. SEISMIC RISK QUANTIFICATION

The quantification of seismically induced failure probabilities and core damage frequencies are accomplished using the computer program SEISIM (Wells, 1983) SEISIM computes the failure probabilities taking into account the dependence between basic events. SEISIM does this by computing the multi-normal integral whose integrand is specified by the means, standard deviations, and correlations of responses and fragilities.

9. RESULTS

The earthquake level definitions and their associated frequencies, based on the LaSalle hazard curve, are given in Table 1. Our analysis resulted in a seismically induced core damage frequency point estimate of 6.0e-7/yr for LaSalle. A large percent of the risk (89%) fell into the lower earthquake levels (lower three of six earthquake levels). We believe that this dominance by the lower level earthquakes can be traced to our assumption of a loss-of-offsite power during an earthquake. Note that this result is different from the Zion analysis in that the bulk of the risk fell into the middle earthquake levels at Zion (earthquake levels 2, 3, 4 and 5). Ninety percent of the core damage frequency is due to only three accident sequences: LOSP-Trans-3, LOSP-Trans-4, and LOSP-Trans-1. All three of these accident sequences are LOSP (loss of offsite power) sequences. The failed systems contained in these accident sequence included HPCS (High Pressure Core-Spray), RCIC (Reactor Core Isolation Cooling), LPCI (Low Pressure Coolant Injection), LPCS (Low Pressure Core Spray), ADS (Automatic Depressurization System), SCS (Shutdown Cooling System), SPC (Suppression Pool Cooling), and CSS (Containment Spray System). LOSP-Trans-3 contains failures of the HPCS, RCIC, LPCI, and LPCS. LOSP-Trans-4 contains failures of the HPCS, RCIC, and ADS. LOSP-Trans-1 contains failures the SCS, SPC, CSS, LPCI, and LPCS. The most dominant sequence is LOSP-Trans-3 and it contributes 42% to the total core damage frequency.

Table 1. Seismic Occurrence Data for the LaSalle Site.

Earthquake Level	Rock Out-crop Acceleration Range	Frequency per year
1	.18-.27	1.1E-4
2	.27-.36	2.9E-5
3	.36-.46	1.1E-5
4	.46-.58	4.7E-6
5	.58-.73	2.1E-6
6	>.73	1.0E-6

The dominant components included diesel generator failures (failure to swing, failure to start, failure to run after started) and failure of the condensate storage tank. In this analysis, the assumption was made that failure of the CST would cause failure of both the HPCS and RCIC pumps because of loss of suction.

10. ADDITIONAL ANALYSIS

During this project a second set of event trees and fault trees was generated. These trees were generated independently of the RMIEP event and fault trees. The fault trees in the second set were less detailed in that fewer systems were modeled and the systems modeled were modeled in less detail than the RMIEP fault trees. Two sets of trees were generated so that a comparison might be made between them to indicate whether or not highly detailed fault trees are warranted for a simplified seismic risk analysis. This second analysis resulted in a seismically induced core damage frequency point estimate of 8.4e-6/yr. Ninety five percent of the risk fell in the lower these earthquake levels with 70% of the risk in earthquake level 1 (lowest acceleration).

These results differ from the other analysis mainly because of modeling differences in the event and fault trees. Both analysis did, however, indicate that LOSP accident sequences dominated, random events dominated, and failure of the diesel generators was a large contributor to the overall risk.

REFERENCES

Shieh, L. C., J. J. Johnson, J. E. Wells, J. C. Chen, and P. D. Smith, "Simplified Seismic Probabilistic Risk Assessment: Procedures and Limitations," Lawrence Livermore National Laboratory, Livermore, California, UCID-20468, NUREG/CR-4331, 1985.

Smith, P. D., R. G. Dong, D. L. Bernreuter, M. P. Bohn, T. Y. Chuang, G. E. Cummings, J. J. Johnson, R. W. Mensing, and J. E. Wells, "Seismic Safety Margins Research - Phase I Final Report," Lawrence Livermore National Laboratory, Livermore, California, UCRL-53021, NUREG/CR-2015, Vols. 1-10, 1981.

Bohn, M. P., L. C. Shieh, J. E. Wells, L. C. Cover, D. L. Bernreuter, J. C. Chen, J. J. Johnson, S. E. Bumpus, R. W. Mensing, W. J. O'Connell, and D. A. Lappa, "Applications of the SSMRP Methodology to the Seismic Risk at the Zion Nuclear Power Plant," Lawrence Livermore National Laboratory, Livermore, California, UCRL-53483, NUREG/CR-3428, 1984.

LaSalle County Station, Final Safety Analysis Report

Bernreuter, D. L., et al. (1985), "Seismic Hazard Characterization of the Eastern United States, Volume 1: Methodology and Results for Ten Sites and Volume 2: Questionnaires," Lawrence Livermore National Laboratory Report UCID-20421.

Johnson, J. J., G. L. Goudreau, S. E. Bumpus, and O. R. Maslenikov, "SSMRP Phase I Final Report - SMACS - Seismic Methodology Analysis Chain with Statistics (Project VIII)," Lawrence Livermore National Laboratory, Livermore, CA, UCRL-53021, Vol. 9, also NUREG/CR-2015, Vol. 9, September, 1981.

Wells, J. E., and D. A. Lappa, "Probabilistic Culling in Fault Tree Evaluation,"Proceedings of the 1983 Reliability and Maintainability Symposium, IEEE, New York, NY, 1983.

Wells, J. E., "SEISIM: A Probabilistic Risk Assessment Tool Used in Evaluating Seismic Risk", Proceedings of the Conference on Seismic Risk and Heavy Industrial Facilities, San Francisco, California, 1983.

Probabilistic analysis of nuclear reactor pressure vessel reliability*

Y.Sun, H.Wu & D.Dong

Institute of Nuclear Energy Technology, Tsinghua University, Beijing, People's Republic of China

SUMMARY

This paper presents a method for estimating the failure probability of nuclear reactor pressure vessels(NRPV) in service.Probability theory and fracture mechanics are used to simmulate the evolution of initial crack size distribution in NRPV' lifetime.The failure probability can be calculated out on the knowledge of crack size distribution,loading condition and fracture toughness distribution.

Crack size,fatique growth characteristics and fracture tough-ness are treated as random variables.They are simulated by using Monte Carlo method.The effect of neutron radiation on fracture toughness because of the presence of copper and phosphorus in the vessel material's composition is considered.

Failure is defined as either of the occurences of the following two events;

    a. crack penetration through the vessel wall because of
       fatique crack growth;
    b. unstable propagaton of a crack.

The occuring probabilities of the two events are calculated respectively.Calculation shows that the first event occures at very small probability under all circumstances and is negligible. For the occuring probability of the second event,sensitivity studes are made to see how the failure probability changes with various influencing factors.The failure probabilities calculated out are more meaningful in relative sense.

1. Introduction

The assessing of reliability of pressure vessels was conven-tionally done by analyzing large amount of statistical data collected from real failures and operational experiences.This is especially true for non-nuclear pressure vessels.This is statistical method as we call it.But as NRPV reliability is concerned statistical method encounters the difficulty of data shortage.The fatal weakness of the method is that it conceals the details of failure that might be important information when people want to improve the perfor-mance of nuclear pressure vessels.

Used in this paper to evaluate NRPV reliability is a different method which we call probabilistic method.The method assumes that all the factors that influence NRPV reliability are of statistical or stochastic nature.Respective probability density functions (P.D.F) are assigned to those factors which are taken to be random variables.Failure criteria are established on the basis of linear

* This work is financially supported by scientific fund of
  Chines Academy of Science.

343

elastic fracture mechanics(LEFM).Sensitivity studes of some factors
are made.Conclusions are drawn from those studes.

2. Modeling in this paper

2.1 Assumptions and conservatism
    Analyzed in this paper is a PWR pressure vessel of Westinghouse
product.It is shown in Fig.1 .Following considerations are made
for simplicity and conservatism.
        a. only the beltline region is considered,where radiation
           dose is the largest;
        b. only surface flaws are analysed,with semi-elliptical shape
           and definite lengh-to-depth ratio;
        c. normal,upset and test transients occures with constant
           frequencies;
        d. fatique crack growth rate(FCGR) is not affected by radiation,
           and no threshold values of ΔKic are considered in the
           calculation of FCGR;
        e. residual stress is not taken into consideration,since it is
           usually negative in the inner part of the vessel wall.
Obviously the assumptions made above can only increase the conservatism
of the calculation results.
2.2 Failure model
    Fracture mechanics has proved to be a very good approch for analysing
thick-walled pressure vessels.In this paper failure criteria are established on the basis of LEFM.
    Failure is defined as either of occurences of the following two events,
        a. crack penetration through the vessel wall because of fatique
           crack growth;
        b. unstable propagation of a crack.
    The occuring probability PF1 of the first event is the integration
of f(a) over region (t,∞),where f(a) and t are respectively the P.D.F
of crack size and vessel wall thickness.

$$FP1 = \int_t^\infty f(a)\, da \qquad\qquad (1)$$

    The occuring probability PF2 of the second event is expressed by

$$FP2 = \int_{-\infty}^\infty \left[f(K) \int_{-\infty}^K f(K_{Ic})\, dK_{Ic} \right] dK \qquad (2)$$

where, K -- sress intensity factor(S.I.F);
        Kic -- fracture toughness;
        f(K) -- P.D.F of S.I.F;
        f(Kic) -- P.D.F of fracture toughness.
2.3 Fatique evaluation
    Before operation the vessel should undergo hydrostatic tests.
During operation the vessel are subjected to many cyclic loads produced
by transients of normal,upset and test conditions(see Table 1.).The
fatique growth of existing cracks due to those cyclic loads is an
important process which affects failure and consequently is taken into
account.
    It should be made clear here that only normal,upset and test condi-
tions are considered in the fatique analysis.Emergency and faulted
conditions are not taken into account in this paper.
    The famous Paris' low is used for evaluating fatique crack growth
rate,
$$\frac{da}{dN} = C(\Delta K)^n \qquad\qquad (3)$$

Where, da/dN -- FCGR;
        Kic -- S.I.F change;
        C,n -- coefficients.
In calculation the coefficient C is treated as a random variable to
take into account the fact that the fatique crack growth rate data
are dispersive.
2.4 Distributions of parameters
    As having been stated above,the random variables considered in this
article are initial crack size,fracture toughness and fatique crack
growth coefficient C.

344

Theoretically the initial crack size distribution is dependent on two things: manufacturing process and the reliability of non-destructive test.So it is a complex job to find exact flaw shapes, sizes,locations and to process them into proper distribution.Therefor for simplicity a simple P.D.F of initial crack size is used.It takes exponential form,

$$f(a)=R \times EXP[-R(a-a0)] \qquad (a > a0) \qquad (4)$$

Where,R and a0 are constants,a is crack depth.

Two forms of probability density functions have been chosen for fracture toughness.One is normal distribution,the other is Weibull distribution.

Normal distribution: $\quad f(Kic)= \dfrac{1}{\sqrt{2\pi}\,\sigma}\, exp\left[\dfrac{(K_{1c}-\bar{K}_{1c})^2}{2\sigma^2}\right]$ $\qquad$ (5)

Weibull distribution:
$$f(Kic)=\begin{cases} 0 & Kic<Ko \\ k(K_{1c}-K_o)^m exp\left\{-\dfrac{k}{m+1}(K_{1c}-K_o)^{m+1}\right\} & Kic>Ko \end{cases} \qquad (6)$$

Where, Kic -- mean of fracture toughness;
$\quad\quad \sigma$ -- standard deviation of fracture toughness;
$\quad\quad$ Ko,k,m -- Weibull distribution parameters.
There exist following relationships between the above parameters,

$$Kic=Ko+(\dfrac{m}{k})^{\frac{1}{m+1}}$$

$$\sigma=(\dfrac{m+1}{k})^{\frac{1}{m+1}}\left\{ \Gamma(\dfrac{m+3}{m+1}) - \Gamma^2(\dfrac{m+2}{m+1}) \right\}^{\frac{1}{2}} \qquad (7)$$

Where, Γ is the Gama function.
The mean value of fracture toughness is evaluated by

$$Kic=A+B \times TANH[(T'-To)/D] \qquad (8)$$

While the effective temperature is defined by

$$T'=T-(RTNDTO+ \Delta RTNDT)$$
$$\Delta RTNDT= min \begin{cases} [40+1000(Cu-0.08)+5000(P-0.008)](F/10^{19})^{0.5} \\ 250(1.67 F/10^{19})^{0.2041} \end{cases} \qquad (9)$$

Where, T -- working temperature;
$\quad\quad$ RTNDTO -- initial reference temperature;
$\quad\quad$ RTNDT -- increase in reference temperature;
$\quad\quad$ Cu -- copper content(%);
$\quad\quad$ P -- phosphorus content(%);
$\quad\quad$ F -- integrated flux.
As for the probability density function of coeficient C in Paris' low,only normal distribution is adopted.
2.5 Calculation method
All the random variables are simulated using Monte Carlo method. The integration expressed by (2) is calculated by Monte Carlo simulation too.A FORTRAN programme is written for VAX computer.A simple block diagram of the programme is given(see Fig.2).

3. Sensitivity studes
One great advantage of probabilistic method is that it allows people to analyse the weight of factors which affect reliability.Sensitivity studes of the following factors are made.
$\quad$ a. length-to-depth ratio of semi-elliptical cracks;
$\quad$ b. initial nil-ductile reference temperature;
$\quad$ c. copper content;
$\quad$ d. relative standard deviation of fracture toughness;
$\quad$ e. constant r in the initial crack size distribution;
$\quad$ f. integrated flux.
Results and conclusions are discussed in the next section.

4. Conclusion
It is concluded that the probability of crack penetration due to

fatique crack growth through out the Vessel's lifetime is so small that it can be neglected.

The total failure probability at the end of lifetime lies between 10^{-4} and 10^7

The failure probability calculated using Weibull distribution for fracture toughness is 2-10 times that when normal distribution is used.

The eccentricity of crackis not a sensitive factor.When lentgh-to-depth ratio increases from 1:1 to 10:1, failure probability increases from $1.45x10^{-5}$ to $8.05x10^{-5}$.

When copper content varies from 0.05% to 0.35%,failure probability changes by 17%.

The two sensitive factors are the relative standard deviation of fracture toughness and the initial crack distribution constant R.When relative standard deviation of fracture toughness varies by 40%,failure probability changes by almost 90%.Similarly when R varies by 40%,failure probability changes by around 80%.

Expanded work is under way to consider more accurate modeling and to quicken the calculation process and make it more accurate.

REFERENCES

1. Becher,P.E., Pedersen,A., "Application of Statistical Linear Elastic Fracture Mechanics to Pressure Vessel Reliability Analysis", Nucl. Eng. Design 27, 413-425 (1974)
2. Lidiard,A.B., Williams,M., "A Simplified Analysis of Pressure Vessel Reliability", J. Br. Nucl. Energy Soc., 1977,16,July, No.3, 207-223
3. Riccardella,P.C., Mager,T.R., "Fatique Crack Growth Analysis of Pressurized Water Reactor Vessels", Stress Analysis and Growth of Cracks, Proceedings of the 1971 National Symposium on Fracture Mechanics, Part I,ASTM STP 513,American Society for Testing and Materials, 1972, 260-279
4. Jouris,G.M., Witt,F.J., "An Application of Probabilistic Fracture Mechanics to Reactor Pressure Vessels Including Multiple Initiation and Arrest Events", Transactions of the 7th SMiRT,M1/3,1983
5. Lucia,A.C., Elbaz,J., Brunnhuber,R., "COVASTOL: A Computer Code for the Estimation of Pressure Vessel Reliability", Transactions of the 5th SMiRT, M8/5, 1979
6. Marshall,W., "An Assessment of the Integrity of PWR Pressure Vessels, A U.K. Study Group Report", Transactions of the 4th SMiRT, G6/1*, 1977
7. Paris,P.C., Bucci,R,J., Wessel,E.T., Clark,W.G., Mager,T.R., "Extensive Study of Low Fatique Crack Growth Rates in A533 and A508 Steels", ASTM STP 513, 1972, 146-176

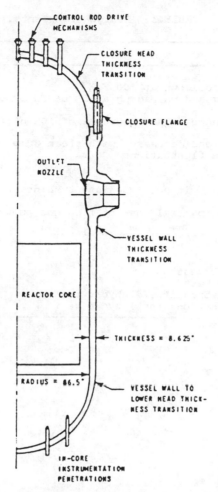

Fig.1　Pressure Vessel

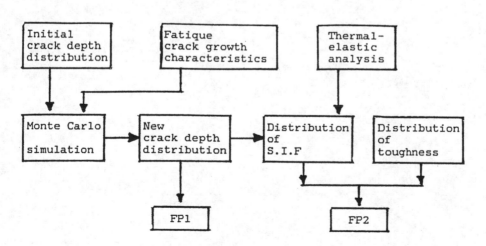

Fig. 2　Block Diagram for Calculation

TABLE 1. Design Transients.

Conditions	Occurences
Normal	
Heatup and cooldown at 100 F/H	200(each)
Unit loading and unloading at 5% of full power/min	18400(each)
Step load increase and decrease of 10% of full power	2000(each)
Large step load decrease, with steam dump	200
Steady state fluctuations	1000000
Upset	
Loss of load, without immediate turbine or reactor trip	80
Loss of flow(partial loss of flow,one pump only)	80
Reactor trip from full power	400
Test	
Turbine roll test	10
Hydrostatic test	
Pre-operational(cold) hydro test	5
Post-operational(hot) hydro test	40

Application of probabilistic methodology to the assessment of pressurised components containing defects

G.W.Gould, P.J.Browne & Z.A.Gralewski
Electrowatt Engineering Services (UK) Ltd, Horsham, Sussex
N.F.Haines & R.Wilson
CEGB, Berkeley Nuclear Laboratories, UK

1. INTRODUCTION

The use of fracture mechanics is now well established in the design and assessment of pressure vessels and piping for nuclear power plants and other industrial facilities. Over the last few years a number of design procedures for assessing defects in such components have been developed specifically for use by the designer rather than the fracture mechanics specialist. One such method is the CEGB R6 procedure (Milne I, et al, 1986) which has been in use for several years and is well validated by a considerable amount of experimental and analytical work (Milne I. et al, 1987).

The deterministic methods have been developed to give conservative assessments of defective structures which in turn leads to a high level of confidence that failure will be avoided. Although the deterministic methods have these in-built safety factors there may still be a finite probability of failure, albeit a very small one, due to uncertainties in material properties, NDE data and loading conditions. The aim of the probabilistic approach is to provide an estimate of the failure probability of a structure. In addition, a clearer indication of the residual life of a component can be obtained using probabilistic analysis. Thus, application of probabilistic methodology should be seen as an extension of the deterministic "safe/unsafe" concept rather than an alternative.

In particular, if the probabilistic concept is combined with a well established deterministic procedure it is hoped a useful assessment tool will result. It was with this objective that a computer code has been developed in which probabilistic methodology has been linked to the CEGB R6 procedure.

2. THE COMPUTER MODEL

The computer code developed by BNL and EWE comprises two main parts. There is the deterministic aspect which considers failure assessment given all the input parameters and there is the probabilistic aspect which handles the sampling and final estimation of the failure probabilities. Such a division has the important result that it is possible to verify separately the deterministic and probabilistic portions of the program. A simplified flow chart of the computer code is shown in Figure 1.

2.1 Deterministic Analysis

The program uses the CEGB R6 procedure (Milne, Ainsworth, Dowling and Stewart 1986) for the criterion for failure by either fracture or plastic collapse. This approach is well established and it is simply necessary to select the appropriate form of the diagram for the materials under consideration. A typical failure assessment diagram is shown in Figure 2. The failure criterion for creep rupture is taken from the appropriate diagram of the ISO Data 1 Report (1978).

The relevant assessment parameters are calculated by the program using the available fracture mechanics and creep rupture models as selected by the user.

2.2 Probabilistic Analysis

Monte Carlo simulation is used to estimate the distributions of the various failure parameters given the probability distributions of the stochastic input variables such as fracture toughness, yield strength and initial defect size. In effect a large number of assessments are carried out for each case using different values for the stochastic variables in each assessment. These values are selected from the appropriate distributions using a random number generator. A count is kept of both the total number of assessments performed and those for which failure is predicted (or at least where the assessment does not fall within the 'failure avoidance' region on the diagram). The failure probability is simply the ratio of the number of predicted failures to the total number of assessments. A major difficulty with the small expected failure probabilities is that a very large number of samples must be taken for the uncertainty in the probability to be suitably small. This may require excessive computer time and therefore the computing costs may be prohibitive. Fortunately techniques are available which (under suitable conditions) greatly reduce the time to obtain a result of the required accuracy. Two such techniques, stratified and importance sampling are employed in the program and further details are given below.

3. SAMPLING TECHNIQUES

3.1 Stratified Sampling

The purpose of a so-called stratified sampling scheme is to ensure that specified intervals in the distribution of sample values contain the appropriate number of samples. This will ensure that the 'tails' of the distribution (i.e. the regions which often have the greatest effect on the output parameters) are not empty or severely under-filled due to chance. Instead each interval contains the correct number of samples (rounded to the nearest integer). Sampling is then carried out using the standard method first of generating a pseudo-random number in the range 0 - 1 and taking this to be the cumulative probability for the sample variate. The value of the sample is then calculated according to the distribution assumed.

After each set of sample variates have been generated (one value for each of the variables) the deterministic assessment is carried out using these values. The whole process is repeated until the required number of cases has been run (for example 1000 runs). The distributions included in the program at present are normal, log-normal, uniform, log-uniform, Weibull and Gumbel. Others could be added if necessary.

350

3.2 Importance Sampling

Importance sampling can reduce the variance of an output variable for a given number of samples by concentrating the sampling on those regions of the distribution which have the greatest effect on the relevant output parameter (e.g. in the tails of the distributions).

The biasing of certain parts of the distribution is achieved by applying suitable weighting factors (>1.0) to the relevant intervals in the stratified sampling scheme. The numbers in each interval and the cumulative probabilities corresponding to each interval boundary are modified accordingly. Sampling followed by a deterministic calculation is then carried out as before for the required number of samples. It is of course necessary to process the results to remove the effect of the biasing before the failure probability can be determined and this is done within the program. The values of the weighting factors to be used are determined from an initial calculation without weights. This analysis is facilitated by selecting an option in the program which will calculate the necessary functions which can then be plotted and analysed to give the weights.

The effectiveness of importance sampling is illustrated in Figure 3. This shows, for a hypothetical load case, the cumulative probability of failure plotted against the number of runs using a log scale. It is clear that IS shows an oscillation that is very much smaller than that shown by RS. This is supported by Figure 4 which shows, for the same load case, the standard deviation of the cumulative failure probability plotted against the number of runs using a log scale. The increased efficiency of importance sampling over random sampling may be seen directly.

4. APPLICATION OF COMPUTER CODE

Consider a longitudinal weld in a large diameter pipe. The component has been in service for some time and results from a large number of non-destructive examinations of the weld are available. A finite element analysis has been carried out to determine the stresses in the weld under safety valve pressure conditions. The owner wishes to operate the plant for at least a further five years so the probability of failure of the pipe at that time needs to be determined.

4.1 Loading Data and Material Thickness

Loading data required by the computer code are, applied stresses (divided into membrane and bending), operating temperature and pressure, and the safety-valve lift-off pressure. Other loading conditions can be introduced so that fault conditions and transients may be examined. Probability distributions may be associated with these parameters but, for this example, constant values have been used as follows:

        Applied membrane stress = 100 MPa
        Operating Pressure   = 2.1 MPa
        Safety Valve Pressure   = 2.5 MPa
        Nominal Operating Temperature = 370°C.

4.2 Material Properties and Defect Depth Distribution

The material properties required are yield stress, ultimate tensile strength and fracture toughness. For this example, normal

351

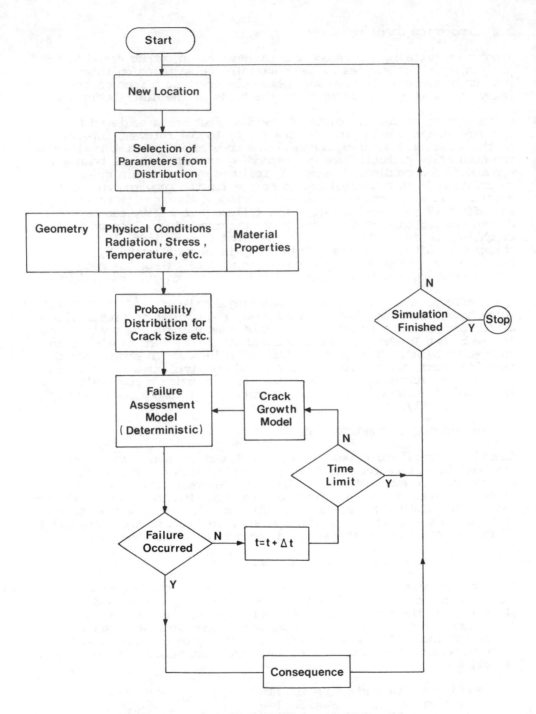

Figure 1: Flowchart for Combined Deterministic and Statistical Model

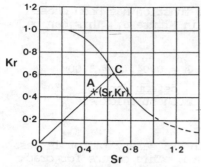

Figure 2: Failure Assessment Diagram

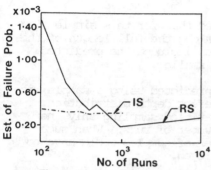

Figure 3: IS vs RS Estimation of Cumulative Probability

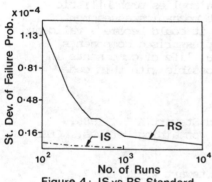

Figure 4: IS vs RS Standard Deviations

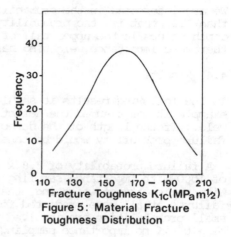

Figure 5: Material Fracture Toughness Distribution

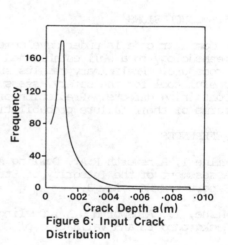

Figure 6: Input Crack Distribution

distributions are used to reflect the variability of these properties, defined as follows:

Yield Stress = 175 MPa with a standard deviation of 16.53 MPa
UTS = 433 MPa with a standard deviation of 15.81 MPa
Fracture
Toughness = 164 MPa m$^{1/2}$ with a standard deviation of 26.51 MPa m$^{1/2}$

These distributions are curtailed at $\pm 2\sigma$. The input distribution for fracture toughness, K_{1c}, is shown in Figure 5.

353

For this example the Gumbel extreme value function has been chosen for the defect depth distribution as shown in Figure 6. This can be defined as follows:

$$PDF(X) = BA^B \exp\left[-\frac{x}{A}^{-B}\right] X^{-(B+1)} \quad ; \quad A = 0.04, \; B = 3.00$$

where, X = normalised defect depth = $\dfrac{\text{defect depth}}{\text{Component thickness}}$

4.3 Importance Weighting

We are interested in the probability that the R6 safety factor is less than 1.0, that is, the probability of failure. This occurs for crack depth values in the upper tail of the Gumbel distribution and therefore importance weighting has been applied to this region.

4.4 Results

In the following results it should be noted that for this simple example it was assumed the defect exists along the full length of the weld. If the length of the defect had been included the predicted failure probability would be considerably smaller.

A failure probability of 6.8×10^{-4} is predicted using a sample size of 1000 with importance sampling of the defect depth distribution. With no importance weighting, that is, stratified sampling only, no failures at all were predicted for this number of runs. When such small probabilities are involved the number of samples would have to be $>10^4$ if no importance sampling was employed.

5. CONCLUSIONS

A computer code is under development which applies probabilistic methodology to a well established fracture mechanics assessment procedure. Preliminary results show that it could become a valid and useful tool for the safety assessment of pressurised components containing defects. Prediction of residual life of components in terms of their failure probabilities is possible with this code.

REFERENCES

Milne I, Ainsworth R.A., Dowling A.R., and Stewart A.T., 1986. Assessment of the Integrity of Structures Containing Defects. CEGB Report R/H/R6 - Rev 3.

Milne, I., Ainsworth R.A., Dowling A.R., Stewart A. T., 1987. Background to and validation of CEGB Report R/H/R6-Rev.3.

ACKNOWLEDGEMENT

This work was performed under a contract with the C.E.G.B.

List of abbreviations

ABWR	Advanced Boiling Water Reactor
ACI	American Concrete Institute
ACRS	Advisory Committee on Reactor Safeguards
ACS	Above Core Structure
AE	Acoustic Emission
AEB	See OBE
AEOD	Analysis and Evaluation of Operational Data
AFNOR	Association Française de Normalisation
AFP	Auxiliary Feedwater Pump
AFPT	Auxiliary Feedwater Pump Turbine
AGR	Advanced Gas-Cooled Reactor
AI	Artificial Intelligence
AICE	American Institute of Chemical Engineers
AISC	American Institute of Steel Construction
AISI	American Iron and Steel Institute
ALE	Arbitrary Lagrange Euler
ANL	Argonne National Laboratory (Argonne, Illinois, USA)
ANS	American Nuclear Society
ANSI	American National Standards Institute
ARS	Amplified Response Spectra

ASCE	American Society of Civil Engineers
ASME	American Society of Mechanical Engineers
ASTM	American Society for Testing and Materials
ATWS	Anticipated Transient Without Scram
BAM	Bundesanstalt für Materialprüfung und -forschung
BARC	Bhabha Atomic Research Centre
BDI	Bundle Duct Interaction
BEM	Boundary Element Method
BFR	Binomial Failure Rate
BM	Base Material
BNL	Brookhaven National Laboratory
BOL	Beginning-of-Life
BOT	Breeder at of Tube
BP	Basic Parameter
BS	British Standard
BWR	Boiling Water Reactor
CAD	Computer Aided Design
CAE	Computer Aided Engineering
CANDU	Canada Deuterium Uranium
CCF	Common Cause Failure
CCF-RBE	Common Cause Failure Reliability Benchmark Exercise
CCG	Creep Crack Growth
CCP	Centre Cracked Plate Specimen
CDM	Charge/Discharge Machine
CEA	Commissariat à l'Energie Atomique
CEB	Comité Européen du Béton

356

CEC	Commission of the European Communities
CEE	See CEC
CEGB	Central Electricity Generating Board, Berkeley, UK
CEIT	Centro de Estudios e Investigaciones Tecnicas de Giupuzcoa (San Sebastian, Spain)
CEN	Centre d'Etudes Nucléaires
CF	Corrosion Fatigue
CFR	Code of Federal Regulations
CIT	Compact Ignition Tokamak
CNRS	Centre National de la Recherche Scientifique (National Centre for Scientific Research, France)
COD	Crack Opening Displacement
CODF	Cristallite Orientation Distribution Function
COSA	Project for Comparison of Computer Codes for Salt
COVA	Code Validation
CPD	Cumulative Plastic Deformation
CPU	Central Processing Unit
CRD	Control Rod Drive
CRDL	Control Rod Drive Line
CRDM	Control Rod Drive Mechanism
CRIEPI	Central Research Institute of the Electric Power Industry, Japan
CSR	Constant Strain Rate
CSS	Core Support Structure
CT	Calandria Tube
CTL	Construction Technology Laboratories, Skokie, Illinois, USA
CTOD	Crack Tip Opening Displacement
CW	Cavity Wall
DCI	Dynamic Cross-Interaction

DEGB	Double-Ended Guillotine Break
DEMO	Demonstration Reactor
DENS	Diffuse Elastic Neutron Scattering
DFA	Dependent Failure Analysis
DIB	Dynamic Intelligent Building System
DIM	Direct Integration Method
DIP	Defect Identification Program
DN	Double Null
DOE	US Department of Energy
DOF	Degree of Freedom
DRS	Design Response Spectra
DSI	Dynamic Substructure Iteration
EAS	Energy Absorbing Support
E/B	Enclosure Building
E/C	Erosion/Corrosion
ECCS	Emergency Core Cooling System
ECT	Eddy-Current Testing
EDF	Electricité de France
EDG	Emergency Diesel Generator
EF	Emergency Feed
EHS	Elastic Half Space
EIR	Swiss Federal Institute for Reactor Research
EMPA	Swiss Federal Institute for Materials Testing and -Research
ENEA	Comitato Nazzionale per la Ricerca e per lo Sviluppo dell'Energia Nucleare e delle Energie Alternative (National Committee for the Research and Developement of Nuclear and Alternative Energy)
ENEL	Italian Electricity Generating Board
ENS	European Nuclear Society

EPFL	Ecole Polytechnique Fédérale de Lausanne, Switzerland
EPRI	Electric Power Research Institute
EPFM	Elastic Plastic Fracture Mechanics
ESG-PA	Equivalent Static-g, Plastic Analysis
ETA	Event Tree Analysis
ETHZ	Eidgenössische Technische Hochschule Zürich, Switzerland

FBR	Fast Breeder Reactor
FBTR	Fast Breeding Test Reactor
F/C	Fuel Channel
FCGR	Fatigue Crack Growth Rate
FDM	Finite Difference Method
FEM	Finite Element Method
FF	Force-Flow
FFT	Fast Fourier Transform
FHC	Fuel Handling Crane
F/M	Fuelling Machine
FMEA	Failure Modes and Effects Analysis
FORM	First Order Second Moment
FRPS	Fuel Rod Process Simulator
FRS	Floor Response Spectrum
FSI	Fluid Structure Interaction
FSTF	Full Scale Test Facility
FTA	Fault Tree Analysis
FW	First Wall

| GCHR | Gas Cooled Heating Reactor |
| GKS | Graphical Kernel System |

GSU   Gulf State Utilities

HAZ   Heat Affected Zone

HAZOP   Hazard and Operability Analysis

HCDA   Hypothetical Core Disruptive Accident

HCLPF   High Confidence Low Probability of Failure

HCP   Hardened Cement Paste

HCU   Hydraulic Control Unit

HDR   Heissdampfreaktor (Superheated Steam Reactor)

HFTD   Hybrid Frequency - Time Domain

HPI   High Pressure Injection

HSST   Heavy Section Steel Technology

HTGR   High Temperature Gas Cooled Reactor

HTR   High Temperature Reactor

HWC   Hydrogen Water Chemistry

IAEA   International Atomic Energy Agency

IAR   Irradiated Annealed Reirradiated

IASCC   Irradiation Assisted Stress Corrosion Cracking

IASMIRT   International Association for Structural Mechanics in
                       Reactor Technology e.V.

I/C   Inner Concrete Structure

ICCGR   International Cooperative Group on Cyclic Crack Growth
                       Rate Testing and Evaluation

ICF   Inertial Confinement Fusion

ICIS   In Core Instrumentation System

IDCOR   Industry Degraded Core Rulemaking Program

IDVP   Independent Design Verification Program

IEA   International Energy Agency

360

IFA	Instrumental Fuel Assembly
IFR	Integral Fast Reactor
IFSMTF	International Fusion Superconducting Magnet Test Facility
IGSCC	Intergranular Stress Corrosion Cracking
IHCP	Inverse Heat Conduction Problems
IHSI	Induction Heating Stress Improvement
IHX	Intermediate Heat Exchanger
II	Inertial Interaction
ILRT	Integrated Leak Rate Test
INPO	Institute for Nuclear Power Operations
IPSN	Institute of Nuclear Protection and Safety
ISI	Inservice Inspection
JAERI	Japan Atomic Energy Research Institute
JAPEIC	Japan Power Engineering and Inspection Corporation
JCF	Japanese Commission on Corrosion Fatigue
JET	Joint European Torus
JINS	Japan Institute of Nuclear Engineering
JIS	Japanese Industrial Standard
JPDR	Japan Power Demonstration Reactor
JRC	Joint Research Center, ISPRA, Italy
JRR	Japan Research Reactor
KI	Kinematic Interaction
KKW	Kalkar Nuclear Power Plant
LBB	Leak Before Break
L.B.LOCA	Large Break Loss of Coolant Accident

LCF	Low Cycle Fatigue
LCT	Large Coil Tasks
LEBP	Laminated Elastomer Bearing Pad
LEFM	Linear Elastic Fracture Mechanics
LESED	Local Equivalent Strain Energy Density
LHGR	Linear Heat Generating Rate
LLNL	Lawrence Livermore National Laboratory
LMC	Laboratory for Building Materials, EPF Lausanne, Switzerland
LMFBR	Liquid Metal Fast Breeder Reactor
LMR	Liquid Metal Cooled Reactor
LOCA	Loss of Coolant Accident
LOF	Loss of Flow
LOFWS	Loss of Flow Without Scram
LSM	Line Spring Model
LVDT	Linear Voltage Displacement Transducer
LWR	Light Water Reactor
MAM	Mode Acceleration Method
MDF	Multi-Degree of Freedom
MDSI	Multilevel Dynamic Substructure Iteration
MFCM	Microfriction Creep Model
MFM	Microfriction Model
MFR	Multinomial Failure Rate
MFTF	Mirror Fusion Test Facility
MFW	Main Feedwater System
MGL	Multiple Greek Letter
MHD	Magnetohydrodynamics
MIMD	Multiple Instruction - Multiple Data Stream Processor

MIPS	Mega Instructions Per Second
MITI	Ministry of International Trade and Industry, Japan
MLRPV	Multilayer Reactor Pressure Vessel
MPA	Materialprüfungsanstalt, Stuttgart, Germany
MPE	See OBE
MSBR	Molten Salt Breeder Reactor
MSIP	Mechanical Stress Improvement Process
MSIV	Main Steam Isolation Valve
μSR	Muon Spin Rotation Spectroscopy

NAGRA	Nationale Genossenschaft für die Lagerung radioaktiver Abfälle
NBS	National Bureau of Standards
NDC	Numerical Dissipation Coefficient
NDE	Non-Destructive Examination
NDT	Non-Destructive Testing
NEFF	Swiss National Foundation for Energy Research
NET	Next European Torus
NGS	Nuclear Generating Station
NIB	Nuclear Island Building
NLSM	Non-Linear Line-Spring Model
NPAR	Nuclear Plant Aging Research
NPP	Nuclear Power Plant
NPRDS	Nuclear Plant Reliability Data System
NRC	U.S. Nuclear Regulatory Commission
NRPV	Nuclear Reactor Pressure Vessel
NSSS	Nuclear Steam Supply System
NUPEC	Nuclear Power Engineering Test Center

OBE	Operating Basis Earthquake
ODE	Ordinary Differential Equation
ODF	Orientation Distribution Function
ONS	Office of Nuclear Safety
OPC	Ordinary Portland Cement
ORNL	Oak Ridge National Laboratory
PB	Pool-Boiling
PCCV	Prestressed Concrete Containment Vessel
PCPV	Prestressed Concrete Pressure Vessel
PCRV	Prestressed Concrete Reactor Vessel
PCS	Prestressed Concrete Structure
PCV	Pressure Containment Vessel
PDE	Partial Differential Equation
PDF	Probability Density Function
PDS	Plant Damage State
PFBR	Prototype Fast Breeder Reactor
PFDR	Piping and Fitting Dynamic Reliability
PFM	Probabilistic Fracture Mechanics
PHTS	Primary Heat Transport System
PHWR	Pressurized Heavy Water Reactor
PISC	Plate Inspection Steering Committee
PLEX	Plant-Life Extension
PLR	Primary Loop Recirculation System
PNET	Probabilistic Network Evaluation Technique
PNP	German Project Nuclear Process Heat
PNSMM	Positive-Negative Sequence Modification Method
PRA	Probabilistic Risk Assessments

PRISM	Power Reactor Inherently Safe Module
PRS	Pressure Relief Structure
PSA	Probabilistic Safety Assessment
PSD	Power Spectral Density
PSDF	Power Spectral Density Functions
PSI	Pre-Service Inspection
PT	Pressure Tube
PTS	Pressurized Thermal Shock
PVRC	Pressure Vessel Research Council
PWR	Pressurized Water Reactor

R/C	Reinforced Concrete
RCC	Règles de Conception et de Construction
RCCA	Rod Cluster Control Assembly
RCCV	Reinforced Concrete Containment Vessel
RCP	Reactor Coolant Pump
REB	Reactor External Building
RFP	Reversed Field Pinch
RILEM	International Union of Testing and Research Laboratories for Materials and Structures
RMIEP	Risk Methods Integration and Evaluation Program
RMS	Root-Mean-Square
RPV	Reactor Pressure Vessel
RSMA	Response Spectrum Modal Analysis
RSO	Response Spectrum of Oscillator
RV	Reactor Vessel

S/A	Subassembly

SANS	Small Angle Neutron Scattering
SC	Superconducting Coil
SCC	Stress Corrosion Crack
SCE	Southern California Edison Company
SDOF	Single Degree of Freedom
SEB	See SSE
SEM	Scanning Electron Microscopy
SEP	Systematic Evaluation Program
SFC	Stress Fatigue Crack
SFCGR	Strain Fatigue Crack Growth Rate
SFEM	Stochastic Finite Element Method
SFRC	Steel Fibre Reinforced Concrete
SHB	Split Hopkinson Bar
SICC	Strain Induced Corrosion Cracking
SIF	Stress-Intensity Factor
SLD	Self-Locking Device
SMA	Seismic Margin Assessment
SMACNA	Sheet Metal and Air Conditioning Contractors National Association
SME	Seismic Margin Earthquake
SMIRT	Structural Mechanics in Reactor Technology
SOK	State-of-Knowledge
SPRA	Seismic Probabilistic Risk Assessment
SPSS	Simplified Piping Support System
SRV	Safety Relief Valve
SQUG	Seismic Qualification Utility Group
SRSS	Square Root Sum of Squares
SRP	Standard Review Plan
SRT	Seismic Review Team

SSCV	Spherical Steel Containment Vessel
SSE	Safe Shutdown Earthquake
SSI	Soil-Structure Interaction
SSMRP	Seismic Safety Margins Research Program
SSR	Slow Strain Rate Testing
SSRC	Structural Stability Research Council
SSY	Small-Scale Yielding
STS	Stainless Steel
TES	Top Entry System
TF	Toroidal Field
TFTR	Tokamak Fusion Test Reactor
TOP	Transient Overpower
TOPWS	Transient Overpower Without Scram
TREX	Tube Reduced Extrusions
TSP	Tube Support Plate
TÜV	Technischer Überwachungs-Verein
UBC	Uniform Building Code
UIS	Upper Internal Structure
UKAEA	UK Atomic Energy Authority
USE	Upper-Shelf Energy
USI	Unresolved Safety Issue
USNRC	See NRC
UT	Ultrasonic Testing
VB	Vacuum Building
VCE	Virtual Crack Extension

VTT	Valtion Teknillinen Tutkimuskeskus, Technical Research Centre of Finland
V.V.	Vacuum Vessel
WLM	Weakest Link Method
WM	Weld Material
WOR	Weld Overlay Repair
ZPA	Zero Period Acceleration

Table of contents to Volumes B–N

Volume B

General and fusion structural mechanics methods development

Contact-impact problems

Computer codes and methods

CAD and optimization

Fluid-structure interaction I

Model methods

Nonlinear statics and dynamics

Computational fracture mechanics

Instability and buckling

Creep and plasticity analysis of structures

Volume C

Fast reactor fuels and subassemblies

Thermal reactor fuels and assemblies

Volume D

In-service inspection and monitoring

Diesel rotating machinery

Volume E

Design of LMFBR components

Fluid structure interaction

Volume F

Environmentally assisted cracking I: Data evaluation

390

Vibration and piping dynamics

Environmentally assisted cracking II: Time domain assessment and stress corrosion

Components I: Stress analysis

Components II: Materials selection and stress analysis

Volume G

Fracture mechanics – Testing aspects

Fracture assessment and surface cracks

Non destructive examination

Failure assessment of piping leak before break aspects

Thermal mixing and thermal shock I

Volume H

402

Dynamic material properties and modelling

Behaviour of structural elements

Measurements, observations and related methods

Special problems

Volume J

Effects of high temperature and fire

Sandia 1/6th scale model test

Containment overpressure and hydrodynamic effects

Aircraft impact: Testing, analysis and design

Volume K1

Testing of structures I

Testing of structures II

Volume K2

Isolation concepts I

Isolation concepts II

Fluid-structure interaction/piping failures

Secondary input and response

Response of piping systems I

Volume L

427

Experiments (structures)

Creep fatigue, damage

Computational aspects

Volume M

432

Volume N

Structural, mechanical and thermal design issues of blankets

443

444

445

446

447

Authors' index to the complete work

450

451

453

457

459